Test Bank

to accompany

Tortora/Grabowski

PRINCIPLES OF ANATOMY AND PHYSIOLOGY
Eighth Edition

Pamela Langley
New Hampshire Technical Institute

 HarperCollins*CollegePublishers*

Test Bank to accompany Tortora/Grabowski, *Principles of Anatomy and Phsyiology, 8/e*

Table of Contents

For their thoughtful commentary and sharp eyes, thanks to the reviewers listed below:

Joan Barber - Delaware Technical and Community College
Ralph Bowers - Leeward Community College
Sandra R. Grabowski - Purdue University
Elden Martin - Bowling Green State University
Robert Morris - Widener University
Elizabeth Murray - College of Mount St. Joseph
Ted Namm - University of Massachusetts Lowell
Michael Vitale - Daytona Beach Community College
College Charles Weitze - Mount Wachusett Community

For their support and patience over the past year, thanks to Cyndy Taylor and Bonnie Roesch.

For their confidence in me and for their unwavering enthusiasm for my involvement in this project, special thanks to Brett and Grace Sullivan.

Pamela Langley

Chapter 1   An Introduction to the Human Body

Multiple-Choice

Choose the one alternative that best completes the statement or answers the question.

Pages: 4
1. Which of the following correctly lists the levels of organization from **least** complex to **most** complex?
   A) cellular, tissue, chemical, system, organ, organism
   B) chemical, cellular, tissue, organ, system, organism
   C) tissue, cellular, chemical, organ, system, organism
   D) chemical, tissue, cellular, system, organ, organism
   E) organism, system, organ, tissue, cellular, chemical
   Answer: B

Pages: 4
2. A structure that has a specific function and that is composed of two or more different types of tissues is called a(n):
   A) cell
   B) system
   C) organelle
   D) organ
   E) organism
   Answer: D

Pages: 4
3. The sum of all chemical reactions that occur in the body is known as:
   A) anabolism
   B) catabolism
   C) metabolism
   D) differentiation
   E) responsiveness
   Answer: C

Pages: 7
4. The change a cell undergoes to develop from an unspecialized one to a specialized one is called:
   A) reproduction
   B) anabolism
   C) responsiveness
   D) differentiation
   E) growth
   Answer: D

Pages: 14
5. The brain is found in the:
    A) mediastinum
    B) ventral body cavity (only)
    C) thoracic cavity (only)
    D) dorsal body cavity
    E) both B and C
    Answer: D

Pages: 13
6. A plane or section that divides an organ such that you would be looking at an **inferior** surface of the section of that organ would be a:
    A) coronal section
    B) medial section
    C) sagittal section
    D) transverse section
    E) oblique section
    Answer: D

Pages: 14
7. The mediastinum is part of the:
    A) thoracic cavity (only)
    B) ventral body cavity (only)
    C) cranial cavity
    D) spinal cavity
    E) both A and B
    Answer: E

Pages: 4
8. The breakdown of large, complex molecules into smaller, simpler ones is called:
    A) anabolism
    B) catabolism
    C) growth
    D) differentiation
    E) homeostasis
    Answer: B

Pages: 7
9. Interstitial fluid is the fluid:
    A) inside blood vessels
    B) inside cells
    C) between the cells in a tissue
    D) inside lymph vessels
    E) that is consumed as part of the diet
    Answer: C

Pages: 13

10. A midsagittal plane divides the body into:
    A) superior and inferior portions
    B) equal right and left halves
    C) equal anterior and posterior halves
    D) ventral and dorsal body cavities
    E) quadrants
    Answer: B

Pages: 14

11. The organs found in the pleural cavities are the:
    A) heart and large blood vessels
    B) liver and stomach
    C) brain and spinal cord
    D) kidneys
    E) lungs
    Answer: E

Pages: 8

12. Homeostasis is regulated by the nervous system and the:
    A) cardiovascular system
    B) muscular system
    C) respiratory system
    D) reproductive system
    E) endocrine system
    Answer: E

Pages: 11

13. The word **plantar** refers to the:
    A) groin
    B) armpit
    C) sole of foot
    D) palm of hand
    E) wrist
    Answer: C

Pages: 11

14. The word **inguinal** refers to the:
    A) groin
    B) hip
    C) buttocks
    D) armpit
    E) back of elbow
    Answer: A

Pages: 12
15. Which of the following best describes the relationship between the liver and the heart?
    A) The liver is medial to the heart.
    B) The liver is superior to the heart.
    C) The liver is distal to the heart.
    D) The liver is inferior to the heart.
    E) The liver is posterior to the heart.
    Answer: D

Pages: 12
16. Which of the following best describes the relationship between the carpal region and the brachial region?
    A) The carpal region is proximal to the brachial region.
    B) The carpal region is distal to the brachial region.
    C) The carpal region is inferior to the brachial region.
    D) The carpal region is anterior to the brachial region.
    E) The carpal region is lateral to the brachial region.
    Answer: B

Pages: 18
17. The region of the abdominopelvic cavity that is superior and lateral to the umbilical region is the:
    A) hypogastric region
    B) epigastric region
    C) left hypochondriac region
    D) right iliac (inguinal) region
    E) both C and D fit the description
    Answer: C

Pages: 18
18. Which of the following regions of the abdominopelvic cavity is the most inferior?
    A) epigastric
    B) right lumbar
    C) left hypochondriac
    D) hypogastric
    E) umbilical
    Answer: D

Pages: 6
19. The body system that distributes oxygen and nutrients to cells and carries carbon dioxide and wastes away from cells is the:
    A) respiratory system
    B) cardiovascular system
    C) endocrine system
    D) urinary system
    E) integumentary system
    Answer: B

20. The system that regulates the volume and chemical composition of blood, eliminates wastes, and regulates fluid and electrolyte balance is the:
    A) respiratory system
    B) cardiovascular system
    C) endocrine system
    D) urinary system
    E) integumentary system
    Answer: D

21. Which of the following most correctly describes the relationship between the ribs and the heart?
    A) The ribs are anterior to the heart.
    B) The ribs are posterior to the heart.
    C) The ribs are deep to the heart.
    D) The ribs are superficial to the heart.
    E) The ribs are medial to the heart.
    Answer: D

22. Which of the following best describes the relationship of the heart to the spine?
    A) The heart is medial to the spine.
    B) The heart is lateral to the spine.
    C) The heart is dorsal to the spine.
    D) The heart is ventral to the spine.
    E) The heart is superior to the spine.
    Answer: D

23. A plane or section that divides an organ such that you would be looking at a **medial** surface of the section of that organ would be a(n):
    A) coronal section
    B) horizontal section
    C) sagittal section
    D) transverse section
    E) oblique section
    Answer: C

24. The abdominopelvic region that is immediately superior to the right lumbar region is the:
    A) epigastric region
    B) right hypochondriac region
    C) right iliac (inguinal) region
    D) left lumbar region
    E) hypogastric region
    Answer: B

Pages: 11
25. The term **brachial** refers to the:
    A) armpit
    B) arm
    C) neck
    D) wrist
    E) thigh
    Answer: B

Pages: 12
26. Which of the following best describes the relationship between the ears and the tip of the nose?
    A) The ears are medial and posterior to the tip of the nose.
    B) The ears are lateral and anterior to the tip of the nose.
    C) The ears are medial and anterior to the tip of the nose.
    D) The ears are lateral and posterior to the tip of the nose.
    E) The ears are superior and medial to the tip of the nose.
    Answer: D

Pages: 12
27. Which of the following best describes the relationship between the right shoulder and the navel?
    A) The right shoulder is inferior and medial to the navel.
    B) The right shoulder is inferior and lateral to the navel.
    C) The right shoulder is superior and medial to the navel.
    D) The right shoulder is superior and lateral to the navel.
    E) The right shoulder is superior and proximal to the navel.
    Answer: D

Pages: 6
28. Which of the following organ systems is made up of glands that secrete hormones?
    A) cardiovascular
    B) endocrine
    C) muscular
    D) skeletal
    E) urinary
    Answer: B

Pages: 8
29. A feedback system consists of three basic components: a control center, a receptor, and a(n):
    A) regulator
    B) monitor
    C) integrator
    D) amplifier
    E) effector
    Answer: E

30. A feedback loop in which the response of the effector enhances or amplifies the original stimulus (stress) is called a(n):
    A) positive feedback loop
    B) polarized feedback loop
    C) negative feedback loop
    D) neutral feedback loop
    E) anabolic feedback loop
    Answer: A

31. **ALL** of the following are part of the ventral body cavity **EXCEPT** the:
    A) thoracic cavity
    B) abdominal cavity
    C) pelvic cavity
    D) cranial cavity
    E) pleural cavities
    Answer: D

32. Chemical reactions that break the molecule $C_6H_{12}O_6$ into CO_2 and H_2O would be examples of:
    A) anabolism
    B) catabolism
    C) differentiation
    D) synthesis
    E) ionization
    Answer: B

33. During the process of development of the skeletal system, embryonic cells, known as mesenchyme cells, may develop into either osteoblasts or chondroblasts, which, in turn, may develop into osteocytes and chondrocytes (respectively). This process is an example of:
    A) anabolism
    B) catabolism
    C) differentiation
    D) responsiveness
    E) carcinogenesis
    Answer: C

34. Which of the following body parts would be considered **contralateral** to each other?
    A) heart and diaphragm
    B) right arm and right leg
    C) left wrist and left elbow
    D) collar bones and shoulder blades
    E) right lung and left lung
    Answer: E

Pages: 12
35. Which of the following is **distal** to the thigh bone (femur)?
    A) shin bone (tibia)
    B) hip bone
    C) elbow
    D) plantar region
    E) both A and D are correct
    Answer: E

Pages: 3
36. **ALL** of the following are primarily studies of anatomy (as opposed to physiology) **EXCEPT**:
    A) observing the arrangement of cells in the adrenal gland
    B) describing the process by which nerve impulses are transmitted
    C) exploring the embryonic origins of endocrine glands
    D) finding the location of the biceps femoris muscle
    E) identifying the types of tissues present in the walls of the intestinal tract
    Answer: B

Pages: 3
37. **ALL** of the following are primarily studies of physiology (as opposed to anatomy) **EXCEPT**:
    A) describing the process by which glucose is catabolized
    B) explaining how substances are secreted from cells
    C) describing the process by which nerve impulses are transmitted
    D) identifying the types of tissues present in the walls of the intestinal tract
    E) identifying the factors that affect blood pressure
    Answer: D

Pages: 4
38. Amino acids are bonded together in the body to form larger protein molecules. This is an example of:
    A) anabolism
    B) catabolism
    C) differentiation
    D) growth
    E) reproduction
    Answer: A

Pages: 6
39. Skin, hair, and sweat glands are part of the:
    A) muscular system
    B) endocrine system
    C) cardiovascular system
    D) integumentary system
    E) nervous system
    Answer: D

Pages: 6

40. Organs such as the liver, gallbladder, and pancreas, which are associated with the gastrointestinal tract, are part of the:
    A) digestive system
    B) cardiovascular system
    C) integumentary system
    D) nervous system
    E) endocrine system
    Answer: A

Pages: 8

41. Osmometer cells sense changes in the concentration of blood plasma; therefore they must be:
    A) receptors
    B) control centers
    C) stress inducers
    D) part of the cardiovascular system
    E) effectors
    Answer: A

Pages: 8

42. Osmometer cells in the brain sense an increase in the concentration of the blood plasma. They then notify the pituitary gland to release the hormone, ADH. This hormone causes the kidney to save water, which lowers the concentration of the plasma. **ALL** of the following are **TRUE** for this scenario **EXCEPT**:
    A) The kidney acts as an effector in this feedback loop.
    B) The osmometer cells act as receptors in this feedback loop.
    C) The stress in this feedback loop is an increase in plasma concentration.
    D) The aspect of homeostasis (i.e., the controlled condition) regulated by this feedback loop is constant ADH secretion.
    E) This is an example of a negative feedback loop.
    Answer: D

Pages: 8

43. Which of the following is an example of a **positive** feedback loop?
    A) A neuron is stimulated, thus opening membrane channels to allow sodium ions to leak from the extracellular fluid to the intracellular fluid. This causes more membrane channels to open, thus allowing more sodium ions to enter the intracellular fluid.
    B) Baroreceptors notify the brain that the blood pressure has increased. The brain then notifies the blood vessels to dilate, thus lowering the blood pressure.
    C) Low levels of glucose in the blood cause the pancreas to release less insulin (a hormone that lowers blood glucose).
    D) Elevated body temperature is sensed by cells in the brain. As a result, sweat is produced, and heat is lost as the water in the sweat evaporates.
    E) An auto factory produces 1000 cars per week. The sales office could sell 1200 cars per week. Extra production personnel are added at the factory to meet the sales demand.
    Answer: A

Pages: 8

44. You are eating a hot fudge sundae. The pleasant taste information is sensed by your taste buds, which notify your brain. Your brain releases endorphins, which make you feel very good. You now associate the good feeling with hot fudge sundaes, so you eat another hot fudge sundae. Now you feel even better.

Which of the following statements is **TRUE** regarding this scenario?
A) This is a negative feedback loop because two hot fudge sundaes will make you sick.
B) This is a positive feedback loop because the results make you feel good.
C) This is a negative feedback loop because you were doing something bad for your health in the first place, and the result makes the situation worse.
D) This is a positive feedback loop because the stress (eating a hot fudge sundae) and the effect (eating another hot fudge sundae) are the same.
E) This is a negative feedback loop because the stress (eating a hot fudge sundae) and the effect (eating another hot fudge sundae) are the same.
Answer: D

Pages: 4

45. Which of the following levels of organization is **most** complex?
A) chemical
B) organ
C) tissue
D) system
E) cellular
Answer: D

Pages: 4

46. Which of the following levels of organization is **least** complex?
A) chemical
B) organ
C) tissue
D) system
E) cellular
Answer: A

Pages: 4

47. "A group of similarly specialized cells plus their intercellular substance" best describes a(n):
A) cell
B) system
C) organ
D) tissue
E) organelle
Answer: D

Pages: 4

48. A triglyceride is broken down into a molecule of glycerol and three molecules of fatty acids. This is an example of:
    A) anabolism
    B) catabolism
    C) differentiation
    D) reproduction
    E) synthesis
    Answer: B

Pages: 10

49. **ALL** of the following would be considered **signs** of an infection **EXCEPT**:
    A) skin lesions of chickenpox
    B) elevated body temperature
    C) swollen lymph nodes
    D) dull pain localized in the back of the neck
    E) enlargement of the liver
    Answer: D

Pages: 3

50. Renal physiology is the study of the function of the:
    A) heart
    B) lungs
    C) endocrine glands
    D) liver
    E) kidneys
    Answer: E

Pages: 6

51. Which of the following body systems provides protection against disease and returns proteins and plasma to the cardiovascular system?
    A) respiratory
    B) urinary
    C) endocrine
    D) lymphatic
    E) integumentary
    Answer: D

Pages: 6

52. Which of the following body systems provides support and protection, stores minerals, and assists in body movements?
    A) digestive
    B) endocrine
    C) integumentary
    D) muscular
    E) skeletal
    Answer: E

Pages: 14
53. **ALL** of the following are located in the mediastinum **EXCEPT** the:
A) heart
B) lungs
C) esophagus
D) trachea
E) thymus gland
Answer: B

Pages: 14
54. **ALL** of the following are located in the abdominal cavity **EXCEPT** the:
A) urinary bladder
B) spleen
C) liver
D) stomach
E) gallbladder
Answer: A

Pages: 3
55. Embryology is the anatomical study of development from the point of fertilization of the egg through:
A) birth
B) implantation of the fertilized egg in the uterine wall one week later
C) the first eight weeks of development in the uterus
D) the first six months of development in the uterus
E) the development of all adult structures
Answer: C

Pages: 16
56. **ALL** of the following are found in the pelvic cavity **EXCEPT** the:
A) large intestine
B) urinary bladder
C) uterus
D) kidneys
E) ureters
Answer: D

Pages: 4
57. Organelles are:
A) structures containing two or more different tissues having a specific function and recognizable shape
B) group of specialized cells plus the substance surrounding them
C) specialized structures found within cells
D) organs without a recognizable shape
E) organs that have not developed fully
Answer: C

Pages: 3
58. The microscopic study of tissues is known as:
    A) embryology
    B) cytology
    C) pathology
    D) histology
    E) gross anatomy
    Answer: D

Pages: 10
59. Which of the following would be considered a **symptom** of disease?
    A) crusty lesions on the skin
    B) elevated body temperature
    C) swollen lymph nodes
    D) increased secretion of mucus
    E) dull pain localized in the back of the neck
    Answer: E

Pages: 11
60. A person in **anatomical position** will exhibit **ALL** of the following **EXCEPT**:
    A) standing erect
    B) facing observer
    C) feet flat on floor
    D) arms at sides
    E) palms against the lateral sides of the thighs
    Answer: E

Pages: 12
61. An anatomical term that pertains to the outer wall of a body cavity is:
    A) proximal
    B) distal
    C) superficial
    D) parietal
    E) visceral
    Answer: D

Pages: 14
62. The vertebral (spinal) canal is part of the:
    A) dorsal body cavity
    B) thoracic body cavity (only)
    C) abdominal cavity (only)
    D) pelvic cavity (only)
    E) thoracic, abdominal, and pelvic cavities
    Answer: A

63. The pleura is a serous membrane associated with the:
    A) brain
    B) heart
    C) lungs
    D) abdominal viscera
    E) spinal cord
    Answer: C

64. A sonogram is produced by:
    A) the response of protons to a pulse of radio waves while they are being magnetized
    B) comparison of an x-ray of a body organ before and after a contrast dye has been injected into a blood vessel
    C) an x-ray beam moving in an arc around the body
    D) high-frequency sound waves transmitted to a video monitor
    E) computer interpretation of radioactive emissions from injected substances
    Answer: D

65. Which of the following employs a form of radiation thought to be harmless to all except possibly pregnant women?
    A) positron emission tomography
    B) magnetic resonance imaging
    C) conventional radiography
    D) computed tomography
    E) digital subtraction angiography
    Answer: B

True-False

Write T if the statement is true and F if the statement is false.

1. A chemical reaction that breaks down a large protein molecule into smaller amino acid molecules would be an anabolic reaction.
   Answer: False

2. Interstitial fluid is an extracellular fluid filling the spaces between the cells of tissues.
   Answer: True

3. The effector is the part of a feedback loop that notifies the control center of changes in the environment.
   Answer: False

4. Negative feedback loops are so-called because they cause harm to the body.
   Answer: False

Pages: 8

5. A feedback loop in which the response of the effectors enhances or amplifies the original stress (stimulus) is referred to as a positive feedback loop.
    Answer: True

Pages: 10

6. The skin lesions of chickenpox are considered signs of the infection.
    Answer: True

Pages: 12, 14

7. The pleural cavities are lateral to the mediastinum.
    Answer: True

Pages: 14

8. The diaphragm separates the abdominal cavity from the pelvic cavity.
    Answer: False

Pages: 18

9. The left lumbar region is immediately superior to the left iliac (inguinal) region.
    Answer: True

Pages: 13

10. A sagittal plane divides the body (or an organ) into anterior and posterior portions.
    Answer: False

Pages: 14

11. The dorsal body cavities are the cranial cavity and the pelvic caivty.
    Answer: False

Pages: 14

12. The mediastinum is located between the lungs, and includes all the contents of the thoracic cavity except the lungs.
    Answer: True

Pages: 14

13. The pericardial cavity contains the lungs.
    Answer: False

Pages: 7

14. Plasma is a type of intracellular fluid.
    Answer: False

Pages: 4

15. The basic structural and functional units of an organism are known as tissues.
    Answer: False

Short Answer

Write the word or phrase that best completes each statement or answers the question.

Pages: 4
1. Groups of cells (and the substance surrounding them) that usually arise from common ancestor cells and work together to perform a particular function comprise a _____.
Answer: tissue

Pages: 3
2. Renal physiology is the study of the function of the _____.
Answer: kidneys

Pages: 4
3. The chemical reactions that break down large, complex molecules into smaller, simpler ones are referred to as _____.
Answer: catabolism

Pages: 6
4. _____ is the ability of an organism to detect and react to changes in the external or internal environment.
Answer: responsiveness

Pages: 7
5. _____ is a condition in which the body's internal environment remains within certain physiological limits.
Answer: homeostasis

Pages: 8
6. Any disturbance in homeostasis is referred to as _____.
Answer: stress

Pages: 6
7. The chemicals produced by the endocrine system that help regulate homeostasis are called

_____.
Answer: hormones

Pages: 8
8. In a feedback loop, the control center provides output to and elicits a response from a(n)

_____.
Answer: effector

Pages: 8
9. The stimulus (stress) in a feedback loop is an increase in blood sugar. If this feedback loop is a negative feedback loop, then the effector will cause the blood sugar to _____.
Answer: decrease

Pages: 8
10. The stimulus (stress) in a feedback loop is an increase in blood sugar. If this is positive feedback loop, then the effector will cause blood sugar to _____.
Answer: increase

Pages: 12
11. The anatomical term that means "on the same side of the body" is _____.
Answer: ipsilateral

Pages: 16
12. The serous membrane associated with the heart is the _____.
Answer: pericardium

Pages: 11
13. Description of any region of the body by means of directional terms and body planes assumes that the body is in _____ position.
Answer: anatomical

Pages: 12
14. The anatomical term that means "away from the head or toward the lower part of a structure" is

_____.
Answer: inferior

Pages: 6
15. The _____ system is composed of a series of glands that secrete hormones.
Answer: endocrine

Pages: 8
16. A feedback system consists of three basic components: a control center, a receptor, and a(n)

_____.
Answer: effector

Pages: 4
17. Specialized structures within a cell that function in the overall cell's anatomy and physiology are known as _____.
Answer: organelles

Pages: 3
18. The study of structure and the relationships between stuctures is _____.
Answer: anatomy

Pages: 3
19. The study of the function of body parts is _____.
Answer: physiology

Pages: 4
20. The basic structural and functional unit of an organism is the _____.
Answer: cell

Pages: 4
21. _____ are structures that are composed of two or more different tissues, have specific functions, and usually have recognizable shapes.
Answer: organs

Pages: 4
22. The level of structural organization of the body consisting of several related organs that have a common function is the _____.
Answer: system

Pages: 3
23. The microscopic study of the structure of cells is _____.
Answer: cytology

Pages: 3
24. The microscopic study of the structure of tissues is _____.
Answer: histology

Pages: 14
25. A plane or section that divides an organ such that you would be looking at a medial surface of a section of that organ would be a _____.
Answer: sagittal section

Pages: 8
26. Homeostasis is regulated by the endocrine system and the _____ system.
Answer: nervous

Pages: 12
27. The component of a feedback loop that senses changes in the environment and notifies the control center of the changes is called the _____.
Answer: receptor

Pages: 12
28. The wrist is _____ to the elbow.
Answer: distal

Pages: 12
29. The intestines are _____ to the heart.
Answer: inferior

Pages: 12
30. The muscles are _____ to the skin.
Answer: deep

Matching

Choose the item from Column 2 that best matches each item in Column 1.

1. brachial   arm

2. cervical   neck

3. otic   ear   eye

4. crural   lower leg

5. cephalic   head

6. carpal   wrist   hand

7. calcaneal   heel of foot

8. plantar   sole of foot

9. popliteal   hollow behind knee   knee cap

10. axillary   armpit

Essay

Write your answer in the space provided or on a separate sheet of paper.

Pages: 8
1. Compare and contrast the mechanisms by which the nervous system and endocrine system regula homeostasis.
   Answer: The neurons of the nervous system transmit information about deviations from homeostasis and changes to be made by effectors via nerve impulses (electrical impulses). Information is transmitted relatively rapidly. The endocrine system operates via the use of chemicals (hormones) released by endocrine glands. Changes initiated by the endocrine system occur relatively slowly.

Pages: 14
2. The human body can be planed (sectioned) by cutting it from various directions. Name and describe the planes (sections) used in human anatomical studies.
   Answer: A sagittal section cuts the body into right and left portions. A midsagittal plane cuts the body into equal right and left portions, while a parasagittal plane separates the body into unequal right and left portions. A frontal (coronal) plane separates the body into anterior and posterior portions. A transverse (horizontal, cross-sectional) plane separates the body into superior and inferior portions. An oblique plane sections the body at an angle between a horizontal plane and either a frontal or sagittal plane.

Pages: 4

3. Outline the organizational levels of the human body in order from least complex to most complex. Define/discuss the components of each level.

   Answer: · Chemical - interactions of atoms and molecules
   · Cellular - cells = basic structural and functional units of organism; smallest units that can survive on own
   · Tissue - two or more cell types plus intercellular substance
   · Organ - two or more tissue types working for common purpose
   · System - two or more organs working for common purpose
   · Organism - eleven body systems working together

Pages: 8

4. Define the term **homeostasis**. Identify the components of a typical feedback loop, and describe the role of each.

   Answer: Homeostasis is the maintenance of a relatively constant internal environment within certain physiological limits. Homeostasis is regulated by feedback loops, which typically consist of a receptor, a control (integrating) center, and an effector. The receptor monitors changes in a controlled condition, and sends this information to a control (integrating) center. The control center compares this input with other information, and notifies an effector to make an appropriate change. The effector makes the appropriate response, as dictated by the control center.

Pages: 8

5. Explain how a positive feedback loop differs from a negative feedback loop.

   Answer: In a positive feedback loop, the response of the effector enhances or amplifies the original stress, that is, the condition is moved further away from homeostasis. In a negative feedback loop, the response of the effector is the opposite of the orginal stress and tends to move the controlled condition back toward homeostasis.

Pages: 11

6. Describe the position of a body in anatomical position. Explain the usefulness of the concept of anatomical position to the study of anatomy.

   Answer: A body in anatomical position is standing erect, with feet flat on the floor, facing the observer. The arms are at the sides with the palms facing the observer. Having a standard anatomical position allows directional terms to be clear. Any body part or region can be described in relation to any other body part or region, regardless of the actual position of the body.

Pages: 14

7. Name the two major body cavities and the subdivisions of each. Describe the contents of each subdivision.

   Answer: · Dorsal cavity = cranial cavity (brain) plus the spinal (vertebral) canal (spinal cord)
   · Ventral body cavity = thoracic and abdominopelvic cavities, separated by the diaphragm; thoracic cavity = two pleural cavities (lungs) lateral to mediastinum (heart, trachea, esophagus, etc.); abdominopelvic cavity = abdominal cavity (most digestive organs) superior to pelvic cavity (part of large intestine, urinary bladder, and internal reproductive organs)

Pages: 8

8. Consider the following situation as a feedback loop: Tuition at a small college has been $1000 per semester for many years, and the student enrollment has remained constant at 1000 students for an equal number of years. This income exactly covers the expenses of faculty/staff salaries. The college administrators voted this year to fund a raise for the faculty and staff by raising tuition to $2000 per semester. Now only 300 students are enrolled at the college. Continue this story as a negative feedback loop.

   Answer: There are many possible answers to this question, but the usual answer is to forget the raise and lower tuition so that the students come back.

Pages: 8

9. Consider the following situation as a feedback loop: Tuition at a small college has been $1000 per semester for many years, and enrollment at the college has remained constant at 1000 students for an equal number of years. This income exactly covers the expenses of faculty/staff salaries. The college administrators voted this year to fund a raise for the faculty and staff by raising tuition to $2000 per semester. Now only 300 students are enrolled at the college. Continue this story as a positive feedback loop.

   Answer: There are several correct answers to this question, but the usual answer is that the administrators raise tuition again, enrollment drops further, and the school closes.

Chapter 2   The Chemical Level of Organization

Multiple-Choice

Choose the one alternative that best completes the statement or answers the question.

Pages: 36
1. Which of the following is **true** regarding this situation: Solution A has a pH of 7.38 and Solution B has a pH of 7.42:
    A) Solution B is more acidic than Solution A.
    B) The pH of Solution A falls within the homeostatic pH range for extracellular body fluids, but the pH of Solution B does not.
    C) Solution A contains a higher concentration of hydrogen ions than Solution B.
    D) Solution B contains a higher concentration of hydrogen ions than Solution A.
    E) both B and C are correct
    Answer: C

Pages: 38
2. Which of the following is a carbohydrate?
    A) sucrase
    B) ATP
    C) cholesterol
    D) raffinose
    E) There is no way of knowing without seeing the chemical structures.
    Answer: D

Pages: 43
3. Which of the following is a protein?
    A) sucrase
    B) ATP
    C) cholesterol
    D) raffinose
    E) There is no way of knowing without seeing the chemical structures.
    Answer: A

Pages: 38
4. Glycogen is an example of a:
    A) polysaccharide (only)
    B) carbohydrate (only)
    C) peptide
    D) lipid
    E) both a polysaccharide and a carbohydrate
    Answer: E

5. A triglyceride is a:
   A) simple sugar
   B) lipid
   C) protein
   D) nucleic acid
   E) polysaccharide
   Answer: B

6. Peptide bonds are found in:
   A) carbohydrates
   B) lipids
   C) proteins
   D) inorganic compounds
   E) Any type of molecule can contain a peptide bond.
   Answer: C

7. Glycerol is the backbone molecule for:
   A) disaccharides
   B) DNA
   C) peptides
   D) triglycerides
   E) ATP
   Answer: D

8. When a protein undergoes a hydrolysis reaction, the end-products are:
   A) amino acids
   B) monosaccharides
   C) fatty acids
   D) nucleic acids
   E) nucleotides
   Answer: A

9. To produce lactose:
   A) two amino acids must form a peptide bond
   B) pairing of nitrogenous bases must occur between nucleotides
   C) glucose and galactose must undergo a dehydration reaction
   D) glucose and fructose must undergo a hydrolysis reaction
   E) at least two fatty acids must bind to glycerol
   Answer: C

Pages: 46
10. The biological function of a protein is determined by its:
    A) primary structure
    B) secondary structure
    C) tertiary structure
    D) quaternary structure
    E) denatured structure
    Answer: C

Pages: 43
11. Enzymes are:
    A) polysaccharides
    B) proteins
    C) steroids
    D) triglycerides
    E) phospholipids
    Answer: B

Pages: 49
12. The function of ATP is to:
    A) act as a template for production of proteins
    B) store energy
    C) act as a catalyst
    D) determine the function of the cell
    E) hold amino acids together in a protein
    Answer: B

Pages: 27
13. An element required for blood clotting, muscle contraction, and hardening the matrix of bone is:
    A) calcium
    B) hydrogen
    C) iron
    D) nitrogen
    E) glucose
    Answer: A

Pages: 38
14. **All** organic compounds contain the elements:
    A) iron and oxygen
    B) carbon and oxygen
    C) carbon and hydrogen
    D) iron and hydrogen
    E) carbon and nitrogen
    Answer: C

Pages: 34
15. **All** of the following are organic compounds **except**:
    A) ATP
    B) glucose
    C) DNA
    D) enzymes
    E) water
    Answer: E

Pages: 34
16. The most abundant inorganic substance in the human body is:
    A) glucose
    B) fat
    C) ATP
    D) water
    E) iron
    Answer: D

Pages: 36
17. Which of the following is considered to be neutral on the pH scale?
    A) urine
    B) pure water
    C) blood plasma
    D) cytoplasm
    E) interstitial fluid
    Answer: B

Pages: 40
18. Steroids are classified as:
    A) carbohydrates
    B) lipids
    C) proteins
    D) nucleic acids
    E) inorganic compounds
    Answer: B

Pages: 41
19. A "saturated" fat is saturated with:
    A) nitrogen
    B) carbon
    C) hydrogen
    D) ATP
    E) oxygen
    Answer: C

Pages: 43
20. A hormone is an example of which functional class of proteins?
    A) contractile
    B) structural
    C) transport
    D) regulatory
    E) catalytic
    Answer: D

Pages: 43
21. Hemoglobin is an example of which functional class of proteins?
    A) contractile
    B) structural
    C) transport
    D) regulatory
    E) catalytic
    Answer: C

Pages: 38
22. The sugar found in RNA is:
    A) fructose
    B) galactose
    C) deoxyribose
    D) ribose
    E) sucrose
    Answer: D

Pages: 26
23. 96% of the human body is composed of the elements carbon, hydrogen, oxygen, and:
    A) aluminum
    B) iron
    C) calcium
    D) nitrogen
    E) sodium
    Answer: D

Pages: 34
24. An inorganic acid dissociates in water into:
    A) one or more hydroxide ions and one or more cations
    B) one or more hydrogen ions and one or more anions
    C) one or more hydroxide ions and one or more anions
    D) one or more hydrogen ions and one or more cations
    E) cations and anions other than hydroxide and hydrogen ions
    Answer: B

Pages: 34

25. An inorganic base dissociates in water into:
    A) one or more hydrogen ions and one or more anions
    B) one or more hydroxide ions and one or more cations
    C) one or more hydrogen ions and one or more cations
    D) one or more hydroxide ions and one or more anions
    E) cations and anions other than hydroxide and hydrogen ions
    Answer: B

Pages: 34

26. An inorganic salt dissociates in water into:
    A) one or more hydrogen ions and one or more anions
    B) one or more hydroxide ions and one or more cations
    C) one or more hydrogen ions and one or more cations
    D) one or more hydroxide ions and one or more anions
    E) cations and anions other than hydroxide and hydrogen ions
    Answer: E

Pages: 50

27. In the presence of oxygen, glucose is completely broken down into:
    A) water and hydrogen
    B) hydrogen and oxygen
    C) carbon dioxide and oxygen
    D) oxygen and water
    E) carbon dioxide and water
    Answer: E

Pages: 49

28. In RNA, the nitrogenous base that takes the place of thymine is:
    A) adenine
    B) cytosine
    C) guanine
    D) uracil
    E) glycine
    Answer: D

Pages: 49

29. Which of the following represents accurate base-pairing in DNA molecules?
    A) adenine to adenine and guanine to guanine
    B) adenine to uracil and cytosine to guanine
    C) adenine to cytosine and guanine to thymine
    D) adenine to thymine and cytosine to guanine
    E) adenine to guanine and cytosine to thymine
    Answer: D

30. The suffix that denotes a sugar is:
    A) -ase
    B) -ose
    C) -ide
    D) -amine
    E) -oid
    Answer: B

31. Energy that travels in waves, such as x-rays or visible light, is known as:
    A) heat energy
    B) chemical energy
    C) radiant energy
    D) mechanical energy
    E) electrical energy
    Answer: C

32. The type of energy present in the bonds between atoms is:
    A) heat energy
    B) chemical energy
    C) radiant energy
    D) mechanical energy
    E) electrical energy
    Answer: B

33. An atom of one element is distinguished from an atom of another element by the number of:
    A) neutrons in the nucleus
    B) electrons in the nucleus
    C) protons in the nucleus
    D) electrons orbiting the nucleus
    E) electrons it can lose when bonding
    Answer: C

34. The **atomic number** of an atom is the:
    A) sum of the numbers of subatomic particles
    B) number of electrons in the outer orbital shell
    C) number of neutrons in the nucleus
    D) number of protons in the nucleus
    E) sum of the numbers of protons and neutrons only
    Answer: D

Pages: 28
35. The **mass number** of an atom is the:
    A) sum of the numbers of subatomic particles
    B) number of electrons in the outer orbital shell
    C) number of neutrons in the nucleus
    D) number of protons in the nucleus
    E) sum of the number of protons and neutrons only
    Answer: E

Pages: 29
36. The electron shell nearest the nucleus of an atom holds a maximum of how many electrons?
    A) one
    B) two
    C) eight
    D) eighteen
    E) all of an atom's electrons, regardless of number
    Answer: B

Pages: 30
37. A molecule that can be broken down into two or more elements by chemical means is a(n):
    A) isomer
    B) cation
    C) compound
    D) anion
    E) inert element
    Answer: C

Pages: 28
38. Two different atoms of an element that have the same number of protons, but different numbers of neutrons are called:
    A) isotopes
    B) compounds
    C) anions
    D) electrolytes
    E) cations
    Answer: A

Pages: 28
39. The half-life of a radioactive isotope is the time it takes for:
    A) death to occur in half of humans who are exposed
    B) half of the atoms in a solid gram of the substance to become a gas
    C) half of the non-radioactive atoms in a sample to become radioactive
    D) half of radioactive atoms in a sample to decay into non-radioactive atoms
    E) atoms in a radioactive substance to become incorporated into half of the cells in the body
    Answer: D

Pages: 28
40. An atom of oxygen has 8 protons, 8 neutrons, and 8 electrons. What is its atomic number?
    A) 8, because it has 8 protons
    B) 8, because it has 8 neutrons
    C) 8, because it has 8 electrons
    D) 16, because it has 8 protons plus 8 electrons
    E) 16, because it has 8 protons plus 8 neutrons
    Answer: A

Pages: 28
41. An atom of oxygen has 8 protons, 8 electrons, and 8 neutrons. Which of the following is the mass number of oxygen?
    A) 8, because it has 8 protons
    B) 8, because it has 8 electrons
    C) 8, because it has 8 neutrons
    D) 16, because it has 8 protons plus 8 electrons
    E) 16, because it has 8 protons plus 8 neutrons
    Answer: E

Pages: 29
42. An atom of oxygen has 8 protons, 8 electrons, and 8 neutrons. How would you expect the electrons to be "arranged" in this atom?
    A) all electrons are located in the nucleus of the atom
    B) all 8 electrons are located in the outermost electron shell
    C) all 8 electrons are located in the electron shell nearest the nucleus
    D) 2 electrons are in the electron shell nearest the nucleus, and 6 are in the second shell
    E) 6 electrons are in the electron shell nearest the nucleus, and 2 are in the second shell
    Answer: D

Pages: 29
43. An atom of oxygen has 8 protons, 8 electrons, and 8 neutrons. In order to achieve stability, the oxygen atom will:
    A) donate two electrons from its outer shell to another atom
    B) accept two electrons into its outer electron shell from another atom
    C) donate two protons to another atom
    D) accept two protons from another atom
    E) do nothing, because its outer electron shell is already stable
    Answer: B

Pages: 30
44. An atom becomes a **cation** by:
    A) accepting electrons into its outermost electron shell
    B) giving up electrons from its outermost electron shell
    C) donating protons to another atom
    D) accepting protons from another atom
    E) accepting neutrons from another atom
    Answer: B

Pages: 30
45. An atom becomes an **anion** by:
    A) accepting electrons into its outermost electron shell
    B) giving up electrons from its outermost electron shell
    C) donating protons to another atom
    D) accepting protons from another atom
    E) accepting neutrons from another atom
    Answer: A

Pages: 31
46. In a covalent bond, atoms are bonded by:
    A) attraction of unlike electrical charges
    B) sharing pairs of electrons
    C) direct connections between the atomic nuclei
    D) radiant energy
    E) sharing pairs of protons
    Answer: B

Pages: 33
47. Which of the following is **true** about the products of a chemical reaction?
    A) In a synthesis reaction, the products contain more atoms than the reactants.
    B) In a decomposition reaction, the products contain fewer atoms than the reactants.
    C) In any chemical reaction, the products may contain more or fewer atoms than the reactants,
       because the reaction is a random event.
    D) The products of a chemical reaction always contain the same number of atoms as the reactants.
    E) both A and B are correct
    Answer: D

Pages: 33
48. Anabolic reactions are:
    A) synthesis reactions
    B) nuclear reactions
    C) exchange reactions
    D) decomposition reactions
    E) hydrolysis reactions
    Answer: A

Pages: 33
49. Catabolic reactions are:
    A) synthesis reactions
    B) nuclear reactions
    C) exchange reactions
    D) decomposition reactions
    E) both B and D are correct
    Answer: D

Pages: 33
50. Buffer reactions, which help maintain normal acid-base balance, are examples of:
    A) synthesis reactions
    B) nuclear reactions
    C) exchange reactions
    D) decomposition reactions
    E) hydrogen bonding
    Answer: C

Pages: 33
51. In an oxidation reaction, a molecule is oxidized if it:
    A) gains electrons
    B) loses electrons
    C) loses hydrogen atoms
    D) loses oxygen atoms
    E) both B and C are correct
    Answer: E

Pages: 33
52. In a reduction reaction, a molecule is reduced if it:
    A) undergoes a decomposition reaction
    B) loses oxygen atoms
    C) gains hydrogen atoms
    D) loses electrons
    E) both C and D are correct
    Answer: C

Pages: 33
53. Which of the following is most likely to be **true** in a redox reaction occuring in the body?
    A) One molecule gains oxygen atoms, while another loses oxygen atoms.
    B) The reaction must be anabolic.
    C) One molecule gains hydrogen atoms, while another loses hydrogen atoms.
    D) Oxygen atoms are removed from all molecules involved for use inside mitochondria.
    E) Hydrogen atoms are bonded to all molecules involved to maintain proper pH balance.
    Answer: C

Pages: 32
54. In an exergonic reaction occurring in the body, energy is:
    A) released as bonds are broken and reformed
    B) consumed as new bonds are formed
    C) released as atomic nuclei are broken and reformed
    D) consumed as subatomic particles are rearranged within the nucleus
    E) converted from chemical into mechanical energy
    Answer: A

Pages: 36
55. Which of the following compounds is an acid? (Assume all compunds will ionize in water.)
    A) H_2CO_3
    B) $Ca_2(PO_4)_2$
    C) NaOH
    D) KCl
    E) There is no way of knowing.
    Answer: A

Pages: 36
56. An electrolyte whose pH was measured at 11 could be a(n):
    A) base
    B) hydrogen ion acceptor
    C) molecule containing a hydroxide ion
    D) ionic compound
    E) all of these
    Answer: E

Pages: 46
57. Denaturing of a protein always results in:
    A) loss of biological function
    B) addition of new amino acids to the molecule
    C) destruction of the primary structure
    D) formation of new hydrogen bonds
    E) stimulation of new protein synthesis
    Answer: A

Pages: 35
58. Water is a good solvent for **ionic** compounds because water:
    A) is a universal solvent
    B) neutralizes the electrical differences between the atoms because it has a neutral pH
    C) contributes electrons to the bond between the atoms
    D) contains polar covalent bonds, which separate the ions by means of electrical attraction
    E) is a bigger molecule than most ionic compounds, so it breaks ionic bonds during atomic collisions
    Answer: D

Pages: 36
59. The pH scale measures:
    A) total electrolyte concentration
    B) ATP levels
    C) the level of enzyme activity
    D) hydrogen ion concentration
    E) total hydrogen content of all organic compounds in the body
    Answer: D

Pages: 43
60. The element that is always found in a protein that is not necessarily present in carbohydrates and fats is:
    A) carbon
    B) hydrogen
    C) oxygen
    D) iron
    E) nitrogen
    Answer: E

Pages: 50
61. When adenosine triphosphate is hydrolyzed by the addition of a water molecule:
    A) energy is released
    B) ADP is a product of the reaction
    C) energy is stored
    D) another phosphate group is added to the molecule
    E) both A and B are correct
    Answer: E

Pages: 50
62. The energy needed to attach a third phosphate group to ADP is supplied mainly by the catabolism of:
    A) DNA
    B) steroids
    C) glucose
    D) electrolytes
    E) vitamins
    Answer: C

Pages: 47
63. Purines and pyrimidines are part of:
    A) DNA
    B) steroids
    C) glycogen
    D) triglycerides
    E) phospholipids
    Answer: A

Pages: 46
64. The protein portion of an enzyme is known as the:
    A) apoenzyme
    B) holoenzyme
    C) cofactor
    D) substrate
    E) active site
    Answer: A

Pages: 49
65. Deoxyribose is:
    A) found in DNA
    B) a carbohydrate
    C) a monosaccharide
    D) a five-carbon sugar
    E) all of these
    Answer: E

True-False

Write T if the statement is true and F if the statement is false.

Pages: 47
1. The monomers of nucleic acids are nucleotides.
   Answer: True

Pages: 47
2. DNA is an example of a carbohydrate.
   Answer: False

Pages: 36
3. A solution with a pH of 7.35 contains more hydrogen ions than a solution with a pH of 8.00.
   Answer: True

Pages: 33
4. All decomposition reactions that occur in your body are collectively called anabolic reactions.
   Answer: False

Pages: 27
5. The electrons of an atom are located within the atom's nucleus.
   Answer: False

Pages: 28
6. The number of protons in an atom always equals the number of neutrons.
   Answer: False

Pages: 31
7. When a covalent bond forms, neither of the combining atoms loses or gains electrons.
   Answer: True

Pages: 50
8. Energy is released when a phosphate group is added to ADP.
   Answer: False

Pages: 36
9. The pH scale of 0 to 14 is based on the concentration of hydrogen ions in a solution.
   Answer: True

10. A solution with a hydrogen ion concentration of 0.001 moles/liter has a pH of 3.
Answer: True

Pages: 36
11. A solution that has more hydrogen ions than hydroxide ions is known as an alkaline solution.
Answer: False

Pages: 28
12. The atomic number of an atom represents the number of protons in the nucleus of the atom.
Answer: True

Pages: 40
13. Triglycerides are major energy storage molecules in the body.
Answer: True

Pages: 49
14. Thymine is found in both DNA and RNA.
Answer: False

Pages: 49
15. DNA has a double helix structure, but RNA does not have a double helix structure.
Answer: True

Short Answer

Write the word or phrase that best completes each statement or answers the question.

Pages: 30
1. The most plentiful extracellular cation is _____.
Answer: sodium

Pages: 26
2. Anything living or nonliving that occupies space and has mass is known as _____.
Answer: matter

Pages: 28
3. Different atoms of an element that have the same number of protons, but different numbers of neutrons are called _____.
Answer: isotopes

Pages: 30
4. A substance that can be chemically broken down into two or more different elements is a

_____.
Answer: compound

Pages: 44
5. The covalent bond that forms between a pair of amino acids is called a _____.
Answer: peptide bond

6. Substances that can speed up chemical reactions without being altered themselves are known as _____.
Answer: catalysts

7. The two purines found in DNA nucleotides are _____.
Answer: adenine and guanine

8. The complete hydrolysis of proteins would yield _____.
Answer: amino acids

9. The term **double helix** describes the structure of _____.
Answer: DNA

10. The major lipid component of cell membranes is _____.
Answer: phospholipids

11. Cholesterol, bile salts, and sex hormones are all examples of a class of lipids known as _____.
Answer: steroids

12. Homeostatic mechanisms maintain the pH of blood between 7.35 and _____.
Answer: 7.45

13. A molecule that gains hydrogen atoms during chemical reactions in the body is said to be _____.
Answer: reduced

14. The collision energy needed for a chemical reaction to occur is called the _____.
Answer: activation energy

15. The two factors that most influence the chance that a collision will occur between atoms are the concentration and the _____.
Answer: temperature

16. A positively charged ion is called a _____.
Answer: cation

17. The most abundant inorganic compound in the human body is _____.
Answer: water

18. Organic compounds are held together mostly or entirely by _____ bonds.
    Answer: covalent

19. In a solution, the substance that is dissolved is the _____.
    Answer: solute

20. In the formation of macromolecules, monomers are joined together by a reaction called a
    _____, which involves the elimination of a water molecule from the reactants.
    Answer: dehydration synthesis

21. Macromolecules are broken down into monomers by the addition of water in a reaction known as
    _____.
    Answer: hydrolysis

22. Sugars and starches are examples of _____.
    Answer: carbohydrates

23. The principal function of carbohydrates is to provide _____.
    Answer: (a readily available) energy (source)

24. Lipids are said to be **hydrophobic**, which means that they are _____.
    Answer: water-fearing or insoluble in water

25. Triglycerides are made up of fatty acids and _____.
    Answer: glycerol

26. Fats whose fatty acids contain multiple double bonds between their carbon atoms are said to be
    _____.
    Answer: polyunsaturated

27. Prostaglandins and leukotrienes are examples of a class of lipids known as _____.
    Answer: eicosanoids

28. Catalytic proteins are called _____.
    Answer: enzymes

29. The biological function of a protein is determined by its _____ structure.
    Answer: tertiary

30. The number of substrate molecules converted to product per enzyme molecule in one second is called the _____.
    Answer: turnover number

Matching

Choose the item from Column 2 that best matches each item in Column 1.

1. made up of amino acids   proteins

2. glucose, fructose, and   monosaccharides
    ribose are examples

3. double helix structure   DNA

4. the cell's "energy   ATP
    currency"

5. starch and glycogen are   polysaccharides
    examples

6. sucrose and lactose are   disaccharides
    examples

7. steroids and triglycerides   lipids
    are examples

8. electrolytes that   acids
    dissociate into one or more
    hydrogen ions

9. electrolytes that   bases
    dissociate into one or more
    hydroxide ions

10. acts as solvent for ionic   water
    compounds in body fluids

Essay

Write your answer in the space provided or on a separate sheet of paper.

Pages: 27
1. Describe/discuss the structural arrangement of the subatomic particles of an atom.
    Answer: The positively charged protons and the uncharged neutrons are located in the nucleus of the atom. Negatively charged electrons, equal in number to the protons, orbit the nucleus in various energy levels, depending on the total number of electrons.

Pages: 30-31

2. Compare and contrast ionic vs. covalent chemical bonds.
   Answer: Chemical bonding involves either donating/receiving electrons or sharing electrons between atoms. In an ionic bond, atoms are held together by the opposite electrical charges created by donating or receiving electrons. In a covalent bond, two atoms share electrons to stabilize their outer electron orbitals without loss or gain of electrons. Most organic compounds are held together by covalent bonds. Electrolytes are held together by ionic bonds.

Pages: 35

3. List and briefly describe the functions of water in the human body.
   Answer: 1. Solvent & suspending medium - due to polar covalent bonds, tends to pull ionic compounds apart
   2. Participant in chemical reactions - digestive breakdown of molecules via hydrolysis or formation of larger molecules via dehydration synthesis
   3. Maintenance of body temperature - great heat-absorbing capacity without large increases in temperature; high heat of vaporization provides excellent cooling mechanism as sweat evaporates
   4. Acts as lubricant - as major part of mucus and other lubricating fluids to reduce friction between body parts

Pages: 37

4. Buffer systems convert strong acids into weak acids and strong bases into weak bases. What does this mean in terms of maintaining the homeostasis of pH in extracellular fluids?
   Answer: Strong acids and bases ionize completely in water, while weak acids and bases ionize only partially. Thus, weak acids and bases affect the pH less than strong acids and bases do. Buffer systems, therefore, help minimize the effect of strong acids and bases on extracellular fluids.

Pages: 46

5. What are enzymes? What is a substrate? How do enzymes work?
   Answer: Enzymes are biological catalysts. Substrates are the molecules upon which enzymes work. Enzymes speed up chemical reactions by lowering the activation energy and properly orienting colliding molecules. There is a specific relationship between enzymes and their substrates. The enzyme has an active site that fits the substrate in a manner similar to a lock and key mechanism. The type of reaction catalyzed is also specific.

Pages: 48-49

6. Compare and contrast the structures of DNA and RNA.
   Answer: DNA and RNA are both formed from smaller units, known as nucleotides. DNA nucleotides contain a phosphate group, deoxyribose, and a nitrogenous base - either adenine, thymine, guanine, or cytosine. RNA nucleotides contain a phosphate group, ribose, and a nitrogenous base - either adenine, uracil, guanine, or cytosine. DNA molecules are made of two strands of nucleotides joined by complementary base pairing (A-T, and C-G). RNA molecules are a single strand of nucleotides, and are much shorter than DNA molecules (they are formed from a fraction of a DNA molecule).

Pages: 43

7. List and briefly describe the functional types of proteins.
    Answer: 1. Structural - form structural framework of body
              2. Regulatory - regulate physiological processes, control growth & development, and mediate responses of nervous system
              3. Contractile - allow production of movement
              4. Immunological - for protection from invading microorganisms and foreign proteins
              5. Transport - carry substances throughout body
              6. Catalytic - act as enzymes to regulate biochemical reactions

Pages: 38

8. What is hydrolysis, and what is its role in human metabolism?
    Answer: Hydrolysis is the breakdown of larger molecules into smaller units by the addition of water. The major groups of biochemicals (e.g., carbohydrates, fats, proteins, etc.) are digested via hydrolysis reactions in the intestinal tract. These smaller units are more easily absorbed than the larger parent compounds.

Pages: 33

9. What are redox reactions?
    Answer: Redox reactions are chemical reduction and oxidation reactions. Molecules that are reduced either gain electrons (or 2H), with a resulting increase in energy content of the molecule. Molecules that are oxidized lose electrons (or 2H) with a resulting decrease in the energy content of the molecule. In a cell these reactions are always coupled.

Pages: 36

10. Consider the following situation: At 8:00 PM, Patient X, a diabetic, is found to have a blood pH of 7.33. His brain tells his diaphragm to contract more frequently so that his respiratory (breathing) rate increases. At midnight, his blood pH is found to be 7.38, and his breathing rate is normal.
    A) State the range of pH that is considered to be homeostatic balance for extracellular fluids.
    B) Did this patient have more hydrogen ions in his blood at 8:00 PM or at midnight? How can you tell?
    C) Does this illustrate a positive or a negative feedback loop? How can you tell?
    Answer: A) 7.35-7.45
              B) more at 8:00 PM - lower pH indicates higher hydrogen ion concentration (more acidic)
              C) negative - homeostasis is restored

Chapter 3   The Cellular Level of Organization

Multiple-Choice

Choose the one alternative that best completes the statement or answers the question.

Pages: 69
1. Ribosomes are made of:
   A) phospholipids
   B) DNA
   C) polysaccharides
   D) RNA
   E) steroids
   Answer: D

Pages: 69
2. The term **chromatin** refers to:
   A) the color of certain cells
   B) uncoiled DNA
   C) highly coiled DNA
   D) the fluid within the nucleus
   E) the protein that makes up the mitotic spindle fibers
   Answer: B

Pages: 73
3. Organelles that contain enzymes that destroy materials engulfed by phagocytes are:
   A) mitochondria
   B) lysosomes
   C) ribosomes
   D) nucleoli
   E) centrioles
   Answer: B

Pages: 73
4. Organelles that contain the enzymes for production of ATP are:
   A) mitochondria
   B) lysosomes
   C) ribosomes
   D) nucleoli
   E) centrioles
   Answer: A

Pages: 71
5. Endoplasmic reticulum is either **smooth** or **rough** depending on the presence of what structures on its outer membrane?
   A) mitochondria
   B) phospholipids
   C) ribosomes
   D) Golgi bodies
   E) centrioles
   Answer: C

Pages: 71
6. Cells that are active in exocytosis would likely contain many:
   A) nuclei
   B) lysosomes
   C) centrioles
   D) peroxisomes
   E) Golgi bodies
   Answer: E

Pages: 61
7. Materials move passively down a hydrostatic pressure gradient in:
   A) simple diffusion
   B) facilitated diffusion
   C) osmosis
   D) filtration
   E) pinocytosis
   Answer: D

Pages: 58
8. The cell does not need to expend energy (ATP) in order to perform:
   A) osmosis
   B) pinocytosis
   C) facilitated diffusion
   D) active transport
   E) both A and C are correct
   Answer: E

Pages: 61
9. Carrier molecules within the cell membrane are required in order to transport a substance across a membrane via:
   A) osmosis
   B) filtration
   C) facilitated diffusion
   D) simple diffusion
   E) exocytosis
   Answer: C

Pages: 64
10. Solid particles are engulfed into a vesicle and brought inside a cell in:
    A) phagocytosis
    B) exocytosis
    C) pinocytosis
    D) filtration
    E) primary active transport
    Answer: A

Pages: 58
11. Solutes move "down" a concentration gradient in:
    A) primary active transport
    B) secondary active transport
    C) osmosis
    D) facilitated diffusion
    E) phagocytosis
    Answer: D

Pages: 67
12. The cytoplasm is mostly:
    A) microtubules
    B) water
    C) suspended proteins
    D) phospholipids
    E) DNA
    Answer: B

Pages: 62
13. Cells try to move sodium ions from the cytoplasm to the outside of the cell, where the sodium concentration is 14 times higher than in the cytoplasm. This means sodium ions are moved out of the cell by:
    A) simple diffusion
    B) facilitated diffusion
    C) osmosis
    D) active transport
    E) filtration
    Answer: D

Pages: 64
14. Pinocytosis is:
    A) a passive means of transporting materials across a membrane
    B) a means by which cells move around
    C) engulfment of liquid particles by cells
    D) a means by which cells can self-destruct
    E) both A and C are correct
    Answer: C

15. The function of the nucleolus is thought to be:
    A) generation of ATP
    B) production of ribosomal RNA
    C) production of vesicles for exocytosis
    D) repair of cell membrane damage
    E) as a carrier of large molecules across the nuclear membrane
    Answer: B

16. Which of the following is **not true** about RNA?
    A) It is shorter than DNA.
    B) It contains some of the same nitrogenous bases as DNA.
    C) It contains the pentose sugar ribose.
    D) It is made of two strands of nucleotides.
    E) There are three types.
    Answer: D

17. The one nitrogenous base found in RNA that is **not** found in DNA is:
    A) adenine
    B) cytosine
    C) thymine
    D) guanine
    E) uracil
    Answer: E

18. The cell structure that controls movements of materials into and out of the cell is the:
    A) endoplasmic reticulum
    B) mitochondria
    C) cell membrane
    D) nucleic acids
    E) centrioles
    Answer: C

19. **Cytokinesis** is the:
    A) movement of a part of a phagocytic cell as it moves toward a solid particle
    B) separation of chromosomes during mitosis
    C) movement of materials across a semipermeable membrane
    D) production of the spindle apparatus during mitosis
    E) division of cell contents other than DNA during cell division
    Answer: E

Pages: 81
20. In mitosis, separation of chromatid pairs takes place during:
    A) interphase
    B) prophase
    C) metaphase
    D) anaphase
    E) telophase
    Answer: D

Pages: 80
21. The mitotic apparatus appears during:
    A) interphase
    B) prophase
    C) metaphase
    D) anaphase
    E) telophase
    Answer: B

Pages: 80
22. The end result of mitosis is:
    A) two diploid cells identical to the parent cell
    B) two haploid cells identical to the parent cell
    C) sperm cells or egg cells
    D) two cells with twice as much DNA as the parent cell
    E) one cell with twice as many organelles as the parent cell
    Answer: A

Pages: 81
23. Chromosomes line up on the equator of the spindle apparatus during:
    A) interphase
    B) prophase
    C) metaphase
    D) anaphase
    E) telophase
    Answer: C

Pages: 81
24. The spindle apparatus and centrioles are made of:
    A) microfilaments
    B) DNA
    C) ribosomal RNA
    D) messenger RNA
    E) microtubules
    Answer: E

Pages: 80
25. Chromosomes attach to the spindle apparatus by:
    A) matching complementary nitrogenous bases
    B) their centromeres
    C) their centrioles
    D) cytokinesis
    E) linking with transfer RNA
    Answer: B

Pages: 55
26. Most of the lipids found in human cell membranes are:
    A) cholesterol
    B) glycolipids
    C) phospholipids
    D) waxes
    E) eicosanoids
    Answer: C

Pages: 77
27. Transcription of DNA is catalyzed by the enzyme:
    A) DNA polymerase
    B) DNA decarboxylase
    C) RNA lipase
    D) RNA dehydrogenase
    E) RNA polymerase
    Answer: E

Pages: 83
28. Which of the following lists the phases of the cell cycle in the correct sequence?
    A) S-phase, G_1-phase, G_2-phase, cytokinesis, S-phase
    B) mitosis, cytokinesis, G_1-phase, G_2-phase, mitosis
    C) cytokinesis, G_1-phase, S-phase, G_2-phase, mitosis
    D) G_1-phase, G_2-phase, mitosis, S-phase, cytokinesis
    E) G_1-phase, S-phase, G_2-phase, mitosis, cytokinesis
    Answer: E

Pages: 88
29. Cancer of blood-forming organs that is characterized by abnormally developed white blood cells
    is known as:
    A) carcinoma
    B) leukemia
    C) melanoma
    D) sarcoma
    E) hematoma
    Answer: B

Pages: 83

30. Cytokinesis usually begins during late:
    A) interphase
    B) anaphase
    C) metaphase
    D) telophase
    E) prophase
    Answer: B

Pages: 77

31. The process of transcription involves production of:
    A) mRNA from a DNA template
    B) two new DNA strands from two original strands
    C) DNA from an mRNA template
    D) an amino acid chain from an mRNA template
    E) new amino acids
    Answer: A

Pages: 78

32. Protein synthesis takes place:
    A) in the cell membrane
    B) on the cristae of mitochondria
    C) on ribosomes
    D) inside lysosomes
    E) inside the nucleus
    Answer: C

Pages: 77

33. What can you tell about the following nucleotide (nitrogenous base) sequence:
    ADENINE-URACIL-GUANINE
    A) It could be part of DNA.
    B) It could be part of RNA.
    C) A complementary strand of RNA nucleotides (nitrogenous bases) would be
       THYMINE-ADENINE-CYTOSINE
    D) The bases are linked by peptide bonds.
    E) both A and C are correct
    Answer: B

Pages: 78

34. During the process of translation, the code carried by mRNA is:
    A) turned into DNA
    B) decoded into a protein
    C) decoded into a pentose
    D) produced by copying the DNA template
    E) used to manufacture tRNA
    Answer: B

Pages: 77
35. If this nitrogenous base sequence (CYTOSINE-CYTOSINE-ADENINE) represents the nucleotides of a codon, then:
    A) the anticodon would be GUANINE-GUANINE-URACIL
    B) the complementary sequence on the sense strand of DNA is the same
    C) it is part of tRNA
    D) it can code for any amino acid
    E) both A and C are correct
    Answer: A

Pages: 55
36. Phospholipid molecules are **amphipathic**, which means that they:
    A) can act as either acids or bases
    B) are water soluble
    C) have polar and nonpolar regions
    D) may be either ionic or covalent compounds
    E) may be either organic or inorganic compounds
    Answer: C

Pages: 56
37. The principal function of cholesterol within the cell membrane is to:
    A) provide an energy source
    B) strengthen the membrane
    C) provide a means of communication among cells
    D) help cells adhere to each other
    E) act as transporters
    Answer: B

Pages: 56
38. Cell membrane receptors are usually what type of molecule?
    A) phospholipid
    B) integral protein
    C) cholesterol
    D) peripheral protein
    E) nucleic acid
    Answer: B

Pages: 57
39. The principle cation in extracellular fluid is:
    A) Na^+
    B) K^+
    C) Cl^-
    D) organic phosphate
    E) Ca^{2+}
    Answer: A

Pages: 57
40. The principle cation in intracellular fluid is:
    A) Na^+
    B) K^+
    C) Cl^-
    D) organic phosphate
    E) Ca^{2+}
    Answer: B

Pages: 57
41. An electrical gradient, or membrane potential, exists across a cell membrane because in most cells the inside surface of the membrane is:
    A) more negatively charged than the outside surface
    B) more positively charged than the outside surface
    C) more hydrophilic than the outside surface
    D) more hydrophobic than the outside surface
    E) richer in lipid molecules
    Answer: A

Pages: 58
42. The selective permeability of a cell membrane to different substances depends on **ALL** of the following **EXCEPT**:
    A) lipid solubility of the substance crossing the membrane
    B) total number of phospholipid molecules in the membrane
    C) electrical charge of the molecule crossing the membrane
    D) presence of transporters to assist substances across the membrane
    E) size of the molecule crossing the membrane
    Answer: B

Pages: 59
43. Solution A and Solution B are separated by a selectively membrane. Solution A contains 5% sodium chloride dissolved in water. Solution B contains 10% sodium chloride dissolved in water. Which of the following statements is **TRUE** regarding this situation?
    A) Solution A has the higher osmotic pressure, because it has a greater concentration of solute.
    B) Solution B has the higher osmotic pressure, because it has a greater concentration of solute.
    C) Solution A has the higher osmotic pressure because it has a greater concentration of solvent.
    D) Solution B has the higher osmotic pressue because it has a greater concentration of solvent.
    E) There is no way of knowing the relative osmotic pressures of these solutions.
    Answer: B

Pages: 60
44. Solution A contains 5% NaCl dissolved in water. Solution B contains 10% NaCl dissolved in water. Which of the following best describes the relative concentrations of these solutions?
    A) Solution A is hypertonic to Solution B.
    B) Solution B is hypertonic to Solution A.
    C) Solutions A and B are isotonic to each other.
    D) Solution B is hypotonic to Solution A.
    E) both A and D are correct
    Answer: B

Pages: 60
45. Red blood cell membranes are not normally permeable to NaCl, and maintain an intracellular concentration of NaCl of 0.9%. If these cells are placed in a solution containing 9% NaCl, what would happen?
    A) Nothing, because the membrane is not permeable to NaCl.
    B) Water will enter the cell because the intracellular fluid has a higher osmotic pressure.
    C) The cell will undergo hemolysis due to membrane damage from the 9% NaCl solution.
    D) The cell will undergo crenation because the extracellular solution has a higher osmotic pressure.
    E) The cell will start to generate NaCl to make up the difference in concentrations.
    Answer: D

Pages: 61
46. The rate of facilitated diffusion is determined primarily by the number of transporters available and the:
    A) steepness of the pressure gradient across the membrane
    B) steepness of the electrical gradient across the membrane
    C) steepness of the concentration gradient across the membrane
    D) pH of the membrane compared with the substance crossing the membrane
    E) amount of ATP available
    Answer: C

Pages: 62
47. Cells maintain the proper gradient to favor facilitated diffusion of glucose into cells by:
    A) breaking down the transporters when the internal concentration of glucose rises too high
    B) inserting insulin into the membrane channel until internal glucose concentrations fall to the proper levels
    C) attaching a phosphate group onto the glucose, thus converting it to a different compound
    D) making more ATP to help the transporters
    E) reversing the pressure gradient to close the membrane channels until glucose levels fall
    Answer: C

Pages: 62
48. During primary active transport, ATP is:
    A) not used at all
    B) split so that energy can be used to change the shape of the transport proteins in the cell membrane
    C) used indirectly to set up ion gradients, which in turn fuel the transport
    D) split so that the energy can be used to convert cations into anions
    E) attached to molecules crossing the membrane to increase their lipid solubility
    Answer: B

Pages: 62
49. When the sodium pump is used to drive the secondary active transport of a substance in the same direction that the sodium is moving, the process is called:
    A) antiport (countertransport)
    B) symport (cotransport)
    C) endocytosis
    D) exocytosis
    E) facilitated diffusion
    Answer: B

Pages: 64
50. Pseudopods are involved in the process of:
    A) phagocytosis
    B) receptor-mediated endocytosis
    C) pinocytosis
    D) secondary active transport
    E) both A and C are correct
    Answer: A

Pages: 65
51. In receptor-mediated endocytosis, the molecule that binds to the receptor is called the:
    A) phagosome
    B) symport
    C) endosome
    D) ligand
    E) antiport
    Answer: D

Pages: 71
52. Fatty acids, phospholipids and steroids are synthesized primarily on:
    A) ribosomes
    B) rough ER
    C) smooth ER
    D) the Golgi complex
    E) mitochondria
    Answer: C

Pages: 73
53. The membranes of lysosomes contain active transport pumps to pump hydrogen ions into the lysosomes because:
    A) hydrogen ions need to be destroyed
    B) the hydrogen ions will raise the pH inside the lysosome so that lysosomal enzymes can work more efficiently
    C) the hydrogen ions will lower the pH inside the lysosome so that the lysosomal enzymes can work more efficiently
    D) they need to pump sodium ions out by secondary active transport
    E) the lysosomal enzymes need to be alternately reduced and oxidized to function
    Answer: C

Pages: 73
54. The function of the enzyme **catalase**, found in peroxisomes, is to:
    A) help proteins enter the Golgi complex
    B) provide a framework upon which DNA can coil
    C) form part of the cytoskeleton of the cell
    D) add a third phosphate onto ADP
    E) use H_2O_2 to oxidize toxic substances
    Answer: E

Pages: 73
55. An organelle that contains DNA and can replicate is the:
   A) ribosome
   B) lysosome
   C) endoplasmic reticulum
   D) Golgi complex
   E) mitochondrion
   Answer: E

Pages: 80
56. Replication of DNA is catalyzed by the enzyme:
   A) DNA polymerase
   B) RNA polymerase
   C) catalase
   D) superoxide dismutase
   E) RNA dehydrogenase
   Answer: A

Pages: 80
57. Cells are engaged in growth, metabolism, and production of substances required for division during:
   A) the G_1-phase of interphase
   B) the S-phase of interphase
   C) telophase of mitosis
   D) prophase of mitosis
   E) cytokinesis
   Answer: A

Pages: 80
58. DNA is replicated during:
   A) the G_1-phase of interphase
   B) the S-phase of interphase
   C) anaphase of mitosis
   D) prophase of mitosis
   E) cytokinesis
   Answer: B

Pages: 81
59. The longest of the phases of mitosis is:
   A) telophase
   B) metaphase
   C) anaphase
   D) prophase
   E) all phases are the same length
   Answer: D

Pages: 80
60. Somatic cells contain:
    A) 22 autosomes and a sex chromosome
    B) 22 pairs of autosomes and a pair of sex chromosomes
    C) 22 pairs of sex chromosomes and a pair of autosomes
    D) 22 sex chromosomes and one autosome
    E) 22 autosomes and a pair of sex chromosomes
    Answer: B

Pages: 84
61. The process of meiosis results in:
    A) gametes
    B) haploid cells
    C) sperm cells or oocytes
    D) cells with 23 chromosomes
    E) all of these
    Answer: E

Pages: 84
62. The process of crossing-over occurs during:
    A) cytokinesis
    B) anaphase of mitosis
    C) prophase I of meiosis
    D) prophase II of meiosis
    E) transcription
    Answer: C

Pages: 88
63. Tumor angiogenesis factors promote:
    A) metastasis
    B) mutation of cellular DNA
    C) activation of oncogenes
    D) growth of new blood vessels in tumors
    E) loss of contact inhibition
    Answer: D

Pages: 88
64. An increase in the number of cells due to an increase in the frequency of cell division is
    called:
    A) atrophy
    B) dysplasia
    C) hyperplasia
    D) hypertrophy
    E) necrosis
    Answer: C

65. Death of a group of cells is called:
    A) atrophy
    B) dysplasia
    C) hyperplasia
    D) hypertrophy
    E) necrosis
    Answer: E

True-False

Write T if the statement is true and F if the statement is false.

Pages: 61
1. The rupture of red blood cells due to their placement into a hypotonic solution is known as **crenation**.
    Answer: False

Pages: 77
2. Transcription is the process by which the genetic information found in DNA is copied onto messenger RNA.
    Answer: True

Pages: 77
3. The DNA strand which serves as a template for RNA synthesis is referred to as the antisense strand.
    Answer: False

Pages: 77
4. Each codon in the messenger RNA molecule specifies either one amino acid or a procedural direction (e.g., stop, skip).
    Answer: True

Pages: 89
5. A permanent structural change in a gene is known as a mutation.
    Answer: True

Pages: 73
6. The chemical reactions of cellular respiration occur on ribosomes.
    Answer: False

Pages: 80
7. Haploid cells result from the process of mitosis.
    Answer: False

Pages: 78
8. The function of tRNA is to bring specific amino acids to the growing peptide chain during translation.
    Answer: True

Pages: 68
9. Cells that do not have a nucleus cannot reproduce.
Answer: True

Pages: 80
10. DNA is replicated during prophase of mitosis.
Answer: False

Pages: 80
11. The mitotic apparatus appears during prophase of mitosis.
Answer: True

Pages: 61
12. Splitting of ATP molecules is required for facilitated diffusion.
Answer: False

Pages: 59
13. Osmosis is the movement of water through a selectively permeable membrane from an area of higher water concentration to an area of lower water concentration.
Answer: True

Pages: 71
14. Cis, medial, and trans cisterns are part of rough endoplasmic reticulum.
Answer: False

Pages: 80
15. Uninucleated somatic cells contain the diploid (2n) chromosome number.
Answer: True

Short Answer

Write the word or phrase that best completes each statement or answers the question.

Pages: 61
1. If Solution A has more solutes and less water than Solution B, then Solution A is considered to be _____ to Solution B.
Answer: hypertonic

Pages: 80
2. A group of nucleotides on a DNA molecule whose purpose is to serve as the "directions" for manufacturing a specific protein is a _____.
Answer: gene

Pages: 80
3. Distribution of two sets of chromosomes into two separate and equal nuclei is known as

_____.

Answer: mitosis

Pages: 83
4. Cytokinesis begins with formation of a _____.
   Answer: cleavage furrow

Pages: 83
5. The union and fusion of gametes is called _____.
   Answer: fertilization

Pages: 84
6. In a diploid cell, the two chromosomes that belong to a pair are called _____ chromosomes.
   Answer: homologous

Pages: 83
7. Chromosome number does not double with each generation of cell division because of a special nuclear divison called _____.
   Answer: meiosis

Pages: 84
8. The process of formation of haploid sperm cells in the male testes is known as _____.
   Answer: spermatogenesis

Pages: 84
9. The process of formation of haploid ova in the female ovaries is known as _____.
   Answer: oogenesis

Pages: 84
10. The homologous pairing of chromosomes in prophase I of meiosis is known as _____.
    Answer: synapsis

Pages: 86
11. The haploid cells in the female that do not function as gametes are known as _____.
    Answer: polar bodies

Pages: 55
12. The fluid mosaic model describes the structure of the _____.
    Answer: cell (plasma) membrane

Pages: 55
13. Temporary structures in the cytoplasm that contain secretions and storage products of the cell are known as _____.
    Answer: inclusions

Pages: 55
14. The part of a phospholipid molecule that lines up facing the intracellular and extracellular fluids is the _____.
    Answer: polar (hydrophilic) part (head)

Pages: 56
15. As a result of the electrochemical gradient across the cell membrane, a voltage called the _____ exists.
    Answer: membrane potential

16. The property of a cell membrane that permits passage of certain substances and restricts passage of others is known as _____.
Answer: selective permeability

17. If two solutions have different concentrations and they are separated by a membrane that is permeable to the solute, then there will be net movement of solute molecules until _____ is reached.
Answer: equilibrium

18. The net movement of water across a selectively membrane by passive means is known as _____.
Answer: osmosis

19. There is no net movement of water molecules across a membrane separating solutions that are _____ to each other.
Answer: isotonic

20. Red blood cells with in intracellular concentration of 0.9% NaCl will undergo _____ when they are placed in a solution of 5% NaCl.
Answer: crenation

21. A type of passive transport across a cell membrane that requires special transporters (carriers) is _____.
Answer: facilitated diffusion

22. Droplets of extracellular fluid flow into vesicles during the process of _____.
Answer: pinocytosis

23. Export of substance from the cell in which vesicles fuse with the plasma membrane and release thair contents into the extracellular fluid is known as _____.
Answer: exocytosis

24. Large organic compounds, such as proteins and glycogen, that remain suspended in the cytoplasm rather than dissolved, are known as _____.
Answer: colloids

25. The proteins around which DNA wraps in a chromatin fiber are called _____.
Answer: histones

26. Organelles that are membrane-enclosed vesicles filled with digestive enzymes are _____ .
Answer: lysosomes

27. Microfilaments, microtubules, and intermediate filaments comprise the _____ .
Answer: cytoskeleton

28. Regions within DNA strands that do not code for synthesis of part of a protein are called

_____ .
Answer: introns

29. Each set of three consecutive nucleotide bases on messenger RNA that specifies one amino acid is called a _____ .
Answer: codon

30. Division of a parent cell's cytoplasm and organelles is called _____ .
Answer: cytokinesis

Matching

Choose the item from Column 2 that best matches each item in Column 1.

1. contains an anticodon   tRNA

2. the end result of   protein
   translation

3. the part of a ribosome made   rRNA
   in the nucleolus

4. template for RNA synthesis   sense strand of DNA

5. parts of sense strand that   introns
   do not code for synthesis
   of part of a protein

6. parts of sense strand that   exons
   do code for parts of
   proteins

7. contains codons   mRNA

8. part of DNA that connects   linker DNA
   nucleosomes

9. double-stranded DNA wrapped   nucleosome
    around histones

10. proteins that help organize   histones
    the coiling and folding
    ability of DNA

Essay

Write your answer in the space provided or on a separate sheet of paper.

Pages: 54
1. Name and briefly describe the four principal parts of a cell.
    Answer: 1. Plasma (cell) membrane - separates internal components from external environment
             2. Cytosol - the semifluid intracellular fluid containing many dissolved and suspended substances
             3. Organelles - highly organized structures with characteristic shapes and highly specialized functions within the cell
             4. Inclusions - temporary storage and secretory structures

Pages: 55
2. Discuss the chemical composition of the cell membrane.
    Answer: The cell membrane is a 50:50 mix of proteins and lipids by weight, but there are many more lipid molecules than protein molecules (due to difference in molecular weight). 75% of the lipids are phospholipids, 5% are glycolipids, and 20% is cholesterol. Membrane proteins are either integral or peripheral proteins, and may be in the form of glycoproteins.

Pages: 55-56
3. Discuss the Fluid Mosaic Model of cell membrane structure.
    Answer: Proteins "float" among phospholipids; phospholipid bilayer, with hydrophobic ends facing each other, and hydrophilic ends facing either ECF or ICF; phospholipids can move sideways within layer; glycolipids (cellular identity markers) face ECF; cholesterol molecules (for stability) among the phospholipids of both layers of the bilayer; integral proteins (channels, transporters, etc.) extend across phospholipid bilayer; peripheral proteins (enzymes, cytoskeleton anchors, etc.) loosely attached to either the inner or outer surface of the membrane

Pages: 81-83
4. Name and describe the events of the four phases of mitosis in sequence.
    Answer: 1. prophase - chromatin fibers shorten and coil into chromosomes; nucleoli and nuclear envelope disappear; centrosomes with centrioles move to opposite poles of cell; mitotic spindle appears
             2. metaphase - centromeres of chromatid pairs line up on metaphase plate of cell
             3. anaphase - centromeres divide; identical sets of chromosomes move to opposite poles of cell
             4. telophase - nuclear envelope reappears to enclose chromosomes; chromosomes revert to chromatin; nucleoli reappear; mitotic spindle disppears

Pages: 81-86
5. Compare and contrast the events and outcomes of mitosis and meiosis.

   Answer: Mitosis results in two daughter cells with the same number and kind of chromosomes as the parent cell (diploid cells, 2n). Meiosis results in four daughter cells with half the number of chromosomes as the parent cell (haploid cells, 1n). Gametes are the result of meiosis. Meiosis occurs in two successive nuclear divisions - a reduction division and an equatorial division. [More detail can be required as desired.]

Pages: 77-78
6. Name the three types of RNA and describe the role of each in protein synthesis.

   Answer: 1. messenger RNA (mRNA) - produced from sense strand of DNA via transcription; carries code for making a particular protein from the nucleus to the ribosome

   2. ribosomal RNA (rRNA) - makes up the ribosome; moves along the mRNA strand to "read" directions for making protein

   3. transfer RNA (tRNA) - transports specific amino acids from the cytoplasm to the growing peptide chain; places amino acids in proper sequence by matching its anticodon with an appropriate codon on the mRNA strand

Pages: 58-59
7. Describe three means by which the rate of diffusion across a semipermeable membrane might be increased.

   Answer: Diffusion might be increased by increasing the temperature, increasing the surface area of the membrane, increasing the steepness of the concentration gradient, or decreasing the distance travelled (thickness of the membrane). [note: Other answers may be acceptable.]

Pages: 58-59
8. Compare and contrast simple diffusion and osmosis.

   Answer: Both are passive types of movements. Simple diffusion involves the net movement of solute molecules from an area of higher solute concentration to an area of lower solute concentration. Osmosis involves the movement of water (solvent) across a semipermeable membrane from an area of higher water concentration to an area of lower water concentration.

Pages: 64-65
9. Compare and contrast phagocytosis and receptor-mediated endocytosis.

   Answer: Both are active means of bringing materials inside the cell. Phagocytosis is performed only by special cells, which extend pseudopods toward an object, and engulf that object in a vesicle to bring it inside. The object is usually then destroyed by the phagocyte. In receptor-mediated endocytosis, a ligand attaches to the cell membrane via specific receptors. A vesicle is formed around the ligand-receptor complex and the complex is brought inside the cell. Ultimately, the ligand is usually used for some cellular process, and the vesicle and receptor are recycled.

Pages: 59-60

10. Consider this situation:

    1) A cell contains 3 Na^+, 7 K^+, and 10 H_2O molecules.

    2) The extracellular fluid contains 7 Na^+, 3 K^+, and 10 H_2O molecules.

    3) All ions create equal osmotic pressure.

    4) The cell membrane is impermeable to the diffusion of both Na^+ and K^+, but is completely permeable to H_2O.

    If 3 Na^+ move from the ICF out to the ECF, then:

    A) Is the ECF now isotonic, hypotonic, or hypertonic to the ICF? Why?

    B) How must the Na^+ have gotten outside? Explain your answer.

    C) If osmosis is also going to occur, in which direction will the net movement of water be? Why?

    Answer: A) hypertonic - more solute outside (therefore, greater osmotic pressure)

    B) active transport - sodium moving up gradient (you may choose to accept other answers)

    C) net movement of water to outside of cell - water must follow the ions to maintain osmotic balance (otherwise, there will be a "water gradient")

Chapter 4   The Tissue Level of Organization

Multiple-Choice

Choose the one alternative that best completes the statement or answers the question.

Pages: 96
1. Which of the following tissues has the **least** amount of matrix (intercellular substance)?
   A) stratified squamous epithelium
   B) areolar connective tissue
   C) osseous tissue
   D) dense irregular (fibrous) connective tissue
   E) blood
   Answer: A

Pages: 98
2. Microvilli and goblet cells are typical modifications of:
   A) skeletal muscle tissue
   B) simple columnar epithelium
   C) osseous tissue
   D) hyaline cartilage
   E) nervous tissue
   Answer: B

Pages: 96
3. The tissue found lining body cavities that do not open to the exterior and that secretes serous fluid is called:
   A) endothelium
   B) mesothelium
   C) transitional epithelium
   D) adipose tissue
   E) hyaline cartilage
   Answer: B

Pages: 104
4. The mast cells that are found in areolar connective tissue produce:
   A) collagen
   B) red blood cells
   C) antibodies
   D) keratin
   E) histamine
   Answer: E

5. Covering and lining are basic functions of:
   A) epithelial tissue (only)
   B) connective tissue
   C) muscle tissue (only)
   D) nervous tissue
   E) both epithelial and muscular tissues
   Answer: A

6. Fibroblasts are the typical cells of:
   A) simple squamous epithelium
   B) dense connective tissue
   C) cardiac muscle tissue
   D) skeletal muscle tissue
   E) nervous tissue
   Answer: B

7. Mucous membranes are:
   A) those that line body cavities that do not open to the exterior
   B) those that line body cavities opening to the exterior
   C) those that secrete serous fluid
   D) made mostly of dense connective tissue
   E) both B and C are correct
   Answer: B

8. The epithelial layer of a mucous membrane could be:
   A) simple columnar epithelium only
   B) stratified squamous epithelium only
   C) mesothelium only
   D) endothelium only
   E) either simple columnar or stratified squamous epithelium
   Answer: E

9. **Stroma** is:
   A) the inorganic part of bone matrix
   B) a membrane lining body cavities that do not open to the exterior
   C) the reticular connective tissue of soft organs
   D) the same thing as intercellular substance
   E) cells that form an organ's functioning part
   Answer: C

Pages: 101
10. Cilia are often found on:
    A) pseudostratified columnar epithelium
    B) stratified squamous epithelium
    C) simple squamous epithelium
    D) elastic cartilage
    E) serous membranes
    Answer: A

Pages: 103
11. Which of the following is typical of an endocrine gland?
    A) It releases hormones.
    B) It is made of epithelial tissue.
    C) It releases its products into ducts that empty onto epithelial surfaces.
    D) both A and B are correct
    E) both B and C are correct
    Answer: D

Pages: 94
12. Nervous tissue originates from:
    A) ectoderm only
    B) endoderm only
    C) mesoderm only
    D) ectoderm and endoderm
    E) endoderm and mesoderm
    Answer: A

Pages: 94
13. Connective tissues and muscle tissues originate from:
    A) ectoderm only
    B) endoderm only
    C) mesoderm only
    D) ectoderm and endoderm
    E) endoderm and mesoderm
    Answer: C

Pages: 94
14. Epithelial tissue originates from:
    A) ectoderm only
    B) endoderm only
    C) mesoderm only
    D) ectoderm and endoderm only
    E) ectoderm, endoderm, and mesoderm
    Answer: E

15. Which of the following connective tissue cells secrete antibodies?
    A) adipocytes
    B) fibroblasts
    C) macrophages
    D) mast cells
    E) plasma cells
    Answer: E

16. Hyaline cartilage is found in **ALL** of the following locations **EXCEPT**:
    A) embryonic skeleton
    B) trachea
    C) bronchial tubes
    D) intervertebral discs
    E) nose
    Answer: D

17. The cells which form the functioning part of an organ are known as the:
    A) granulation tissue
    B) stroma
    C) reticular layers
    D) parenchyma
    E) visceral portion
    Answer: D

18. Which of the following types of muscle tissue is considered voluntary?
    A) cardiac
    B) nonstriated
    C) skeletal
    D) smooth
    E) visceral
    Answer: C

19. Tight junctions are most commonly found in tissues in which:
    A) fluid leakage between cells is a particular risk
    B) cells may be damaged by friction or stretching
    C) fluids pass easily between cells
    D) electrical or chemical signals must pass between cells
    E) one layer of cells needs to move easily over another during growth
    Answer: A

Pages: 94
20. Anchoring junctions are most commonly found in tissues in which:
    A) fluid leakage between cells is a particular risk
    B) cells may be damaged by friction or stretching
    C) fluids pass easily between cells
    D) electrical or chemical signals must pass between cells
    E) all the cells are dead
    Answer: B

Pages: 94
21. Gap junctions are seen in tissues in which:
    A) fluid leakage between cells is a particular risk
    B) cells may be damaged by friction or stretching
    C) fluids pass easily between cells
    D) electrical or chemical signals must pass between cells
    E) adjacent cells are many millimeters apart
    Answer: D

Pages: 95
22. **Connexons** are:
    A) the typical cells of connective tissue
    B) fibers that provide nutrient, waste, and gas exchange for epithelial cells
    C) proteins found in muscle fibers
    D) the routes by which the osteocytes in osseous tissue receive nutrients
    E) proteins that form the channels that provide a means of ion exchange between cells at gap
       junctions
    Answer: E

Pages: 95
23. Epithelial tissue is attached to the underlying connective tissue by:
    A) apical surfaces
    B) a basement membrane
    C) connexons
    D) mesechyme
    E) hyaluronidase
    Answer: B

Pages: 96
24. A special type of simple squamous epithelium that lines the heart and blood vessels is:
    A) transitional epithelium
    B) mesothelium
    C) pseudostratified epithelium
    D) mesenchyme
    E) endothelium
    Answer: E

25. The epithelial portion of serous membranes is called:
    A) transitional epithelium
    B) mesothelium
    C) pseudostratified epithelium
    D) mesenchyme
    E) endothelium
    Answer: B

26. Simple squamous epithelium is found in **ALL** of the following locations **EXCEPT**:
    A) air sacs of lungs
    B) lining heart and blood vessels
    C) serous membranes
    D) urinary bladder
    E) glomerular capsule of kidneys
    Answer: D

27. Transitional epithelium is found in the:
    A) skin
    B) urinary bladder
    C) air sacs of lungs
    D) upper respiratory tract
    E) lining of heart and blood vessels
    Answer: B

28. Most of the upper respiratory tract is lined with:
    A) hyaline cartilage
    B) keratinized stratified squamous epithelium
    C) cilated pseudostratified columnar epithelium
    D) transitional epithelium
    E) skeletal muscle
    Answer: C

29. **ALL** of the following are products of exocrine glands **EXCEPT**:
    A) digestive enzymes
    B) mucus
    C) sweat
    D) hormones
    E) milk
    Answer: D

30. Goblet cells produce:
    A) digestive enzymes
    B) mucus
    C) sweat
    D) hormones
    E) milk
    Answer: B

31. The function of microvilli is to:
    A) increase the surface area of epithelial cell membranes
    B) move materials over the surface of an epithelial tissue
    C) perform phagocytosis
    D) produce mucus
    E) cause contraction of muscle cells
    Answer: A

32. One might expect to see microvilli on epithelial tissues whose principal function is:
    A) protection
    B) movement
    C) absorption
    D) mineral storage
    E) transmission of electrical impulses
    Answer: C

33. Stratified squamous epithelium can be made waterproof and friction resistant by intracellular deposits of:
    A) collagen
    B) hyaluronidase
    C) chondroitin sulfate
    D) mucus
    E) keratin
    Answer: E

34. The function of glandular epithelium is:
    A) support
    B) protection
    C) movement
    D) secretion
    E) sensory reception
    Answer: D

Pages: 103
35. A gland in which the secretory product is discharged from cells via exocytosis is called a(n):
    A) holocrine gland
    B) merocrine gland
    C) apocrine gland
    D) transitional gland
    E) serous gland
    Answer: B

Pages: 103
36. A gland in which the secretory product is accumulated in the apical surfaces of cells, which then pinch off, is called a(n):
    A) holocrine gland
    B) merocrine gland
    C) apocrine gland
    D) transitional gland
    E) serous gland
    Answer: C

Pages: 103
37. The most abundant tissue in the body is:
    A) epithelial tissue
    B) connective tissue
    C) muscle tissue
    D) nervous tissue
    E) all tissues are equally abundant when the body is in homeostasis
    Answer: B

Pages: 103
38. The basic tissue type whose function is particularly related to the form of its matrix (solid, semisolid, or liquid) is:
    A) epithelial tissue
    B) connective tissue
    C) muscle tissue
    D) nervous tissue
    E) the structure of the matrix is unrelated to the functioning of tissues
    Answer: B

Pages: 104
39. The phagocytes found in connective tissues are called:
    A) fibroblasts
    B) macrophages
    C) mast cells
    D) plasma cells
    E) adipocytes
    Answer: B

Pages: 105
40. The most abundant protein in the body is:
    A) collagen
    B) elastin
    C) keratin
    D) hyaluronic acid
    E) connexon
    Answer: A

Pages: 106
41. The embryonic connective tissue from which all other connective tissues arise is:
    A) areolar connective tissue
    B) mucous connective tissue
    C) hyaline cartilage
    D) neuroglia
    E) mesenchyme
    Answer: E

Pages: 105
42. The diffusion of injected drugs can be enhanced by the action of hyaluronidase because it:
    A) increases the rate of ATP synthesis
    B) raises the local temperature by causing inflammation, so molecules move faster
    C) lowers the viscosity of the matrix of areolar connective tissue
    D) forms new, tighter junctions between the cells of connective tissues
    E) increases the rate of matrix secretion by fibroblasts
    Answer: C

Pages: 108
43. A connective tissue that prevents heat loss through the skin by acting as an insulator is:
    A) adipose tissue
    B) osseous tissue
    C) smooth muscle tissue
    D) stratified squamous epithelium
    E) dense regular connective tissue
    Answer: A

Pages: 108
44. The primary function of brown fat is:
    A) energy storage
    B) absorption
    C) protection from friction damage
    D) heat generation in the newborn
    E) protection from bacterial invasion
    Answer: D

Pages: 109
45. A tissue designed to withstand pulling in one direction is:
    A) areolar connective tissue
    B) adipose tissue
    C) dense regular connective tissue
    D) elastic cartilage
    E) stratified squamous epithelium
    Answer: C

Pages: 114
46. **Lacunae** are:
    A) the matrix-producing cells of cartilage
    B) spaces in the matrix of cartilage and bone in which cells are located
    C) fibers in cartilage that provide added strength to the tissue
    D) proteins that make up the ground substance of most connective tissues
    E) the supporting cells in nervous tissue
    Answer: B

Pages: 114
47. The perichondrium is:
    A) a space in the matrix of cartilage where chondrocytes are found
    B) a cell found in mature cartilage
    C) a type of protein found in the matrix of cartilage
    D) dense irregular connective tissue covering the surface of cartilage
    E) an embryonic type of cartilage
    Answer: D

Pages: 114
48. The osteon, or Haversian system, is the basic unit of:
    A) hyaline cartilage
    B) blood
    C) skeletal muscle
    D) compact bone
    E) cancellous bone
    Answer: D

Pages: 114
49. The matrix of osseous tissue is made hard by:
    A) hyaluronic acid
    B) chondroitin sulfate
    C) lacunae
    D) contractile proteins
    E) calcium salts
    Answer: E

Pages: 115
50. The **lamina propria** is:
    A) a space in the matrix of cartilage or bone in which a cell is located
    B) the part of a serous membrane that secretes serous fluid
    C) the dense irregular connective tissue covering cartilage or bone
    D) the connective tissue layer of a mucous membrane
    E) a type of cellular junction seen in muscle tissue
    Answer: D

Pages: 115
51. Membranes that have parietal and visceral portions are:
    A) mucous membranes
    B) serous membranes
    C) cutaneous membranes
    D) synovial membranes
    E) all membranes have parietal and visceral portions
    Answer: B

Pages: 115
52. The space between the parietal and visceral layers of a membrane, such as the pericardium, is normally filled with:
    A) air
    B) blood
    C) serous fluid
    D) synovial fluid
    E) adipose tissue
    Answer: C

Pages: 115
53. The type of membrane found lining freely movable joints is the:
    A) mucous membrane
    B) serous membrane
    C) cutaneous membrane
    D) synovial membrane
    E) stroma
    Answer: D

Pages: 116
54. Both desmosomes and gap junctions are found in the intercalated discs connecting the cells of:
    A) adipose tissue
    B) osseous tissue
    C) skeletal muscle tissue
    D) cardiac muscle tissue
    E) simple squamous epithelium
    Answer: D

55. The basic tissue type with the fewest different basic cell types is:
    A) epithelial tissue
    B) connective tissue
    C) muscle tissue
    D) nervous tissue
    E) all tissues contain the same numbers and types of cells
    Answer: D

56. Dendrites and axons are part of:
    A) chondrocytes
    B) osteocytes
    C) neurons
    D) neuroglia
    E) cardiac muscle cells
    Answer: C

57. If fibrosis occurs during tissue repair, then:
    A) a perfect reconstruction of the injured tissue occurs
    B) rapid replication of parenchymal cells has occurred
    C) the function of the repaired tissue is impaired
    D) only connective tissue was involved in the injury
    E) both B and C are correct
    Answer: C

58. The scab over a wound to the skin is formed primarily by:
    A) keratin in the epidermis
    B) calcium salts
    C) serous fluid as it evaporates
    D) fibrin in blood clots
    E) collagen in torn connective tissue
    Answer: D

59. The granulation tissue of a wound is:
    A) actively growing connective tissue
    B) a scar
    C) a result of bacterial invasion
    D) made from the fibrin in blood clots
    E) dried blood plasma
    Answer: A

Pages: 109
60. Tendons and ligaments are made of:
    A) keratinized stratified squamous epithelium
    B) areolar connective tissue
    C) dense regular connective tissue
    D) hyaline cartilage
    E) osseous tissue
    Answer: C

Pages: 110
61. Elastic connective tissue is found in **ALL** of the following **EXCEPT**:
    A) the walls of large arteries
    B) lung tissue
    C) the true vocal cords
    D) bronchial tubes
    E) tendons
    Answer: E

Pages: 116
62. A long, cylindrical cell that contains many peripheral nuclei is most likely part of:
    A) nervous tissue
    B) skeletal muscle tissue
    C) cardiac muscle tissue
    D) smooth muscle tissue
    E) dense regular connective tissue
    Answer: B

Pages: 114
63. Interstitial growth of cartilage involves:
    A) only cartilage that has been damaged
    B) cartilage in adults
    C) division of existing chondrocytes, and production of matrix within existing cartilage
    D) differentiation of fibroblasts in the perichondrium into chondrocytes
    E) both B and D are correct
    Answer: C

Pages: 111
64. Intervertebral discs and the menisci of the knee are made of:
    A) fibrocartilage
    B) elastic cartilage
    C) hyaline cartilage
    D) osseous tissue
    E) adipose tissue
    Answer: A

65. The epiglottis and the external ear are made of:
   A) fibrocartilage
   B) elastic cartilage
   C) hyaline cartilage
   D) osseous tissue
   E) nervous tissue
   Answer: B

True-False

Write T if the statement is true and F if the statement is false.

Pages: 95
1. Epithelial tissues do not have a direct nerve supply.
   Answer: False

Pages: 96
2. Simple squamous epithelium that forms the epithelial layer of serous membranes is called **endothelium**.
   Answer: False

Pages: 104
3. Connective tissues do not occur on free surfaces, such as the external surface of the body.
   Answer: True

Pages: 105
4. The body's most abundant protein is collagen, representing about 25% of the body's total protein.
   Answer: True

Pages: 116
5. Cardiac muscle tissue is found only in the wall of the heart.
   Answer: True

Pages: 103
6. The cells of merocrine glands must die in order for their products to be secreted.
   Answer: False

Pages: 113
7. Most adipose tissue in adults is white fat.
   Answer: True

Pages: 114
8. Interstitial growth of cartilage occurs primarily during adulthood.
   Answer: False

Pages: 118
9. Neuroglia do not generate or conduct nerve impulses.
   Answer: True

Pages: 118
10. The cardinal factor in the restoration of normal function during tissue repair is the capacity of the parenchymal cells to replicate quickly.
Answer: True

Pages: 115
11. Smooth muscle tissue is both striated and involuntary.
Answer: False

Pages: 115
12. The pleura and the pericardium each have both parietal and visceral layers.
Answer: True

Pages: 114
13. The resilience (ability to assume original shape after deformation) of cartilage is provided by the perichondrium.
Answer: False

Pages: 113
14. Mesenchyme is the embryonic connective tissue from which all other connective tissues arise.
Answer: True

Pages: 95
15. Epithelial tissues contain a large amount of intercellular substance.
Answer: False

Short Answer

Write the word or phrase that best completes each statement or answers the question.

Pages: 102
1. The primary function of stratified squamous epithelium is _____.
Answer: protection

Pages: 96
2. Simple squamous epithelium that lines the heart, blood vessels, lymphatic vessels, and forms the walls of capillaries is known as _____.
Answer: endothelium

Pages: 103
3. _____ glands secrete their products into ducts.
Answer: Exocrine

Pages: 114
4. A connective tissue that is avascular is _____.
Answer: cartilage

Pages: 103
5. The secretions from endocrine glands are called _____.
Answer: hormones

Pages: 106
6. Mucous connective tissue is found primarily in the _____.
   Answer: (umbilical cord of the) fetus

Pages: 114
7. The cells of mature cartilage are called _____.
   Answer: chondrocytes

Pages: 114
8. The most abundant type of cartilage in the human body is _____.
   Answer: hyaline cartilage

Pages: 114
9. _____ growth of cartilage continues throughout life.
   Answer: Appositional

Pages: 115
10. The connective tissue layer of a mucous membrane is called the _____.
    Answer: lamina propria

Pages: 115
11. The type of membrane that lines a body cavity that does not open to the exterior is a

    _____.
    Answer: serous membrane

Pages: 115
12. The serous membrane lining the abdominal cavity and covering the abdominal organs is called the

    _____.
    Answer: peritoneum

Pages: 118
13. _____ conduct nerve impulses away from the cell bodies of neurons.
    Answer: Axons

Pages: 104
14. The type of cell in areolar connective tissue that produces histamine and heparin is the

    _____.
    Answer: mast cell

Pages: 114
15. The type of tissue whose matrix normally contains large amounts of calcium salts is _____.
    Answer: osseous (bone) tissue

Pages: 102
16. The surface area of the apical surfaces of epithelial cell membranes is increased by the
    presence of _____.
    Answer: microvilli

Pages: 102
17. Goblet cells produce _____.
    Answer: mucus

Pages: 110
18. The "gristle" covering the ends of bones at joints is _____ cartilage.
Answer: hyaline

Pages: 111
19. Intervertebral discs and the menisci of the knees are made of _____.
Answer: fibrocartilage

Pages: 94
20. Electrical or chemical signals can pass from cell to cell via connections known as _____.
Answer: gap junctions

Pages: 103
21. The type of epithelium in which the cells may or may not reach the surface, and whose nuclei may lie at different levels giving the appearance of multiple layers is _____.
Answer: pseudostratified epithelium

Pages: 102
22. Keratin may be present in _____ epithelium.
Answer: stratified squamous

Pages: 102
23. Glands in which secretory cells die as their product is discharged are _____ glands.
Answer: holocrine

Pages: 104
24. The immature cells of each major type of connective tissue have names that end in the suffix

_____.
Answer: -blast

Pages: 105
25. The fibers of the connective tissue matrix are embedded in the amorphous _____.
Answer: ground substance

Pages: 118
26. The reticular connective tissue that forms the supporting framework for many soft organs is known as the _____.
Answer: stroma

Pages: 114
27. The basic unit of compact bone is the _____.
Answer: osteon (Haversian system)

Pages: 116
28. A type of tissue that is described as being striated and voluntary is _____.
Answer: skeletal muscle tissue

Pages: 118
29. Propulsion of food through the gastrointestinal tract and contraction of the urinary bladder are functions of _____.
Answer: smooth muscle tissue

Pages: 118
30. Cells that are the structural and functional units of the nervous system are called _____.
Answer: neurons

Matching

Choose the item from Column 2 that best matches each item in Column 1.

1. the embryonic tissue from which all other connective tissues are derived

    mesenchyme

    mucous connective tissue

2. connective tissue specialized for fat storage

    adipose tissue

3. very tough connective tissue containing many parallel collagen fibers

    dense regular connective tissue

4. an especially smooth type of simple squamous epithelium that lines the heart and blood vessels

    endothelium

5. multiple layers of flat cells designed for protection

    stratified squamous epithelium

6. cells may have microvilli and/or cilia

    simple columnar epithelium

7. found in serous membranes and produces serous fluid

    mesothelium

8. most widely distributed tissue in body; located under skin and around all organs and vessels

    areolar connective tissue

9. found only in the wall of the heart; contains intercalated discs

    cardiac muscle tissue

10. covers ends of bones at joints; forms much of embryonic skeleton

    hyaline cartilage

    fibrocartilage

Essay

Write your answer in the space provided or on a separate sheet of paper.

Pages: 94
1. Name the four principal types of human tissues, and briefly describe the function of each.
   Answer: 1. Epithelial tissue - covers body surfaces; lines hollow organs, body cavities, and ducts; forms glands
   2. Connective tissue - protects and supports the body and its organs; binds organs together; stores energy reserves as fat; provides immunity
   3. Muscle tissue - responsible for movement and generation of force
   4. Nervous tissue - initiates amd transmits action potentials (nerve impulses) that help coordinate body activities

Pages: 96
2. Humans are covered with epithelial tissue. What are the advantages of having this type of tissue as a body covering rather than a type of connective tissue, such as cartilage?
   Answer: Epithelial cells are closely packed for a physical barrier, and are attached directly or indirectly to an anchor (basement membrane). The cells of epithelial tissues have a relatively high rate of mitosis. In connective tissues, the cells are usually separated by large amounts of matrix, which is most often semisolid or liquid, and which allows easy diffusion of substances. The cells of connective tissues divide less frequently than those of epithelial tissues.

Pages: 113
3. What is brown fat, and what is its purpose?
   Answer: Brown fat is adipose tissue whose cells have numerous mitochondria and colored cytochrome pigments. This type of fat is seen particularly in the fetus and infant, for whom brown fat helps maintain proper body temperature via its high rate of thermogenesis and rich blood supply.

Pages: 103
4. Compare and contrast exocrine vs. endocrine glands with regard to structure and function.
   Answer: Both types of glands are made up of glandular epithelium. Exocrine glands secrete their products into ducts that empty onto the surface of the tissue. The products of endocrine glands diffuse directly into the blood after being secreted into the extracellular fluid. The products of exocrine glands include sweat, oil, ear wax, breast milk, and digestive secretions. The products of endocrine glands are hormones, which regulate various body activities.

Pages: 113
5. List and briefly describe the functions of adipose tissue.
   Answer: 1. Insulation - reduces heat loss through skin
   2. Serves as important energy reserve - more energy stored per gram of fat than per gram of carbohydrate or protein
   3. Support - holds kidneys in place
   4. Protection - acts as shock absorber around kidneys, eyes, joints, heart
   5. Thermogenesis - brown fat generates heat to help maintain body temperature of newborns

Pages: 104-114
6. Give an example using a connective tissue that illustrates the concept of **differentiation**.
   Answer: There are many correct answers here. All should begin with mesenchyme cells. An
            example would be to continue on to fibroblasts, then to chondroblasts, then to
            chondrocytes.

Pages: 114
7. Compare and contrast interstitial vs. appositional growth of cartilage.
   Answer: During childhood and adolescence, existing chondrocytes reproduce and continuously
            produce new matrix, causing the cartilage to expand from within in interstitial growth
            of cartilage. Appositional growth of cartilage begins later than interstitial growth,
            but continues throughout life. In this type of growth, fibroblasts in the
            perichondrium differentiate into chondroblasts, which surround themselves with
            secreted matrix, and become chondrocytes. The matrix accumulates on the surface of
            existing cartilage only.

Pages: 114
8. Why is the healing of an injury to cartilage so slow?
   Answer: Cartilage is a metabolically inactive tissue. Because cartilage is avascular, all
            substances needed for the repair process must diffuse from blood vessels through the
            perichondrium.

Pages: 105
9. How does the arrangement of collagen fibers affect the nature of the support and strength a
   connective tissue provides? Give examples to support your answer.
   Answer: Collagen fibers that are arranged in parallel bundles, such as is found in dense
            regular connective tissue, provide strength and support in one particular direction.
            When collagen fibers are arranged randomly, such as in areolar connective tissue,
            greater flexibility is allowed, and strength and support are distributed more equally
            in all directions.

Pages: 105
10. Many bacteria produce the enzyme **hyaluronidase**. What advantage does this provide to bacteria
    that invade the human body?
    Answer: The enzyme breaks down hyaluronic acid, an important component of the interstitial
             substance of many tissues. Any bacteria that produce this enzyme can more easily make
             their way between human cells, and spread their infection more quickly.

Chapter 5   The Integumentary System

Multiple-Choice

Choose the one alternative that best completes the statement or answers the question.

Pages: 124
1. The subcutaneous layer of skin is made mostly of:
   A) areolar connective tissue
   B) smooth muscle
   C) stratified squamous epithelium
   D) osseous tissue
   E) simple columnar epithelium
   Answer: A

Pages: 127
2. Nourishment to cells in the epidermis is provided by:
   A) blood vessels running through the stratum basale
   B) keratinocytes
   C) blood vessels in dermal papillae
   D) bacteria that live in sebaceous glands
   E) both A and C are correct
   Answer: C

Pages: 126
3. The reproducing cells of the epidermis are found in the:
   A) stratum basale
   B) stratum spinosum
   C) stratum corneum
   D) stratum lucidum
   E) all of these layers contain reproducing cells
   Answer: A

Pages: 126
4. The outermost layer of the epidermis is the:
   A) stratum lucidum
   B) stratum basale
   C) reticular layer
   D) stratum corneum
   E) superficial fascia
   Answer: D

Pages: 126
5. The function of keratin is to:
   A) make bone hard
   B) make skin tough and waterproof
   C) protect skin from ultraviolet light
   D) provide added pigment to the skin of Asian races
   E) provide nourishment to epidermal cells
   Answer: B

Pages: 128
6. The function of melanin is to:
    A) provide nutrients to dying epidermal cells
    B) make skin tough and waterproof
    C) protect skin from ultraviolet light
    D) provide flexibility to skin
    E) connect the epidermis to the dermis
    Answer: C

Pages: 126
7. The epidermis is made up of:
    A) dense irregular (fibrous) connective tissue
    B) stratified squamous epithelium
    C) areolar connective tissue
    D) smooth muscle
    E) all of these are found in the epidermis
    Answer: B

Pages: 135
8. In the feedback loop involving the skin that helps control body temperature, the effectors are the:
    A) receptors in the dermis
    B) melanocytes
    C) sebaceous glands
    D) sudoriferous glands
    E) keratinocytes
    Answer: D

Pages: 131
9. Sudoriferous glands produce:
    A) hormones
    B) keratin
    C) an oily substance for lubrication of skin
    D) sweat
    E) carotene
    Answer: D

Pages: 131
10. "Goosebumps" occur due to:
    A) over-stimulation of secretion from sudoriferous glands
    B) over-stimulation of secretion from sebaceous glands
    C) separation of the epidermis from the dermis
    D) vasodilation of blood vessels in the skin
    E) the action of arrector pili muscles as they raise hairs to an upright position
    Answer: E

Pages: 138
11. Decubitus ulcers result from:
    A) the varicella-zoster virus
    B) type 1 herpes simplex virus
    C) emotional stress
    D) constant deficiency of blood due to pressure
    E) fungi infecting the epidermis
    Answer: D

Pages: 128
12. Melanin is synthesized by melanocytes from the amino acid:
    A) leucine
    B) histidine
    C) methionine
    D) tyrosine
    E) valine
    Answer: D

Pages: 126
13. Which layer of the epidermis is normally present only in the palms and soles?
    A) stratum basale
    B) stratum corneum
    C) stratum granulosum
    D) stratum lucidum
    E) stratum spinosum
    Answer: D

Pages: 128
14. The substance in skin that protects you from ultraviolet light is:
    A) melanin
    B) sebum
    C) keratin
    D) cerumen
    E) sweat
    Answer: A

Pages: 126
15. The substance in skin that protects you from bacterial invasion and mechanical injury is:
    A) melanin
    B) sebum
    C) keratin
    D) cerumen
    E) sweat
    Answer: C

Pages: 126
16. The stratum corneum is:
    A) the innermost layer of the epidermis
    B) highly vascular
    C) made up of dead cells
    D) seen only on the palms and soles
    E) the layer in which keratin begins to form
    Answer: C

Pages: 129
17. Hair and nails are modifications of the:
    A) epidermis
    B) dermis
    C) hypodermis
    D) melanocytes
    E) sudoriferous glands
    Answer: A

Pages: 126
18. Which of the following statements are **TRUE** regarding the epidermis?
    A) It is keratinized.
    B) Blood vessels travel from the dermis to the outer layers through special channels.
    C) All of the cells in the epidermis reproduce rapidly.
    D) It is made mostly of areolar connective tissue.
    E) both A and C are correct
    Answer: A

Pages: 126
19. The layer of the epidermis that sits on the basement membrane is the:
    A) stratum spinosum
    B) stratum basale
    C) stratum corneum
    D) stratum lucidum
    E) stratum granulosum
    Answer: B

Pages: 126
20. The function of the Langerhans cells found in the skin is to:
    A) produce melanin
    B) produce keratin
    C) provide nourishment to cells isolated in matrix
    D) provide immunity
    E) produce sebum
    Answer: D

Pages: 138
21. The least common, but most deadly, of the skin cancers is:
    A) basal cell carcinoma
    B) squamous cell carcinoma
    C) multiple myeloma
    D) malignant melanoma
    E) hemangioma
    Answer: D

Pages: 136
22. A typical sunburn is an example of a:
    A) first degree burn
    B) second degree burn
    C) third degree burn
    D) hemangioma
    E) squamous cell carcinoma
    Answer: A

Pages: 126
23. Just beneath the stratum basale of the epidermis is the:
    A) stratum corneum of the epidermis
    B) hypodermis
    C) reticular region of the dermis
    D) papillary region of the dermis
    E) skeletal muscle
    Answer: D

Pages: 135
24. In the negative feedback loop involving the integumentary system that helps maintain
    homeostasis by reducing elevated body temperature, the effectors achieve a reduction in body
    temperature by:
    A) raising body hairs to let more heat escape from the body
    B) producing oil to coat and insulate the skin
    C) sending messages to your brain to generate mental images of cool environments
    D) causing constriction of blood vessels in the epidermis
    E) producing sweat, which evaporates, thus taking heat from the skin surface
    Answer: E

Pages: 124
25. The skin can be considered an endocrine gland because it produces:
    A) sweat
    B) sebum
    C) keratin
    D) melanin
    E) a precursor to calcitriol
    Answer: E

Pages: 126
26. Most epidermal cells are:
    A) melanocytes
    B) keratinocytes
    C) Langerhans cells
    D) Merkel cells
    E) fibroblasts
    Answer: B

Pages: 126
27. Cells in the epidermis thought to function in the sensation of touch are the:
    A) melanocytes
    B) keratinocytes
    C) Langerhans cells
    D) Merkel cells
    E) arrector pili
    Answer: D

Pages: 127
28. Nerve endings in the hypodermis that are sensitive to pressure are the:
    A) Meissner's corpuscles
    B) Merkel cells
    C) Langerhans cells
    D) Pacinian (lamellated) corpuscles
    E) keratinocytes
    Answer: D

Pages: 127
29. Superficial fascia is the same thing as the:
    A) epidermis
    B) dermis
    C) hypodermis
    D) melanocytes
    E) arrector pili
    Answer: C

Pages: 125
30. Synthesis of vitamin D begins with the activation of a precursor molecule in the skin by:
    A) melanin
    B) keratin
    C) sebum
    D) UV light
    E) temperatures above 60° F in the external environment
    Answer: D

31. The function of desmosomes is to:
    A) produce melanin
    B) produce keratin
    C) sense changes in pressure
    D) raise hairs in their follicles
    E) join keratinocytes to each other
    Answer: E

32. Which of the following are sensory receptors?
    A) melanocytes
    B) keratinocytes
    C) lamellated (Pacinian) corpuscles
    D) arrector pili
    E) mast cells
    Answer: C

33. Lack of the enzyme tyrosinase results in:
    A) inability to produce melanin
    B) inability to produce keratin
    C) loss of hair
    D) a slower rate of cellular reproduction in the stratum basale
    E) reduced ability to resist bacterial invasion
    Answer: A

34. The cells responsible for the growth of existing hairs and production of new hairs are called the:
    A) cuticle
    B) matrix
    C) shaft
    D) arrector pili
    E) papilla
    Answer: B

35. The function of the arrector pili is to:
    A) generate new hairs
    B) pull hairs into a vertical position
    C) produce sebum
    D) produce sweat
    E) produce the pigment in hairs
    Answer: B

Pages: 131
36. Sebaceous glands are present in the dermis of **ALL** of the following **EXCEPT** the:
    A) torso
    B) external genitalia
    C) palms and soles
    D) face
    E) eyelids
    Answer: C

Pages: 131
37. Eccrine sweat glands are most numerous in the skin of the:
    A) lips
    B) external genitalia
    C) palms and soles
    D) anterior portion of the torso
    E) posterior portion of the torso
    Answer: C

Pages: 132
38. Apocrine sweat glands are most numerous in the skin of the:
    A) pubic region
    B) face
    C) back
    D) palms and soles
    E) eyelids
    Answer: A

Pages: 132
39. Ceruminous glands are modified:
    A) sebaceous glands
    B) endocrine glands
    C) melanocytes
    D) goblet cells
    E) sudoriferous glands
    Answer: E

Pages: 132
40. Ceruminous glands are located in the skin of the:
    A) areolae of the breasts
    B) armpits
    C) eyelids
    D) external auditory meatus (ear canal)
    E) external genitalia
    Answer: D

41. The **eponychium** is the:
    A) cuticle of a nail
    B) the major part of the shaft of a hair
    C) part of a hair follicle where blood vessels are located
    D) muscle that raises hairs into a vertical position
    E) the region of skin in which sudoriferous glands are located
    Answer: A

42. The term **contact inhibition** refers to the:
    A) tendency to avoid touching things that are too hot
    B) inability of blood to flow into the epidermis
    C) social effect of the unpleasant aroma of sweat
    D) end of migration of epidermal cells once in contact with like cells on all sides
    E) extreme sensitivity of exposed dermis
    Answer: D

43. **ALL** of the following events occur during deep wound healing **EXCEPT**:
    A) vasodilation of blood vessels
    B) suspension of the rules of contact inhibition
    C) formation of a blood clot
    D) synthesis of scar tissue by fibroblasts
    E) increased permeability of blood vessels
    Answer: B

44. Granulation tissue is the tissue that forms:
    A) from the debris of phagocytosis during deep wound healing
    B) basal cell carcinomas
    C) a greater than normal amount of melanin
    D) during the migratory phase of deep wound healing
    E) hemangiomas
    Answer: D

45. The process of fibrosis results in:
    A) some form of skin cancer
    B) scar formation
    C) excessive production of skin pigment
    D) excessive production of hair and nails
    E) premature wrinkling of skin
    Answer: B

Pages: 136
46. Embryonically, the epidermis is derived from:
    A) ectoderm only
    B) mesoderm only
    C) endoderm only
    D) mesenchyme
    E) both endoderm and ectoderm
    Answer: A

Pages: 136
47. Embryonically, the dermis is derived from:
    A) ectoderm only
    B) mesoderm only
    C) endoderm only
    D) both ectoderm and endoderm
    E) both mesoderm and endoderm
    Answer: B

Pages: 136
48. Blisters form in second-degree burns because:
    A) the epidermis and dermis separate, and tissue fluid accumulates between the layers
    B) the dermal papillae extend through the damaged epidermis, and are exposed to the external environment
    C) damaged nerve-endings swell
    D) accumulated tissue fluid is necessary for scar formation
    E) the cells of the stratum basale are reproducing at such a rapid rate
    Answer: A

Pages: 136
49. Which of the following occurs in a first-degree burn?
    A) erythema (redness)
    B) blistering
    C) destruction of sensory nerve endings
    D) destruction of epidermal derivatives
    E) all of these
    Answer: A

Pages: 138
50. The most common forms of skin cancer are all caused, at least in part, by:
    A) chronic dryness of skin
    B) over-secretion by sudoriferous glands
    C) chronic exposure to sunlight
    D) over-production of keratin
    E) chronically reduced blood flow in the dermis
    Answer: C

Pages: 131
51. The most potent stimulator of secretion from sebaceous glands is:
    A) testosterone
    B) arrector pili
    C) epidermal growth factor
    D) melanin
    E) nerve impulses from the hypothalamus in the brain
    Answer: A

Pages: 135
52. Which of the following would you expect to happen, if the external temperature is 39° C?
    A) vasoconstriction of blood vessels in the skin
    B) vasodilation of blood vessels in the skin
    C) contraction of arrector pili muscles
    D) both A and C are correct
    E) none of these, because the body is already in homeostasis
    Answer: B

Pages: 135
53. Which of the following would you expect to happen, if the external temperature is 35° C?
    A) vasconstriction of blood vessels in the skin
    B) vasodilation of blood vessels in the skin
    C) increased secretion from sudoriferous glands
    D) both B and C are correct
    E) none of these, because the body is already in homeostasis
    Answer: A

Pages: 125
54. One hour per week in the sunlight with the hands, arms, and face exposed meets the body's needs for activation of:
    A) keratinocytes
    B) sudoriferous glands
    C) the vitamin D precursor molecule
    D) arrector pili
    E) Langerhans cells
    Answer: C

Pages: 126
55. Merkel cells are located in the:
    A) stratum basale of the epidermis
    B) stratum corneum of the epidermis
    C) stratum lucidum of the epidermis
    D) reticular region of the dermis
    E) superficial fascia
    Answer: A

56. The corpuscles of touch (Meissner's corpuscles) are located in the:
   A) hypodermis
   B) stratum basale of the epidermis
   C) basement membrane of the epidermis
   D) papillary region of the dermis
   E) stratum corneum of the epidermis
   Answer: D

57. The differences in the thickness of skin of various regions of the body is due primarily to differences in the thickness of the:
   A) superficial fascia
   B) basement membrane of the epidermis
   C) papillary region of the dermis
   D) epidermis
   E) reticular region of the dermis
   Answer: E

58. Differences in skin color among human races is due primarily to the:
   A) total number of melanocytes
   B) total number of keratinocytes
   C) amount of melanin produced by melanocytes
   D) amount of keratin produced by keratinocytes
   E) amount of iron in hemoglobin molecules
   Answer: C

59. Following an abrasion, the first event of wound healing is:
   A) increased blood flow to the epidermis
   B) separation of cells in the local stratum basale from their basement membrane
   C) stimulation of greater melanin production
   D) secretion of epidermal growth factor from special glands in the epidermis
   E) stimulation of the cells of the stratum basale by neurons of the autonomic nervous system
   Answer: B

60. During deep wound healing, mesenchyme cells that migrate to the site of injury during the inflammatory phase will develop into:
   A) keratinocytes
   B) melanocytes
   C) fibroblasts
   D) phagocytes
   E) collagen fibers
   Answer: C

Pages: 135
61. Heat is lost to the environment from blood in dilated blood vessels in the skin by the mechanism of:
A) radiation
B) evaporation
C) exocytosis
D) active transport
E) vasoconstriction
Answer: A

Pages: 136
62. **Lanugo** is:
A) the whitish, semilunar area at the proximal end of the nail body
B) the region of a hair containing the pigment
C) delicate fetal hair that develops by the fifth or sixth month of development
D) loss of melanocytes over patches of skin
E) a type of yeast infection
Answer: C

Pages: 137
63. The "rule of nines" and the Lund-Brower method are techniques used to estimate the:
A) amount of sweat normally produced at a given envrionmental temperature
B) amount of surface area affected by a burn
C) likelihood that a precancerous lesion will become malignant
D) amount of time it is safe to stay exposed to direct sunlight while wearing sunscreen
E) time it will take to heal a wound of a particular size and depth
Answer: B

Pages: 138
64. Acne is an inflammation of:
A) apocrine sweat glands
B) eccrine sweat glands
C) sebaceous glands
D) the epidermis
E) melanocytes
Answer: C

Pages: 128
65. Lips and nailbeds that are cyanotic look blue because:
A) lack of oxygen changes the color of melanin to deep, purplish blue
B) bilirubin builds up in the skin and epidermally-derived tissues
C) an excess amount of copper compounds has accumulated in the skin and epidermally-derived tissues
D) bacteria have digested the melanin and keratin to blue end-products
E) lack of oxygen changes the color of hemoglobin to deep, purplish blue
Answer: E

True-False

Write T if the statement is true and F if the statement is false.

Pages: 128
1. Homeostatic imbalances in the body may be indicated by changes in skin color.
Answer: True

Pages: 127
2. The dermis is thicker on the dorsal aspects of the body and on the lateral aspects of the extremities than in other regions.
Answer: True

Pages: 131
3. Apocrine sweat glands are much more common that eccrine sweat glands.
Answer: False

Pages: 128
4. The number of melanocytes is approximately the same for all races.
Answer: True

Pages: 137
5. A third-degree burn destroys both the epidermis and the dermis.
Answer: True

Pages: 138
6. Acne is an inflammation of the sudoriferous glands.
Answer: False

Pages: 138
7. Malignant cells conform to the rules of contact inhibition.
Answer: False

Pages: 125
8. People may develop a deficiency of vitamin A due to lack of exposure of skin to sunlight.
Answer: False

Pages: 126
9. The dermis is composed of stratified squamous epithelium.
Answer: False

Pages: 126
10. Most of the cells in the epidermis are keratinocytes.
Answer: True

Pages: 126
11. The cells of the stratum basale are all dead.
Answer: False

12. The epidermis is highly vascular.
    Answer: False

13. The function of melanin is to protect the body from damaging light rays.
    Answer: True

14. The receptors in the negative feedback loop that helps regulate body temperature are the sudoriferous glands.
    Answer: False

15. Vasoconstriction of blood vessels in the skin reduces heat loss from radiation.
    Answer: True

Short Answer

Write the word or phrase that best completes each statement or answers the question.

1. Individuals who do not get enough exposure to sunlight or who do not consume enough fortified milk may develop a deficiency of vitamin _____.
   Answer: D

2. The red-brown-black pigment in skin that absorbs UV light is _____.
   Answer: melanin

3. Most people who have albinism possess melanocytes, but lack the enzyme _____.
   Answer: tyrosinase

4. The deeper region of the dermis is the _____.
   Answer: reticular region

5. Sweat glands found primarily in the skin of the axillae, pubic region, and areolae of the nipples are the _____.
   Answer: apocrine sweat glands

6. The most prevalent life-threatening cancer in young women is _____.
   Answer: malignant melanoma

7. The single layer of continually reproducing cells in the epidermis is called the _____.
   Answer: stratum basale

Pages: 126

8. The most superficial layer of cells in the epidermis is called the _____.
   Answer: stratum corneum

Pages: 126

9. The protein in the outer layer of the epidermis that provides protection against mechanical injury, bacterial invasion, and dehydration is _____.
   Answer: keratin

Pages: 127

10. The region of the skin that is just deep to the stratum basale of the epidermis and that contains loops of capillaries is the _____.
    Answer: papillary region of the dermis

Pages: 127

11. The reticular region of the dermis is attached to underlying organs by the _____.
    Answer: subcutaneous layer (hypodermis, superficial fascia)

Pages: 127

12. The subcutaneous layer contains pressure-sensitive nerve endings known as _____.
    Answer: lamellated (Pacinian) corpuscles

Pages: 128

13. Inability to synthesize the enzyme tyrosinase results in a condition known as _____.
    Answer: albinism

Pages: 128

14. A yellowed appearance of skin and the whites of the eyes due to buildup of bilirubin resulting from liver disease is called _____.
    Answer: jaundice

Pages: 128

15. Redness of the skin due to increased blood flow is known as _____.
    Answer: erythema

Pages: 127

16. Epidermal ridges conform to the contours of the underlying _____.
    Answer: dermal papillae

Pages: 129

17. Hair, nails, and skin glands develop from the embryonic _____.
    Answer: epidermis

Pages: 131

18. The adult pattern of hair growth is determined by _____.
    Answer: sex hormones (androgens and estrogens)

Pages: 131

19. Mammary glands and ceruminous glands are modified _____.
    Answer: sudoriferous glands

Pages: 131

20. Blackheads develop from chemical oxidation of accumulated _____.
    Answer: sebum

Pages: 131

21. The secretion from skin that plays a small role in the elimination of wastes is _____.
    Answer: sweat

Pages: 132

22. The characteristic of epidermal cells that causes them to stop migrating during wound healing once they are touching other epidermal cells on all sides is called _____.
    Answer: contact inhibition

Pages: 134

23. The process of scar tissue formation is called _____.
    Answer: fibrosis

Pages: 135

24. When body temperature begins to fall, to prevent further heat loss blood vessels in the skin will _____.
    Answer: constrict

Pages: 135

25. In the negative feedback loop in which the integumentary system helps regulate body temperature, the effectors are the _____ and the _____.
    Answer: sudoriferous glands; blood vessels in the skin

Pages: 126

26. Cells in the epidermis that work with helper T cells to provide immunity are the _____.
    Answer: Langerhans cells

Pages: 133

27. Formation of a blood clot, vasodilation of blood vessels, and increased permeability of blood vessels for delivery of phagocytes and mesenchyme cells is characteristic of the _____ phase of deep wound healing.
    Answer: inflammatory

Pages: 133

28. Filling of a wound with granulation tissue is characteristic of the _____ phase of deep wound healing.
    Answer: migratory

Pages: 131

29. The more common type of sweat gland is the _____.
    Answer: eccrine sweat gland

Pages: 126

30. The type of tissue that forms the epidermis is _____.
    Answer: stratified squamous epithelium

Matching

Choose the item from Column 2 that best matches each item in Column 1.

1. melanocytes

produce a substance that helps protect the body from UV light

2. keratinocytes

produce a protein that provides protection from mechanical injury, bacterial invasion, and dehydration

3. Langerhans cells

work with helper T cells to provide immunity

4. sudoriferous glands

produce a product that helps regulate body temperature

5. sebaceous glands

produce a product that helps prevent excessive evaporation of water from the skin and keeps skin soft and pliable

6. Merkel cells

function in the sensation of touch

7. lamellated (Pacinian) corpuscles

function in sensations of pressure

8. arrector pili

raises hair to vertical position

9. dermal papillae

provide increased surface area for nutrient, waste, and gas exchange with cells of the stratum basale

10. ceruminous glands

produce earwax

Essay

Write your answer in the space provided or on a separate sheet of paper.

Pages: 124

1. List and briefly discuss the functions of skin.

   Answer: 1) regulation of body temperature via sweat production and changes in blood flow to the skin

   2) protection from mechanical injury, bacterial invasion, and dehydration (via keratin) and from UV light (via melanin)

   3) sensory reception via receptors for temperature, touch, pressure, pain

   4) excretion via sweat

   5) immunity via Langerhans cells

   6) blood reservoir for 8-10% of total blood flow

   7) synthesis of vitamin D from precursors in skin activated by UV light

Pages: 126

2. Name and describe the functions of the types of cells found in the epidermis.

   Answer: 1) keratinocytes - most numerous (90% of all epidermal cells); produce keratin for waterproofing and protection; anchored to each other by desmosomes

   2) melanocytes - produce red-brown-black skin pigment (melanin) from tyrosine; transfer melanin to keratinocytes to protect nuclei from UV light damage

   3) Langerhans cells - work with helper T cell during immune responses

   4) Merkel cells - in contact with sensory neurons called tactile discs; thought to function in sensation of touch

Pages: 126

3. Name the layers of the epidermis in order from deepest to most superficial, and describe the changes that occur in keratinocytes as they move from the deepest to most superficial layer.

   Answer: 1) stratum basale - single layer of cuboidal/columnar cells; undergo continual mitosis

   2) stratum spinosum - 8-10 layers of polyhedral cells very close together

   3) stratum granulosum - 3-5 layers of flattened cells; keratinocytes developing granules of keratohyalin; nuclei degenerating

   4) stratum lucidum - seen only on palms and soles; 3-5 layers of clear, flat, dead cells containing an intermediate substance in keratin production

   5) stratum corneum - 25-30 layers of flat, dead cells filled with keratin; continuously shed and replaced

Pages: 133

4. Describe the process of deep wound healing.

   Answer: 1) inflammatory phase - blood clot forms to loosely unite wound edges; blood vessels dilate and become more permeable; phagocytes and mesenchyme cells to site of injury

   2) migratory phase - clot becomes scab; epithelial cells migrate beneath scab; fibroblasts synthesize scar tissue; begin repair of damaged blood vessels; granulation tissue fills wound

   3) proliferative phase - extensive growth of epithelium beneath scab; fibroblasts deposit collagen fibers in random patterns; further vessel repair

   4) maturation phase - scab falls off; collagen fibers become organized; fibroblasts decrease

Pages: 128

5. Explain the basis of normal skin color.

   Answer: 1) melanin - produced from tyrosine by melanocytes in epidermis; varies skin color from pale yellow to black; amount of melanin produced by melanocytes varies among human races, but number of melanocytes about the same in all races
   2) carotene - produced from vitamin A; a yellow-orange pigment; seen in stratum corneum, dermis, and subcutaneous layer of people of Asian ancestry
   3) hemoglobin - pigment in red blood cells that shows through blood vessels in skin of Caucasian races; gives skin pinkish red color

Pages: 135

6. Describe the changes that occur in the skin during the aging process.

   Answer: After age 40: collagen fibers decrease in number, become tangled; elastic fibers lose elasticity and fray; wrinkles form; fibroblasts and Langerhans cells decrease in number; macrophages less efficient; hair and nails grow more slowly; sebaceous glands decrease in size; sudoriferous glands produce less sweat; melanocytes decrease in number and may enlarge, causing blotching of skin; blood vessels in dermis thicker-walled and less permeable; less subcutaneous fat; skin becomes thinner; slower migration of epidermal cells to surface

Pages: 135

7. Describe in detail the negative feedback mechanisms involving the skin that help relieve the stress of elevated body temperature.

   Answer: 1) controlled condition = constant body temperature
   2) stress = elevation of body temperature
   3) receptors = thermoreceptors in skin and hypothalamus, which detect elevated temperature
   4) control center = brain, which receives input from thermoreceptors and notifies effectors of appropriate response
   5) effectors = sweat glands, which increase secretion of sweat, which evaporates, causing heat loss, and blood vessels in dermis, which dilate, causing heat loss by radiation
   6) output = decreased body temperature; return to homeostasis

Pages: 136

8. Describe the systemic effects of severe burns.

   Answer: 1) dehydration from loss of epidermis, possibly leading to circulatory shock
   2) increased risk of bacterial invasion from loss of epidermis and exposure of vascularized underlying tissues
   3) reduced circulation of blood
   4) decreased urine production
   5) diminished immune response

Pages: 138

9. Identify the risk factors for the development of skin cancer.

   Answer: 1) skin type - higher risk with light-colored, poorly tanning skin
   2) sun exposure - higher risk with higher exposure due to total time in sun or high-altitude
   3) family history - higher risk with family members who have had skin cancer
   4) age - older people at higher risk due to total sun exposure
   5) immunologic status - higher risk if immunosuppressed

Chapter 6   Bone Tissue

Multiple-Choice

Choose the one alternative that best completes the statement or answers the question.

Pages: 143
1. The perichondrium and periosteum are made of:
   A) hyaline cartilage
   B) elastic cartilage
   C) loose connective tissue
   D) dense irregular (fibrous) connective tissue
   E) smooth muscle
   Answer: D

Pages: 143
2. Articular cartilage is:
   A) hyaline cartilage
   B) elastic cartilage
   C) osseous tissue
   D) smooth muscle
   E) fibrocartilage
   Answer: A

Pages: 145
3. Hydroxyapatites are:
   A) the inorganic calcium compounds in bone matrix
   B) the collagen portion of bone matrix
   C) the concentric ring structures seen in compact bone
   D) channels through which nutrients reach osteocytes
   E) the cells of yellow bone marrow
   Answer: A

Pages: 148
4. During endochondral bone formation, the primary center of ossification forms in the:
   A) proximal epiphysis
   B) distal epiphysis
   C) epiphyseal plate
   D) diaphysis
   E) metaphysis
   Answer: D

Pages: 148
5. In endochondral bone formation, the original pattern for the bone is made of:
   A) osseous tissue
   B) keratin
   C) dense irregular (fibrous) connective tissue
   D) elastic cartilage
   E) hyaline cartilage
   Answer: E

Pages: 151

6. The function of the epiphyseal plate is to:
   A) allow more flexibility in a long bone
   B) allow a means by which the bone can increase in diameter
   C) allow a means by which the bone can increase in length
   D) provide nourishment to isolated osteocytes
   E) both B and C are correct
   Answer: C

Pages: 143

7. The diaphysis is:
   A) the end of a long bone
   B) a bone found in a tendon
   C) the covering of a bone
   D) the shaft of a long bone
   E) a plate of bone in spongy bone
   Answer: D

Pages: 144

8. The function of osteoblasts is to:
   A) break down bone
   B) produce blood cells
   C) produce collagen for bone matrix
   D) add new tissue to the periosteum
   E) provide nourishment to the cells of the articular cartilage
   Answer: C

Pages: 147

9. Osteons are typical of the structure of:
   A) dense bone
   B) epiphyseal plates
   C) spongy bone
   D) the endosteum
   E) all of these contain osteons
   Answer: A

Pages: 147

10. Nourishment for osteocytes in cancellous bone is provided by:
    A) blood in vessels in the red marrow
    B) blood in vessels in central canals
    C) blood in yellow marrow
    D) blood in canaliculi
    E) serous fluid in lacunae
    Answer: A

11. During appositional growth of bone and cartilage, new tissue is added:
    A) from within existing tissues
    B) only during embryonic and fetal life
    C) on the outside of existing tissues
    D) only when injury occurs
    E) both A and C are correct
    Answer: C

12. The layers of bone in an osteon are called:
    A) trabeculae
    B) lacunae
    C) canaliculi
    D) lamellae
    E) zones of proliferating bone
    Answer: D

13. Nutrients are provided to osteocytes in dense bone by:
    A) transport through canaliculi from blood in vessels the central canals
    B) blood in the marrow cavity
    C) blood seeping through the matrix of interstitial lamellae
    D) dissolving the matrix around them via enzymes in lacunae
    E) blood in yellow bone marrow
    Answer: A

14. The original pattern for bone that will develop by intramembranous ossification is made of:
    A) hyaline cartilage
    B) osseous tissue
    C) stratified squamous epithelium
    D) elastic cartilage
    E) loose fibrous connective tissue
    Answer: E

15. **Lacunae** are:
    A) channels for blood vessels and nerves extending through dense bone
    B) rings of matrix in osteons
    C) spaces in matrix in which osteocytes and chondrocytes are located
    D) specialized junctions between osteocytes
    E) plates of bone seen in spongy bone
    Answer: C

16. Which of the following cells is most mature (i.e., the most differentiated)?
   A) osteocyte
   B) mesenchyme cell
   C) osteoblast
   D) osteoprogenitor cell
   E) all of the above cells are equally differentiated
   Answer: A

17. **Canaliculi** are:
   A) rings of matrix in osteons
   B) spaces in the matrix in which osteocytes and chondrocytes are located
   C) tiny channels connecting osteocytes with the central canal of osteons
   D) cells that produce the matrix of cartilage
   E) the chambers in spongy bone in which marrow is located
   Answer: C

18. Calcitonin is produced by:
   A) osteoblasts
   B) osteoclasts
   C) the thyroid gland
   D) the parathyroid gland
   E) red bone marrow
   Answer: C

19. The function of calcitonin is to:
   A) lower the level of calcium ions in the blood
   B) raise the level of calcium ions in the blood
   C) activate osteoclasts
   D) force osteoblasts to differentiate into osteoclasts
   E) both B and C are correct
   Answer: A

20. **ALL** of the following happen during the formation of the parietal bone **EXCEPT**:
   A) mesenchyme cells will migrate into the area where ossification will occur
   B) osteoblasts will produce collagen
   C) some osteoprogenitor cells will become chondroblasts
   D) trabeculae of spongy bone will form
   E) outer layers of bone will be remodelled
   Answer: C

Pages: 148

21. During endochondral ossification, calcification of cartilage matrix causes the death of chondrocytes. This leads to:
    A) death of the developing bone
    B) hemopoiesis
    C) closing of the epiphyseal plates
    D) erosion of articular cartilage
    E) development of the marrow cavity
    Answer: E

Pages: 193

22. An exaggerated lumbar curvature of the spine is called:
    A) kyphosis
    B) scoliosis
    C) lordosis
    D) spina bifida
    E) osteoporosis
    Answer: C

Pages: 144

23. The unspecialized cells derived from mesenchyme cells that will develop into osteoblasts are called:
    A) osteoclasts
    B) osteocytes
    C) chondrocytes
    D) fibroblasts
    E) osteoprogenitor cells
    Answer: E

Pages: 145

24. **ALL** of the following are **TRUE** about bone composition **EXCEPT**:
    A) immature bone contains more cells than mature bone
    B) bone matrix, unlike other connective tissues, contains abundant mineral salts
    C) mature bone is generally considered to be completely solid
    D) the diaphysis of a long bone is primarily dense bone
    E) the organic part of bone matrix is primarily collagen
    Answer: C

Pages: 143,149

25. **ALL** of the following are normal sites of hematopoiesis in adults **EXCEPT**:
    A) clavicle (collarbone)
    B) hipbone
    C) rib
    D) sternum
    E) vertebral bodies
    Answer: A

Pages: 143
26. The region between the diaphysis and epiphysis of a long bone, where calcified matrix is replaced by bone, is called the:
    A) periosteal bud
    B) bone collar
    C) primary ossification center
    D) metaphysis
    E) secondary ossification center
    Answer: D

Pages: 151
27. In males, ossification of most bones is usually completed by age:
    A) 2
    B) 6
    C) 14
    D) 18
    E) 25
    Answer: E

Pages: 155
28. **ALL** of the following are **TRUE** about bone homeostasis **EXCEPT**:
    A) bone formation begins before birth
    B) bone constantly remodels and redistributes its matrix along lines of mechanical stress
    C) spongy bone is formed from compact bone
    D) remodeling allows bone to serve as the body's reservoir for calcium
    E) even after bones have reached their adult shapes and sizes, old bone matrix is destroyed and replaced by new bone matrix
    Answer: C

Pages: 157
29. Which of the following statements is **FALSE** regarding aging of bone tissue?
    A) Demineralization usually begins around age 30 in females.
    B) In females, as much as 30% of the calcium in bones is lost by age 70.
    C) Males never lose bone matrix as they age because of the effects of testosterone.
    D) Collagen production generally decreases with age.
    E) Bone remodeling continues throughout life, regardless of age.
    Answer: C

Pages: 154
30. Which of the following best defines a compound (open) fracture?
    A) The bone is broken in more than one place.
    B) The broken ends of the bone protrude through the skin.
    C) The bone is usually twisted apart.
    D) The fracture is at right angles to the long axis of the bone.
    E) One bone fragment is driven into the other.
    Answer: B

Pages: 154
31. Which of the following defines a greenstick fracture?
    A) The anatomical alignment of the bone fragments is preserved.
    B) The bone is usually twisted apart.
    C) One side of the bone breaks, while the other side only bends.
    D) The bone breaks because it has been weakened by disease.
    E) The bone fragments are driven into each other.
    Answer: C

Pages: 154
32. Which of the following defines a simple (closed) fracture?
    A) The bone does not break through the skin.
    B) The broken ends of the bone protrude through the skin.
    C) The bone is usually twisted apart.
    D) One side of the bone breaks, while the other side bends.
    E) The bone fractures because it has been weakened by disease.
    Answer: A

Pages: 154
33. A fracture of the distal end of the radius in which the distal fragment is displaced
    posteriorly is a:
    A) stress fracture
    B) greenstick fracture
    C) Pott's fracture
    D) Paget's fracture
    E) Colles' fracture
    Answer: E

Pages: 159
34. Degeneration of articular cartilage, usually associated with the elderly, in which bones at
    joints may actually touch is referred to as:
    A) osteoarthritis
    B) osteogenic sarcoma
    C) osteomyelitis
    D) osteoporosis
    E) Paget's disease
    Answer: A

Pages: 159
35. A cancer of osteoblasts, which usually affects teenagers during their growth spurt, is:
    A) osteoarthritis
    B) osteogenic sarcoma
    C) osteomyelitis
    D) osteoporosis
    E) Paget's disease
    Answer: B

Pages: 148
36. **ALL** of the following bones would be produced via endochondral ossification **EXCEPT** the:
    A) tibia
    B) femur
    C) frontal
    D) humerus
    E) radius
    Answer: C

Pages: 143
37. The term **hemopoiesis** refers to:
    A) development of compact bone from spongy bone
    B) production of blood cells in red bone marrow
    C) the lack of blood flow to articular cartilage
    D) the formation of a primary ossification center during endochondral bone formation
    E) the growth of a long bone at the epiphyseal plate
    Answer: B

Pages: 145
38. Bone cells whose function is the destruction of bone matrix are the:
    A) osteoprogenitor cells
    B) osteoblasts
    C) osteocytes
    D) osteoclasts
    E) all of these can destroy bone matrix
    Answer: D

Pages: 145
39. The organic part of bone matrix is mostly:
    A) calcium salts
    B) magnesium salts
    C) collagen fibers
    D) osteocytes
    E) keratin
    Answer: C

Pages: 147
40. Which of the following is **TRUE** about the blood supply to a long bone?
    A) Osseous tissue is completely avascular.
    B) The only blood vessels in bone are found in the medullary cavity and spongy bone of the epiphyses.
    C) The smallest blood vessels are found in the canaliculi of osteons.
    D) A nutrient artery passes through the diaphysis and branches into epiphyseal arteries and periosteal arteries.
    E) both B and C are correct
    Answer: D

Pages: 147
41. In both intramembranous and endochondral ossification, the first stage of development of bone is:
A) penetration of the diaphysis by a nutrient artery
B) migration of mesenchyme cells to the area of bone formation
C) formation of a cartilage model of the bone
D) fusion of trabeculae
E) development of a periosteal bud
Answer: B

Pages: 151
42. The area of the epiphyseal plate that serves as an anchor to the bone of the epiphysis is the zone of:
A) resting cartilage
B) proliferating cartilage
C) hypertrophic cartilage
D) calcified cartilage
E) articular cartilage
Answer: A

Pages: 151
43. The region of the epiphyseal plate that consists mostly of dead cells in calcified matrix is the zone of:
A) resting cartilage
B) proliferating cartilage
C) maturing cartilage
D) calcified cartilage
E) articular cartilage
Answer: D

Pages: 152
44. Before puberty, bone growth is stimulated mainly by the hormone:
A) calcitonin
B) testosterone
C) estrogen
D) human growth hormone
E) PTH
Answer: D

Pages: 152
45. Bone constantly remodels and redistributes its matrix along lines of:
A) blood flow
B) nervous stimulation
C) canaliculi
D) mechanical stress
E) overlying muscles
Answer: D

Pages: 152
46. A function of vitamin D in proper bone development is to:
    A) increase absorption of calcium from the intestine
    B) activate osteoclasts
    C) bind hydroxyapatites to collagen fibers
    D) promote absorption of amino acids from the intestine
    E) increase the rate of protein synthesis
    Answer: A

Pages: 155
47. A hormone that increases the number and activity of osteoclasts is:
    A) human growth hormone
    B) testosterone
    C) calcitonin
    D) parathyroid hormone
    E) calcitriol
    Answer: D

Pages: 153
48. The first step in the repair of a bone fracture is:
    A) formation of the bony callus
    B) remodeling of the bone
    C) formation of the fibrocartilaginous callus
    D) development of a fracture hematoma
    E) development of granulation tissue (procallus)
    Answer: D

Pages: 155
49. **ALL** of the following are effects of parathyroid hormone (PTH) **EXCEPT**:
    A) promotes increased activity of osteoblasts
    B) promotes increased activity of osteoclasts
    C) promotes formation of calcitriol
    D) promotes elimination of phosphate in urine
    E) promotes recovery of calcium ions by kidneys
    Answer: A

Pages: 156
50. An effect of mechanical stress on bone is to promote greater secretion of the hormone:
    A) parathyroid hormone (PTH)
    B) insulin
    C) human growth hormone
    D) testosterone
    E) calcitonin
    Answer: E

51. Bones are more brittle in the elderly because:
    A) levels of calcitonin are higher in the elderly
    B) there is less collagen relative to the amount of mineral salts
    C) low levels of growth hormone prevent deposition of calcium in bone
    D) the calcium salts in the bone matrix are of a different type than those in younger people
    E) too much collagen takes up space that should be occupied by hydroxyapatites
    Answer: B

52. Limb buds normally appear on the developing embryo during which week of development?
    A) third
    B) fifth
    C) seventh
    D) tenth
    E) twelfth
    Answer: B

53. Endochondral ossification has usually begun by which week of embryonic development?
    A) second
    B) third
    C) fifth
    D) seventh
    E) There is no specific time that ossification begins.
    Answer: D

54. The notochord is:
    A) the remnants of the primary ossification center of a long bone
    B) what remains after the epiphyseal plate of a long bone has closed
    C) what remains after fontanels have ossified
    D) a rod of tissue located where the vertebral column will develop in the embryo
    E) the marrow cavity of embryonic bones
    Answer: D

55. A condition characterized by a greatly accelerated bone remodeling process that results in thickened and softened bones is:
    A) osteoporosis
    B) osteoarthritis
    C) Paget's disease
    D) osteomyelitis
    E) osteosarcoma
    Answer: C

Pages: 143
56. An important energy reserve is provided by the skeletal system in the form of:
    A) hemopoietic tissue
    B) hydroxyapatites
    C) yellow bone marrow
    D) articular cartilage
    E) the periosteum
    Answer: C

Pages: 143
57. A bone would be unable to repair itself if it lacked a(n):
    A) articular cartilage
    B) periosteum
    C) epiphyseal plate
    D) epiphysis
    E) diaphysis
    Answer: B

Pages: 143
58. The lining of the medullary cavity is called the:
    A) endosteum
    B) periosteum
    C) metaphysis
    D) articular cartilage
    E) epiphyseal plate
    Answer: A

Pages: 144
59. The cells responsible for maintaining the daily cellular activities of bone tissue, such as exchange of nutrients and wastes with the blood, are the:
    A) osteoprogenitor cells
    B) osteoblasts
    C) osteoclasts
    D) osteocytes
    E) chondroblasts
    Answer: D

Pages: 145
60. The crystallization of mineral salts in bone occurs only in the presence of:
    A) collagen
    B) PTH
    C) osteoclasts
    D) hyaline cartilage
    E) yellow bone marrow
    Answer: A

61. The central (Haversian) canal of an osteon contains:
    A) red bone marrow
    B) yellow bone marrow
    C) a blood vessel and nerve
    D) an osteocyte
    E) the remnants of the embryonic cartilage pattern of the bone
    Answer: C

62. Periosteal arteries and nerves enter the diaphysis of a long bone through:
    A) central (Haversian) canals
    B) perforating (Volkmann's) canals
    C) the epiphyseal plate
    D) nutrient foramina
    E) the articular cartilage
    Answer: B

63. During the first stage of bone development, mesenchyme cells differentiate into osteoprogenitor cells. The subsequent differentiation of these cells into either osteoblasts or chondroblasts depends on the presence or absence of:
    A) osteoclasts
    B) parathyroid hormone
    C) capillaries
    D) a periosteum
    E) collagen
    Answer: C

64. The secondary ossification centers of a long bone normally develop:
    A) during the seventh week of embryonic development
    B) during the twelfth week of embryonic development
    C) at the same time as the primary ossification center
    D) around the time of birth
    E) at age two
    Answer: D

65. The "growth spurt" that occurs during the teenage years is due primarily to the action of:
    A) estrogens and testosterone
    B) calcitonin
    C) growth hormone
    D) parathyroid hormone
    E) insulinlike growth factors
    Answer: A

True-False

Write T if the statement is true and F if the statement is false.

Pages: 143
1. The epiphysis is the region where the diaphysis joins the metaphysis.
Answer: False

Pages: 145
2. Osteoclasts are believed to develop from circulating erythrocytes.
Answer: False

Pages: 145
3. There are more cells in immature bone than in mature bone.
Answer: True

Pages: 151
4. The activity of the epiphyseal plate is the only means by which the diaphysis can increase in length.
Answer: True

Pages: 152
5. Magnesium deficiency inhibits the activity of osteoblasts.
Answer: True

Pages: 153
6. Injuries to cartilage heal more slowly than those to bone because the blood vessels that serve cartilage do not extend into the cartilage matrix as they do in bone.
Answer: True

Pages: 143
7. Yellow bone marrow is a hemopoietic tissue.
Answer: False

Pages: 144
8. Osteoprogenitor cells are derived from mesenchyme.
Answer: True

Pages: 144
9. The organic part of bone matrix is produced by osteoblasts.
Answer: True

Pages: 147
10. Osseous tissue is an avascular tissue.
Answer: False

Pages: 147
11. Osteons (Haversian systems) are common in spongy bone.
Answer: False

12. The parietal bones form by intramembranous ossification.
Answer: True

13. Trabeculae are the spaces between plates of bone in spongy bone.
Answer: False

14. The humerus (upper arm bone) forms by intramembranous ossification.
Answer: False

15. Before puberty, bone growth is stimulated primarily by estrogens and testosterone.
Answer: False

Short Answer

Write the word or phrase that best completes each statement or answers the question.

Pages: 143
1. The study of bone structure and treatment of bone disorders is known as _____.
Answer: osteology

Pages: 143
2. The inner layer of the periosteum is called the _____ layer.
Answer: osteogenic

Pages: 143
3. The membrane that lines the medullary cavity of a long bone is called the _____.
Answer: endosteum

Pages: 144
4. Mature bone cells that are completely surrounded by matrix are called _____.
Answer: osteocytes

Pages: 145
5. Bones are less brittle than other calcium-based products, such as oyster shells and egg shells, because they contain more of the protein _____.
Answer: collagen

Pages: 147
6. The channels in osteons that connect lacunae with central canals are called _____.
Answer: canaliculi

Pages: 147
7. Areas between osteons are filled in with _____ lamellae.
Answer: interstitial

Pages: 148
8. In intramembranous ossification, the highly vascularized mesenchyme on the outside of the new bone develops into the _____.
    Answer: periosteum

Pages: 143
9. Yellow bone marrow consists primarily of _____.
    Answer: adipose cells

Pages: 143
10. The part of a long bone that is not covered by periosteum is covered by _____.
    Answer: articular cartilage

Pages: 145
11. Cells whose primary function is bone resorption are the _____.
    Answer: osteoclasts

Pages: 145
12. About 50% of bone matrix consists of _____.
    Answer: mineral salts (hydroxyapatites)

Pages: 145
13. The external layer of all bones and the bulk of the diaphyses of long bones is made up of _____ bone.
    Answer: compact (dense)

Pages: 145
14. Blood vessels run longitudinally through compact bone in _____.
    Answer: central (Haversian) canals

Pages: 147
15. Thin plates of bone in spongy bone are called _____.
    Answer: trabeculae

Pages: 148
16. The artery to the diaphysis of a long bone is called the _____ artery.
    Answer: nutrient

Pages: 148
17. The flat bones of the skull form by _____ ossification.
    Answer: intramembranous

Pages: 148
18. In endochondral ossification, the embryonic pattern for the bone is made of _____.
    Answer: hyaline cartilage

Pages: 148
19. Cells responsible for cartilage formation are called _____.
    Answer: chondroblasts

Pages: 148
20. During endochondral ossification of a long bone, the primary ossification center forms in the

    _____.
    Answer: diaphysis

Pages: 151
21. A long bone increases in length due to the activity of cartilage cells at the _____.
    Answer: epiphyseal plate

Pages: 152
22. Human growth hormone is produced by the _____.
    Answer: anterior pituitary gland

Pages: 152
23. Bone constantly remodels and redistributes matrix along lines of _____.
    Answer: mechanical stress

Pages: 152
24. Hormones that are produced locally by bone and also by the liver that stimulate the uptake of
    amino acids and promote proteins synthesis are _____.
    Answer: insulin-like growth factors

Pages: 155
25. Levels of calcium ions in the blood are increased by the effects of the hormone _____.
    Answer: parathyroid hormone (PTH)

Pages: 153
26. The blood clot that forms in and around the site of a bone fracture is called a _____.
    Answer: fracture hematoma

Pages: 154
27. A fracture in which the bone has splintered at the site of impact, leaving smaller fragments
    between the two main fragments is called a _____ fracture.
    Answer: comminuted

Pages: 155
28. In the negative feedback loop that controls the release of parathyroid hormone, the control
    center is the _____.
    Answer: gene for PTH within the parathyroid gland cell

Pages: 156
29. A hormone that inhibits the activity of osteoclasts and promotes deposition of calcium ions
    into bones is _____.
    Answer: calcitonin

Pages: 159
30. Inflammation of the bone marrow, caused by a pathogenic microorganism, is called _____.
    Answer: osteomyelitis

Matching

Choose the item from Column 2 that best matches each item in Column 1.

1. mesenchyme cells

    stem cells from which more differentiated bone and cartilage cells arise

2. osteoprogenitor cells

    develop into osteoblasts; found in periosteum and endosteum

3. osteoblasts

    do not undergo mitosis; secrete collagen

4. osteocytes

    sit in lacunae in bone matrix; maintain daily cellular activities of bone

5. osteoclasts

    located on bone surfaces; function in bone resorption

6. hydroxyapatites

    the mineral portion of bone matrix

7. osteons

    units of concentric lamellae in compact bone

8. trabeculae

    thin plates of bone in spongy bone

9. canaliculi

    tiny channels that allow communication between osteocytes and blood vessels in central canals

10. chondroblasts

    form the matrix of cartilage

Essay

Write your answer in the space provided or on a separate sheet of paper.

Pages: 143

1. List and briefly discuss the functions of the skeletal system.
   Answer: 1) support - point of attachment for skeletal muscles; supports soft organs
   2) protection - bones surround major organs
   3) assists in movement - skeletal muscles attached to bones, provide leverage
   4) mineral homeostasis - exchange of minerals (especially Ca and P) between bone and blood
   5) hemopoiesis - red bone marrow produces blood cells
   6) energy storage - fat in yellow bone marrow

Pages: 153

2. Describe the process of bone fracture repair.
   Answer: 1) formation of fracture hematoma w/i 6-8 hrs; traumatized tissue removed
   2) formation of procallus (granulation tissue); production of collagen by fibroblasts
   3) development of fibrocartilaginous callus from procallus via action of chondroblasts
   4) conversion of fibrocartilaginous callus into bony callus of spongy bone by osteoblasts
   5) remodeling of bony callus

Pages: 152

3. Describe the roles of human growth hormone, insulin-like growth factors, and sex hormones in maintaining the homeostasis of the skeletal system.
   Answer: All promote skeletal growth. hGH promotes growth of all tissues and stimulates production of IGFs by bone and liver. IGFs stimulate uptake of amino acids and promote protein synthesis - necessary for collagen synthesis. Sex hormones increase the activity of osteoblasts and dictate deposition of matrix by gender.

Pages: 145-147

4. Compare and contrast the structure of compact bone vs. spongy bone.
   Answer: Compact bone arranged in osteons; concentric lamellae of matrix surrounding central canal containing blood vessel and nerve; osteocytes in lacunae at edges of lamellae, communicating with central canal via canaliculi; interstitial lamellae between osteons. Spongy bone has no osteons; thin plates of bone (trabeculae) surround spaces containing red bone marrow; osteocytes in lacunae get nutrients directly from blood in marrow

Pages: 155

5. Describe in detail a negative feedback loop involving the endocrine and skeletal systems that results in an increase in serum calcium levels.
   Answer: Following a decrease in serum calcium levels (stress), the change would be sensed by both the thyroid and parathyroid glands. The thyroid would decrease the release of calcitonin to halt deposition of calcium in bone, and the parathyroids would increase the release of PTH to promote bone resorption by osteoclasts, as well as retention and absorption of calcium from kidneys and intestines.

Pages: 148-149
6. Explain how the medullary cavity forms in a long bone during endochondral ossification.
   Answer: Following development of the primary ossification center in disintegrating calcified cartilage, spongy bone forms in center of bone. Osteoclasts break down new spongy bone, leaving a cavity, which fills with red marrow.

Pages: 155
7. Patient X has a tumor of the parathyroid glands that causes a hypersecretion from these glands. Predict the effect on the skeletal system and on the secretion of calcitonin.
   Answer: High levels of PTH would cause high levels of osteoclast activity, thus removing calcium from bones. Bones would become weak and soft. Excess phosphate would be lost from kidneys. High levels of calcium ions in blood may disrupt nerve and muscle function. Calcitonin levels would probably be high, trying to restore homeostasis by increasing deposition of calcium into bone.

Pages: 152
8. How might prolonged sunlight deprivation affect the skeletal system?
   Answer: Resulting vitamin D deficiency may cause low levels of calcium absorption, thus less deposition in bone, thus weakening and softening of bones, assuming vitamin D is not adequately obtained in diet.

Pages: 152
9. Archaeologists have unearthed the upper arm bones of some ancient humans. They note with interest that these bones have greatly enlarged deltoid tuberosities, which are the points where the deltoid muscles attach to these bones. What might they infer from this finding about the life of these people, and why?
   Answer: Possibly some activity that required the repeated use of this muscle (such as archery or rowing), that caused continual mechanical stress on this point of muscle attachment, thus inducing remodeling via addition of bone.

Pages: 152
10. Janet has heard that taking megadoses of vitamins is beneficial to her health. What would you tell her about the possible effects of these extra vitamins on the health of her skeletal system?
    Answer: Vitamin D helps in the absorption of calcium from the intestines, and vitamin A helps coordinate the activities of osteoblasts and osteoclasts. However, both vitamins A and D may be toxic in large doses. Vitamin B_{12} may inhibit osteoblast activity. Vitamin C helps maintain collagen in bone matrix.

Chapter 7   The Skeletal System: The Axial Skeleton

Multiple-Choice

Choose the one alternative that best completes the statement or answers the question.

Pages: 163
1. Bones whose length and width are nearly equal are classified as:
    A) long bones
    B) short bones
    C) flat bones
    D) irregular bones
    E) Wormian bones
    Answer: B

Pages: 163
2. The most extensive surface area for muscle attachment is provided by:
    A) long bones
    B) short bones
    C) flat bones
    D) irregular bones
    E) Wormian bones
    Answer: C

Pages: 163
3. Long bones are made mostly of:
    A) compact bone
    B) spongy bone
    C) hemopoietic tissue
    D) dense regular (fibrous) connective tissue
    E) yellow bone marrow
    Answer: A

Pages: 163
4. Bones such as the vertebrae and maxillae are classified as:
    A) long bones
    B) short bones
    C) flat bones
    D) irregular bones
    E) Wormian bones
    Answer: D

Pages: 163
5. Bones located in tendons are referred to as:
    A) Wormian bones
    B) tendonous bones
    C) loose bones
    D) muscular bones
    E) sesamoid bones
    Answer: E

Pages: 163

6. Bones located in the joints of cranial bones are classified as:
   A) Wormian bones
   B) sesamoid bones
   C) synovial bones
   D) articular bones
   E) short bones
   Answer: A

Pages: 163

7. **ALL** of the following are part of the axial skeleton **EXCEPT** the:
   A) occipital bone
   B) hyoid bone
   C) vertebrae
   D) coxal bones
   E) sternum
   Answer: D

Pages: 163

8. Which of the following bones is considered to be part of the axial skeleton?
   A) humerus
   B) coxal
   C) hyoid
   D) patella
   E) talus
   Answer: C

Pages: 166

9. Of the 80 bones in the axial skeleton, how many are part of the skull?
   A) 7
   B) 14
   C) 22
   D) 40
   E) all 80
   Answer: C

Pages: 164

10. **ALL** of the following terms refer to processes or projections from bones **EXCEPT**:
    A) tubercle
    B) fossa
    C) trochanter
    D) condyle
    E) spine
    Answer: B

Pages: 164
11. Which of the following terms refers to a hole in a bone through which blood vessels, nerves, or ligaments pass?
    A) fossa
    B) tubercle
    C) epicondyle
    D) sinus
    E) foramen
    Answer: E

Pages: 164
12. A small rounded process on a bone is called a:
    A) spine
    B) tubercle
    C) tuberosity
    D) foramen
    E) sulcus
    Answer: B

Pages: 164
13. A fontanel is:
    A) a hole in a bone through which blood vessels, nerves, or ligaments pass
    B) a bone whose epiphyseal plates have closed prematurely
    C) the space between cranial bones of the newborn that is filled with fibrous connective tissue
    D) a large, rough projection from a bone to which muscles attach
    E) a hole in a bone through which fluid flows
    Answer: C

Pages: 164
14. A condyle is:
    A) a large, flat surface to which muscles can attach extensively
    B) an indentation in a bone
    C) an air-filled space in a bone
    D) a large, rounded articular prominence
    E) a long ridge on a bone
    Answer: D

Pages: 164
15. **ALL** of the following are considered part of the appendicular skeleton **EXCEPT** the:
    A) humerus
    B) coxal bones
    C) fibula
    D) ribs
    E) calcaneus
    Answer: D

16. There are normally **TWO** of **EACH** of the following bones **EXCEPT** the:
    A) vomer
    B) maxilla
    C) nasal
    D) temporal
    E) zygomatic
    Answer: A

17. The lambdoid suture is located between the:
    A) parietal bones
    B) parietal bones and the frontal bone
    C) parietal and temporal bones
    D) maxillae
    E) parietal bones and the occipital bone
    Answer: E

18. The squamous sutures are located between the:
    A) parietal bones
    B) parietal bones and the frontal bone
    C) parietal bones and the temporal bones
    D) maxillae
    E) parietal bones and the occipital bone
    Answer: C

19. The supraorbital foramina are located in the:
    A) sphenoid bone
    B) ethmoid bone
    C) frontal bone
    D) zygomatic bones
    E) lacrimal bones
    Answer: C

20. The zygomatic process is part of the:
    A) zygomatic bone
    B) mandible
    C) frontal bone
    D) sphenoid bone
    E) temporal bone
    Answer: E

Pages: 171
21. The internal and middle ear are housed by the:
    A) external auditory meatus
    B) mastoid process of the temporal bone
    C) sella turcica of the sphenoid bone
    D) petrous portion of the temporal bone
    E) greater wings of the sphenoid bone
    Answer: D

Pages: 171
22. The mandibular fossa of the temporal bone articulates with what part of the mandible?
    A) condylar process
    B) coronoid process
    C) alveolar process
    D) zygomatic process
    E) mental foramen
    Answer: A

Pages: 171
23. The temporal bone articulates with **ALL** of the following **EXCEPT** the:
    A) parietal bone
    B) zygomatic bone
    C) mandible
    D) frontal bone
    E) sphenoid bone
    Answer: D

Pages: 171
24. The part of the temporal bone that is located posterior and inferior to the external auditory meatus is the:
    A) zygomatic process
    B) mastoid process
    C) temporal process
    D) mandibular fossa
    E) petrous portion
    Answer: B

Pages: 181
25. The hyoid bone is suspended from the:
    A) mastoid processes of the temporal bones
    B) occipital condyles
    C) superior nasal conchae of the ethmoid bone
    D) sella turcica of the sphenoid bone
    E) styloid processes of the temporal bones
    Answer: E

Pages: 184
26. The superior articular facets of the atlas articulate with the:
    A) occipital condyles
    B) mastoid processes of the temporal bones
    C) dens of the axis
    D) inferior articular facets of the axis
    E) first ribs
    Answer: A

Pages: 173
27. The superior and inferior nuchal lines are ridges on the:
    A) ethmoid bone
    B) occipital bone
    C) nasal bones
    D) mandible
    E) cervical vertebrae
    Answer: B

Pages: 173
28. The cranial bone that articulates with all other cranial bones is the:
    A) frontal
    B) ethmoid
    C) temporal
    D) sphenoid
    E) parietal
    Answer: D

Pages: 173
29. The pterygoid processes are found on the inferior portion of the:
    A) palatine bones
    B) maxillae
    C) temporal bones
    D) sphenoid bone
    E) ethmoid bone
    Answer: D

Pages: 173
30. The pituitary gland is located in the:
    A) sella turcica of the sphenoid bone
    B) petrous portion of the temporal bone
    C) mastoid process of the temporal bone
    D) hypoglossal canal of the occipital bone
    E) crista galli of the ethmoid bone
    Answer: A

Pages: 173
31. The bone that forms the major supporting structure of the nasal cavity is the:
    A) nasal bone
    B) sphenoid bone
    C) ethmoid bone
    D) zygomatic bone
    E) temporal bone
    Answer: C

Pages: 179
32. The bone whose superior border articulates with the perpendicular plate of the ethmoid to form the nasal septum is the:
    A) lacrimal
    B) nasal
    C) palatine
    D) sphenoid
    E) vomer
    Answer: E

Pages: 173
33. The superior and middle nasal conchae are part of the:
    A) nasal bones
    B) lacrimal bones
    C) maxillae
    D) ethmoid bone
    E) sphenoid bone
    Answer: D

Pages: 173
34. The function of the nasal conchae is to:
    A) provide a surface for muscle attachment
    B) create turbulence in inspired air for cleansing purposes
    C) provide extensive surface area for gas exchange
    D) act as an anchor for periodontal ligaments
    E) protect olfactory nerves as they travel to the brain
    Answer: B

Pages: 176
35. Paranasal sinuses are found in **ALL** of the following bones **EXCEPT** the:
    A) frontal
    B) sphenoid
    C) zygomatic
    D) ethmoid
    E) maxilla
    Answer: C

Pages: 176
36. The hard palate is formed by the:
    A) horizontal plates of the palatine bones and the palatine processes of the maxillae
    B) perpendicular plate of the ethmoid and the vomer
    C) alveolar processes of the maxillae and mandibles
    D) all the nasal conchae
    E) greater and lesser wings of the sphenoid bone
    Answer: A

Pages: 178
37. The tear ducts pass through the:
    A) lacrimal bones
    B) ethmoid bone
    C) maxillae
    D) sphenoid bones
    E) inferior nasal conchae
    Answer: A

Pages: 180
38. The foramen located between the petrous portion of the temporal bone and the occipital bone is the:
    A) optic foramen
    B) jugular foramen
    C) foramen magnum
    D) foramen ovale
    E) foramen rotundum
    Answer: B

Pages: 180
39. The hypoglossal canals are located:
    A) in the petrous portions of the temporal bones
    B) at the base of the sella turcica of the sphenoid bone
    C) in the cribriform plate of the ethmoid bone
    D) superior to the base of the occipital condyles
    E) in the palatine bones
    Answer: D

Pages: 180
40. Cranial nerve I (olfactory) passes through the:
    A) superior orbital fissure
    B) cribriform plate of the ethmoid bone
    C) petrous portion of the temporal bone
    D) jugular foramen
    E) foramen ovale
    Answer: B

Pages: 180
41. The foramen located in the petrous portion of the temporal bone is the:
    A) carotid foramen
    B) foramen magnum
    C) optic foramen
    D) inferior orbital fissure
    E) mental foramen
    Answer: A

Pages: 180
42. The mandibular branch of cranial nerve V (trigeminal) passes through the:
    A) foramen magnum
    B) foramen ovale
    C) mental foramen
    D) incisive foramen
    E) mandibular foramen
    Answer: B

Pages: 180
43. The foramen bounded by parts of the sphenoid, temporal and occipital bones is the:
    A) foramen ovale
    B) foramen magnum
    C) hypoglossal canals
    D) foramen lacerum
    E) foramen rotundum
    Answer: D

Pages: 180
44. **ALL** of the following are foramina in the sphenoid bone **EXCEPT** the:
    A) foramen rotundum
    B) superior orbital fissure
    C) foramen ovale
    D) optic foramen
    E) jugular foramen
    Answer: E

Pages: 180
45. Cranial nerves III, IV, V (branch), and VI all pass through the:
    A) superior orbital fissure
    B) inferior orbital fissure
    C) jugular foramen
    D) foramen magnum
    E) hypoglossal canals
    Answer: A

Pages: 181
46. The hyoid bone is located between the larynx and the:
    A) occipital bone
    B) temporal bone
    C) mandible
    D) ethmoid bone
    E) vomer
    Answer: C

Pages: 181
47. The majority of vertebrae belong to which region of the vertebral column?
    A) cervical
    B) thoracic
    C) lumbar
    D) sacral
    E) coccygeal
    Answer: B

Pages: 181
48. Which of the following lists the regions of the vertebral column in the correct order from superior to inferior?
    A) cervical, lumbar, thoracic, coccygeal, sacral
    B) coccygeal, sacral, lumbar, thoracic, cervical
    C) coccygeal, lumbar, sacral, thoracic, cervical
    D) cervical, thoracic, lumbar, sacral, coccygeal
    E) thoracic, cervical, lumbar, sacral, coccygeal
    Answer: D

Pages: 181
49. The lumbar curve of the vertebral column develops:
    A) only if the intervertebral discs are damaged
    B) when an infant begins to hold its head erect
    C) when a child begins to sit, stand, and walk
    D) during the third month of fetal development
    E) just prior to birth
    Answer: C

Pages: 183
50. The weight-bearing part of a vertebra is the:
    A) spinous process
    B) transverse processes
    C) nucleus pulposus
    D) centrum
    E) vertebral arch
    Answer: D

Pages: 183
51. The spinal cord passes through the:
    A) intervertebral foramen
    B) transverse foramen
    C) vertebral foramen
    D) centrum
    E) both C and D are correct
    Answer: C

Pages: 185
52. When whiplash injuries result in death, the usual cause is damage to the medulla oblongata of the brain by the:
    A) rupturing of intervertebral discs
    B) dens of the axis
    C) spinous processes of the cervical vertebrae
    D) vertebra prominens
    E) lateral masses of the atlas
    Answer: B

Pages: 186
53. The largest and strongest vertebrae belong to the:
    A) cervical region
    B) thoracic region
    C) lumbar region
    D) sacral region
    E) coccygeal region
    Answer: C

Pages: 186
54. An inferior entrance to the vertebral canal is the sacral:
    A) cornua
    B) ala
    C) promontory
    D) canal
    E) hiatus
    Answer: E

Pages: 186
55. An important obstetrical landmark is the sacral:
    A) cornua
    B) ala
    C) promontory
    D) foramina
    E) crests
    Answer: C

Pages: 186
56. The auricular surface of the sacrum articulates with the:
    A) coccyx
    B) fifth lumbar vertebra
    C) ilium of the hipbone
    D) femur
    E) pubis of the hipbone
    Answer: C

Pages: 190
57. The most inferior portion of the sternum is the:
    A) manubrium
    B) body
    C) xiphoid process
    D) suprasternal notch
    E) coccyx
    Answer: C

Pages: 191
58. The articular part of the tubercle of a rib articulates with the:
    A) body of the sternum
    B) thoracic vertebrae
    C) lumbar vertebrae
    D) xiphoid process of the sternum
    E) both B and C are correct
    Answer: B

Pages: 193
59. An abnormal lateral bending of the vertebral column, usually seen in the thoracic region, is called:
    A) scoliosis
    B) kyphosis
    C) lordosis
    D) spina bifida
    E) herniation
    Answer: A

Pages: 193
60. The lamina of the vertebrae fail to fuse in the congenital condition known as:
    A) scoliosis
    B) kyphosis
    C) lordosis
    D) cleft palate
    E) spina bifida
    Answer: E

61. The vertebral artery passes through what part of the cervical vertebrae?
    A) vertebral canals
    B) intervertebral foramina
    C) transverse processes
    D) vertebra prominens
    E) bodies
    Answer: C

62. Paired L-shaped bones that form part of the hard palate, part of the floor and walls of the nasal cavity, and part of the floor of the orbit are the:
    A) lacrimal bones
    B) nasal bones
    C) maxillae
    D) vomers
    E) palatines
    Answer: E

63. The vertebral arch of a vertebra is formed by the pedicles and the:
    A) laminae
    B) centrum
    C) superior articular facets
    D) demifacets
    E) transverse processes
    Answer: A

64. The meninges that cover the brain attach anteriorly to the:
    A) sella turcica of the sphenoid bone
    B) styloid processes of the temporal bones
    C) external occipital protuberance
    D) perpendicular plate of the ethmoid bone
    E) crista galli of the ethmoid bone
    Answer: E

65. The manubrium of the sternum articulates with the:
    A) first and second pairs of ribs
    B) medial ends of the clavicles
    C) eleventh and twelfth pairs of ribs
    D) no other bones
    E) both A and B are correct
    Answer: E

True-False

Write T if the statement is true and F if the statement is false.

Pages: 163
1. Flat bones are generally composed of two plates of spongy bone enclosing a layer of compact bone.
   Answer: False

Pages: 163
2. Wormian bones are small bones found in the tendons at joints.
   Answer: False

Pages: 164
3. A trochanter is a large projection seen only on the femur.
   Answer: True

Pages: 164
4. A tubercle is a small, rounded process on a bone.
   Answer: True

Pages: 173
5. The pterygoid processes are part of the palatine bones.
   Answer: False

Pages: 173
6. The ethmoid bone forms part of the medial wall of the orbit.
   Answer: True

Pages: 185
7. Transverse foramina are seen in the transverse processes of the cervical vertebrae, but not in the transverse processes of the vertebrae of the thoracic and lumbar regions.
   Answer: True

Pages: 164
8. The appendicular skeleton contains more bones than the axial skeleton.
   Answer: True

Pages: 167
9. The coronal suture is located between the two parietal bones.
   Answer: False

Pages: 176
10. The alveoli of the maxillae are air-filled spaces.
    Answer: False

Pages: 171
11. The temporal process is a projection from the zygomatic bone.
    Answer: True

12. The medulla oblongata of the brain passes through the foramen magnum.
Answer: True

Pages: 183
13. The spinal cord passes through the bodies of the vertebrae.
Answer: False

Pages: 191
14. Women normally have only one pair of floating ribs.
Answer: False

Pages: 171
15. Cranial nerve VIII (vestibulocochlear) passes through the external auditory meatus.
Answer: False

Short Answer

Write the word or phrase that best completes each statement or answers the question.

Pages: 180
1. The optic foramina are found in the _____ bone.
Answer: sphenoid

Pages: 190
2. The superior portion of the sternum is called the _____.
Answer: manubrium

Pages: 164
3. There are _____ bones in the axial division of the skeletal system, and _____ bones in the appendicular division.
Answer: 80, 126

Pages: 167
4. An immovable joint found only between skull bones is called a _____.
Answer: suture

Pages: 167
5. The suture located between the two parietal bones is the _____ suture.
Answer: sagittal

Pages: 167
6. The suture located between the parietal bones and the occipital bone is the _____ suture.
Answer: lambdoid

Pages: 176,179
7. The hard palate is formed by the _____ of the maxillae and the _____ of the palatine bones.
Answer: palatine plate; horizontal plate

Pages: 176,178
8. The tooth sockets are called _____.
   Answer: alveoli

Pages: 184
9. The first cervical vertebra is called the _____.
   Answer: atlas

Pages: 183
10. Spinal nerves pass through openings between the vertebrae called _____.
    Answer: intervertebral foramina

Pages: 185
11. The atlas and head rotate around the _____ of the _____.
    Answer: dens (odontoid process); axis

Pages: 191
12. The facets and demifacets on the bodies of the thoracic vertebrae are sites of articulation
    with the _____.
    Answer: heads of the ribs

Pages: 190
13. The sternal angle is formed by the junction of the _____ and the _____.
    Answer: manubrium; body (of sternum)

Pages: 191
14. Ribs that attach anteriorly to the cartilage of other ribs are referred to as false ribs, or
    _____ ribs.
    Answer: vertebrochondral

Pages: 181
15. An intervertebral disc consists of the outer annulus fibrosus and the inner elastic structure
    called the _____.
    Answer: nucleus pulposus

Pages: 181
16. When an infant begins to hold its head up, its vertebral column begins to develop a _____
    curve.
    Answer: cervical

Pages: 171
17. The structures involved in hearing and equilibrium are housed in the _____ of the temporal
    bone.
    Answer: petrous portion

Pages: 172
18. The foramen magnum is a large hole in the _____ bone.
    Answer: occipital

Pages: 173
19. The sella turcica of the sphenoid bone provides protection for the _____ gland.
    Answer: pituitary

Pages: 173
20. The olfactory foramina are located in the _____ of the ethmoid bone.
    Answer: cribriform plate

Pages: 176
21. The bony part of the bridge of the nose is formed by the _____.
    Answer: nasal bones

Pages: 176,178
22. Alveolar processes are seen on two facial bones, the _____ and the _____.
    Answer: mandible and maxillae

Pages: 178
23. Two foramina in the mandible that are important sites for injection of dental anesthetics are
    the mandibular foramen and the _____ foramen.
    Answer: mental

Pages: 179
24. A triangular bone that forms the inferior and posterior part of the nasal septum is the

    _____.

    Answer: vomer

Pages: 181
25. A bone located between the mandible and the larynx is the _____ bone.
    Answer: hyoid

Pages: 181
26. The adult vertebral column consists of _____ cervical vertebrae, _____ thoracic vertebrae,
    _____ lumbar vertebrae, _____ sacral vertebrae fused into one, and _____ coccygeal vertebrae
    fused into one or two bones.
    Answer: 7; 12; 5; 5; 4

Pages: 181
27. The anteriorly concave curves of the vertebral column are the _____ and the _____.
    Answer: thoracic; sacral

Pages: 183
28. The two short, thick processes between the body and laminae of a vertebra are the _____.
    Answer: pedicles

Pages: 184
29. The largest vertebral foramina are found in the vertebrae of the _____ region.
    Answer: cervical

Pages: 193

30. An increase in abdominal weight may cause an exaggeration of the lumbar curvature of the spine known as swayback, or _____.
    Answer: lordosis

Matching

Choose the item from Column 2 that best matches each item in Column 1.

1. foramen	an opening through which blood vessels, nerves, or ligaments pass
2. fossa	a depression in or on a bone
3. sulcus	a groove that accommodates a soft structure
4. meatus	a tubelike passageway running within a bone
5. fissure	a narrow, cleftlike opening between adjacent parts of bones
6. facet	a smooth, flat surface
7. crest	a prominent border or ridge
8. tubercle	a small, rounded process
9. tuberosity	a large, rounded usually roughened process

Essay

Write your answer in the space provided or on a separate sheet of paper.

Pages: 163

1. Briefly describe the four principal classifications of bones based on shape.
    Answer: Long bones: greater length than width; curved for strength Short bones: nearly equal length and width Flat bones: two plates of compact bone surrounding spongy bone Irregular bones: complex shapes Wormian bones: located in sutures of skull Sesamoid bones: located in tendons at joints

Pages: 181-186

2. Describe the anatomy of the adult vertebral column, noting unique characteristics of the vertebrae of each region and the characteristics of each region as a whole.
    Answer: Answer should include names of regions, numbers of vertebrae per region, and normal curvatures in each region. Cervical descriptions should include special names, transverse foramina, bifid spinous processes. Thoracic should include rib facets and unique spinous processes. Lumbar should include descriptions of specializations for support. Sacral should include fusion characteristics; likewise for the coccygeal region.

Pages: 179

3. Describe the function of the orbit, and give the names and locations of the bones of the orbit.
    Answer: Function = protection of eyeball & associated structures Bones:
    1) roof = frontal and sphenoid
    2) lateral wall = sphenoid and zygomatic
    3) floor = maxilla, zygomatic, palatine
    4) medial wall = maxilla, lacrimal, ethmoid, sphenoid

Pages: 171

4. Name the major regions, processes, and foramina of the temporal bone.
    Answer: Describe temporal squama, zygomatic process, petrous portion, carotid foramen, jugular foramen, mandibular fossa, articular tubercle, mastoid portion, external auditory meatus, mastoid air cells, mastoid process, mastoid foramen, internal auditory meatus, styloid process.

Pages: 170

5. Define the term "fontanel," and describe the functions of fontanels.
    Answer: Fontanel = membrane-filled spaces between cranial bones of newborn
    1) enable fetal skull to modify size & shape as it passes through birth canal
    2) permit rapid brain growth in infancy
    3) help determine degree of brain development by state of closure
    4) landmark for blood drawing

Pages: 191

6. Describe the classification of ribs based on articulations with other bones.
    Answer: · True (vertebrosternal) - pairs 1-7; attach directly to sternum via costal cartilages
    · False (vertebrochondral) - pairs 8-10; attach to each other's cartilage then to cartilage of pair 7
    · False (vertebral) - pairs 11-12; no anterior attachment

Pages: 173

7. Describe the anatomical features of the ethmoid bone.

   Answer: General anatomy & location - light, spongelike; anterior to sphenoid, posterior to nasals; major support of nasal cavity Specific anatomy - describe lateral masses, ethmoidal sinuses, perpendicular plate, cribriform plate, olfactory foramina, crista galli, nasal conchae

Pages: 181-186

8. Archeologists have unearthed several vertebrae. Vertebrae A and B are extremely tiny, and appear to be fused. Vertebrae C-F have transverse foramina and bifid spinous processes. Vertebra G has a spinous process that projects straight posteriorly, and has superior articular facets that project medially. How should these vertebrae be arranged to put them in the most realistic order? Why?

   Answer: Vertebrae C-F most superior (cervical); Vertebra G (lumbar); Vertebrae A and B (coccygeal)

Pages: 190-191

9. Describe the anatomical composition of the skeletal portion of the thorax.

   Answer: · Sternum - median line of anterior thoracic wall; 6" long
   · Costal cartilages - hyaline cartilage attaching ribs to sternum
   · Ribs - 12 pairs increasing in length 1-7, decreasing in length to twelfth; each articulates posteriorly w/ bodies of thoracic vertebrae; 1-7 = vertebrosternal (attach to sternum); 8-10 = vertebrochondral (attach to cartilage); 11-12 = vertebral (attach only to vertebrae)
   · Bodies of thoracic vertebrae

Chapter 8   The Skeletal System: The Appendicular Skeleton

Multiple-Choice

Choose the one alternative that best completes the statement or answers the question.

Pages: 196
1. **ALL** of the following are part of the appendicular skeleton **EXCEPT** the:
    A) scapula
    B) coxal bones
    C) sternum
    D) fibula
    E) radius
    Answer: C

Pages: 197
2. The conoid tubercle and costal tuberosity are markings found on the:
    A) scapula
    B) radius
    C) ulna
    D) humerus
    E) clavicle
    Answer: E

Pages: 197
3. At the lateral end of the spine of the scapula is the:
    A) acromion process
    B) vertebral column
    C) coracoid process
    D) subscapular fossa
    E) sternum
    Answer: A

Pages: 198
4. The coronoid and olecranon fossae are depressions found on the:
    A) ulna
    B) radius
    C) scapula
    D) humerus
    E) femur
    Answer: D

Pages: 202
5. Most of the muscles of the forearm attach to the:
    A) greater and lesser tubercles of the humerus
    B) coronoid and olecranon fossae of the humerus
    C) deltoid tuberosity of the humerus
    D) acromion and coracoid processes of the scapula
    E) medial and lateral epicondyles of the humerus
    Answer: E

Pages: 202
6. The biceps brachii muscle attaches to the:
   A) styloid process of the ulna
   B) radial tuberosity of the radius
   C) styloid process of the radius
   D) olecranon process of the ulna
   E) coronoid process of the ulna
   Answer: B

Pages: 201
7. Which of the following is part of the elbow joint?
   A) trochlear notch of the ulna
   B) ulnar notch of the radius
   C) glenoid fossa of the humerus
   D) styloid process of the radius
   E) head of the ulna
   Answer: A

Pages: 202
8. **ALL** of the following are in the proximal row of carpal bones **EXCEPT** the:
   A) hamate
   B) lunate
   C) scaphoid
   D) pisiform
   E) triquetrum
   Answer: A

Pages: 202
9. **ALL** of the following are in the distal row of carpal bones **EXCEPT** the:
   A) trapezium
   B) trapezoid
   C) triquetrum
   D) hamate
   E) capitate
   Answer: C

Pages: 202
10. The most commonly fractured of the carpal bones is the:
    A) hamate
    B) capitate
    C) lunate
    D) pisiform
    E) scaphoid
    Answer: E

11. The head of the radius articulates with the:
    A) ulnar notch
    B) coronoid process
    C) capitulum
    D) trochlea
    E) styloid process
    Answer: C

12. The carpal tunnel is formed by the flexor retinaculum, the pisiform, the scaphoid, the trapezium, and the:
    A) styloid process of the ulna
    B) styloid process of the radius
    C) metacarpals
    D) hamate
    E) triquetrum
    Answer: D

13. The structures commonly known as the "knuckles" are the:
    A) heads of the metacarpals
    B) bases of the metacarpals
    C) heads of the proximal phalanges
    D) heads of the distal phalanges
    E) carpals
    Answer: A

14. The greater sciatic notch, through which the sciatic nerve passes, is part of the:
    A) femur
    B) ischium
    C) pubis
    D) humerus
    E) ilium
    Answer: E

15. The auricular surface is the point of articulation between the ilium and the:
    A) ischium
    B) pubis
    C) femur
    D) coccyx
    E) sacrum
    Answer: E

Pages: 205
16. The pelvic brim is the circumference of the oblique plane extending from the sacral promontory to the:
    A) anterior inferior iliac spines
    B) pubic symphysis
    C) ischial tuberosities
    D) greater trochanters of the femurs
    E) posterior superior iliac spines
    Answer: B

Pages: 203
17. The head of the femur articulates with the:
    A) patella
    B) medial and lateral condyles of the tibia
    C) auricular surface of the ilium
    D) acetabulum
    E) both A and B are correct
    Answer: D

Pages: 208
18. When someone says an elderly person "broke her hip," it is most likely that the fracture occurred in the:
    A) sacroiliac joint
    B) pubic symphysis
    C) neck of the femur
    D) acetabulum
    E) ilium
    Answer: C

Pages: 208
19. The greater and lesser trochanters are projections seen on the:
    A) humerus
    B) scapula
    C) tibia
    D) femur
    E) ischium
    Answer: D

Pages: 208
20. The gluteal tuberosity and the linea aspera are ridges seen on the:
    A) ilium
    B) femur
    C) ischium
    D) tibia
    E) pubis
    Answer: B

Pages: 209
21. The medial malleolus is part of the:
    A) tibia
    B) fibula
    C) femur
    D) humerus
    E) ulna
    Answer: A

Pages: 209
22. The lateral malleolus is part of the:
    A) tibia
    B) fibula
    C) femur
    D) humerus
    E) ulna
    Answer: B

Pages: 209
23. The tibia articulates with the:
    A) calcaneus only
    B) talus only
    C) cuboid only
    D) navicular only
    E) both the calcaneus and the talus
    Answer: B

Pages: 209
24. The largest of the tarsal bones is the:
    A) talus
    B) hamate
    C) capitate
    D) calcaneus
    E) cuboid
    Answer: D

Pages: 197
25. The acromion process of the scapula articulates with the:
    A) greater tubercle of the humerus
    B) head of the humerus
    C) lateral end of the clavicle
    D) medial end of the clavicle
    E) vertebral column
    Answer: C

Pages: 197
26. Which of the following is located on the ventral surface of the scapula?
    A) acromion process
    B) spine
    C) supraspinous fossa
    D) subscapular fossa
    E) all of these
    Answer: D

Pages: 199
27. The head of the humerus articulates with the:
    A) trochlear notch of the ulna
    B) head of the radius
    C) glenoid cavity of the scapula
    D) clavicle
    E) both C and D are correct
    Answer: C

Pages: 199
28. The roughened area on the middle portion of the shaft of the humerus is the:
    A) deltoid tuberosity
    B) anatomical neck
    C) capitulum
    D) trochlea
    E) lesser tubercle
    Answer: A

Pages: 200
29. The ulnar nerve lies on the posterior surface of the:
    A) trochlear notch
    B) medial epicondyle of the humerus
    C) lateral epicondyle of the humerus
    D) deltoid tuberosity
    E) carpals
    Answer: B

Pages: 200
30. The prominence of the elbow is formed by the:
    A) coronoid process of the ulna
    B) olecranon process of the ulna
    C) head of the radius
    D) head of the humerus
    E) ulnar tuberosity
    Answer: B

Pages: 201
31. A styloid process can be found at the distal end of the:
    A) femur
    B) tibia
    C) ulna
    D) humerus
    E) all of these have a styloid process
    Answer: C

Pages: 202
32. Which of the following lists the proximal row of carpal bones in the correct order from lateral to medial?
    A) scaphoid, pisiform, lunate, triquetrum
    B) pisiform, triquetrum, lunate, scaphoid
    C) scaphoid, lunate, triquetrum, pisiform
    D) scaphoid, triquetrum, lunate, pisiform
    E) none of these, because these are distal carpal bones
    Answer: C

Pages: 202
33. Which of the following lists the distal row of carpal bones in the correct order from lateral to medial?
    A) hamate, capitate, trapezoid, trapezium
    B) trapezium, trapezoid, capitate, hamate
    C) capitate, hamate, trapezium, trapezoid
    D) trapezoid, trapezium, capitate, hamate
    E) none of these, because these are all proximal carpal bones
    Answer: B

Pages: 202
34. The carpal bone with a large, hook-shaped projection on its anterior surface is the:
    A) scaphoid
    B) hamate
    C) lunate
    D) capitate
    E) pisiform
    Answer: B

Pages: 202
35. The bones making up the palm of the hand are the:
    A) metatarsals
    B) metacarpals
    C) carpals
    D) tarsal
    E) phalanges
    Answer: B

Pages: 202
36. The bases of the proximal phalanges articulate with the:
    A) heads of the distal phalanges
    B) heads of the middle phalanges
    C) heads of the metacarpals
    D) bases of the middle phalanges
    E) bases of the metacarpals
    Answer: C

Pages: 202
37. The superior border of the most superior of the subdivisions of the hipbone is the:
    A) ischial tuberosity
    B) superior ramus of the pubis
    C) pubic crest
    D) ischial spine
    E) iliac crest
    Answer: E

Pages: 202
38. Which metacarpal is proximal to the little finger?
    A) number I
    B) number II
    C) number III
    D) number IV
    E) number V
    Answer: E

Pages: 202
39. An autopsy reveals a missing thumb on the left hand. How many phalanges remain on the left hand?
    A) 4
    B) 8
    C) 10
    D) 11
    E) 12
    Answer: E

Pages: 202
40. When you sit on a stool, which part of the coxal bones touch the stool first?
    A) ischial spines
    B) ischial tuberosities
    C) iliac crests
    D) pubic symphysis
    E) inferior pubic rami
    Answer: B

Pages: 209
41. The prominence felt on the medial surface of the ankle is the:
    A) fibular notch
    B) medial condyle
    C) medial malleolus
    D) talus
    E) tibial tuberosity
    Answer: C

Pages: 209
42. The only bone of the foot that articulates with the tibia is the:
    A) calcaneus
    B) intermediate cuneiform
    C) navicular
    D) talus
    E) cuboid
    Answer: D

Pages: 202
43. The large hole in the coxal bone, through which blood vessels and nerves pass, is the:
    A) acetabulum
    B) pubic symphysis
    C) obturator foramen
    D) iliac fossa
    E) glenoid cavity
    Answer: C

Pages: 205
44. The posterior border of the lesser (true) pelvis is formed by the:
    A) lumbar vertebrae
    B) ischial tuberosities
    C) pubic symphysis
    D) sacrum and coccyx
    E) posterior iliac spines
    Answer: D

Pages: 209
45. The articular facets of the patella articulate with the:
    A) medial and lateral condyles of the tibia
    B) medial and lateral condyles of the femur
    C) greater and lesser trochanters of the femur
    D) medial and lateral epicondyles of the femur
    E) head of the fibula and the tibial tuberosity
    Answer: B

Pages: 209
46. The patellar ligament attaches to the:
    A) greater trochanter of the femur
    B) lesser trochanter of the femur
    C) linea aspera of the femur
    D) tibial tuberosity of the tibia
    E) head of the fibula
    Answer: D

Pages: 209
47. Since the patella develops in the tendon of the quadriceps femoris muscle, it is classified as a:
    A) short bone
    B) Wormian bone
    C) tendinous bone
    D) sesamoid bone
    E) synovial bone
    Answer: D

Pages: 209
48. During walking, the bone that initially bears the entire weight of the body is the:
    A) calcaneus
    B) talus
    C) fibula
    D) first metatarsal
    E) tibia
    Answer: B

Pages: 202
49. The distal end of the radius articulates with the:
    A) radial notch of the ulna
    B) capitulum of the humerus
    C) scaphoid and lunate
    D) coronoid fossa of the humerus
    E) both A and B are correct
    Answer: C

Pages: 205
50. The plane of the pelvic outlet runs:
    A) from the tip of the coccyx to the pubic symphysis
    B) over the superior borders of the iliac crests
    C) from the sacral promontory directly anteriorly
    D) transversely through the body at the level of the superior iliac spines
    E) midsagittally through the sacrum and pubic symphysis
    Answer: A

Pages: 205
51. A pair of coxal bones unearthed from an unmarked grave have oval-shaped obturator foramina and a pubic arch of greater than 90°. What can you tell from this information?
    A) The person was probably very old because the bones have spread apart from their original position.
    B) The bones are probably those of a male because the pubic angle would be less in a female.
    C) The bones are probably those of a female because the pubic angle would be less in a male.
    D) The bones are probably from someone who suffered from a Vitamin D deficiency, which caused abnormal flexibility in the bones.
    E) The bones are probably from a very young person because the obturator foramen is round in adults.
    Answer: C

Pages: 206
52. Which of the following structures would be most visible on a posterior view of the lower extremity?
    A) tibial tuberosity
    B) gluteal tuberosity
    C) intertrochanteric line
    D) patellar surface
    E) both B and C are correct
    Answer: B

Pages: 209
53. The fibular notch is located at the:
    A) proximal end of the tibia
    B) superior surface of the talus
    C) inferior border of the lateral condyle of the femur
    D) distal end of the tibia
    E) head of the fibula
    Answer: D

Pages: 209
54. The base of the first metatarsal articulates with the:
    A) first cuneiform
    B) third cuneiform
    C) cuboid
    D) proximal phalanx of the great toe
    E) proximal phalanx of the smallest toe
    Answer: A

Pages: 196
55. The pectoral girdle consists of the clavicle and the:
    A) scapula
    B) humerus
    C) ribs
    D) sternum
    E) all of these are part of the pectoral girdle
    Answer: A

Pages: 197
56. The medial end of the clavicle articulates with the:
    A) acromion process of the scapula
    B) head of the humerus
    C) seventh cervical vertebra
    D) sternum
    E) coracoid process of the scapula
    Answer: D

Pages: 197
57. The most frequently broken bone in the body is the:
    A) femur
    B) humerus
    C) fibula
    D) ulna
    E) clavicle
    Answer: E

Pages: 197
58. The high point of the shoulder is the:
    A) acromion process of the scapula
    B) greater tubercle of the humerus
    C) conoid tubercle of the clavicle
    D) costal tuberosity of the clavicle
    E) coracoid process of the scapula
    Answer: A

Pages: 199
59. The former site of the epiphyseal plate of the humerus is the:
    A) anatomical neck
    B) surgical neck
    C) intertubercular sulcus
    D) capitulum
    E) fovea capitis
    Answer: A

Pages: 200-201
60. **ALL** of the following parts of the ulna articulate with the humerus **EXCEPT** the:
    A) coronoid process
    B) olecranon process
    C) trochlear notch
    D) head
    E) neither the head nor the coronoid process articulate with the humerus
    Answer: D

61. The ulnar notch is located:
    A) on the posterior side of the distal end of the humerus
    B) on the medial side of the distal end of the radius
    C) between the coronoid and olecranon processes of the ulna
    D) on the medial side of the head of the radius
    E) just below the coronoid process of the ulna
    Answer: B

62. The median nerve passes through the:
    A) obturator foramen
    B) pubic symphysis
    C) greater sciatic notch
    D) trochlear notch
    E) carpal tunnel
    Answer: E

63. The longest, heaviest, and strongest bone in the body is the:
    A) humerus
    B) femur
    C) tibia
    D) fibula
    E) calcaneus
    Answer: B

64. The medial and lateral condyles of the tibia articulate with the:
    A) talus and lateral malleolus of the fibula
    B) medial and lateral epicondyles of the femur
    C) medial and lateral condyles of the femur
    D) greater and lesser trochanters of the femur
    E) talus and calcaneus
    Answer: C

65. Genu valgum develops when:
    A) the patella slips out of the quadriceps femoris tendon
    B) the neck of the femur is at an abnormally decreased angle
    C) the neck of the femur is at an abnormally increased angle
    D) the shaft of the femur is too laterally angled
    E) the acetabulum becomes misshapen following a difficult childbirth
    Answer: B

True-False

Write T if the statement is true and F if the statement is false.

Pages: 199
1. The humerus articulates proximally with the scapula and distally with the ulna and radius.
Answer: True

Pages: 200
2. The head of the ulna is on its distal end, while the head of the radius is on its proximal end.
Answer: True

Pages: 205
3. The portion of the pelvis above the pelvic brim is called the true pelvis.
Answer: False

Pages: 208
4. The patella develops in the tendon of the biceps femoris muscle.
Answer: False

Pages: 212
5. The bones of the female are generally heavier than the bones of the male.
Answer: False

Pages: 207
6. The distal end of the femur articulates with both the tibia and the fibula.
Answer: False

Pages: 208
7. The linea aspera is a vertical ridge on the posterior aspect of the femur.
Answer: True

Pages: 204
8. The lesser sciatic notch is located just below the ischial spine.
Answer: True

Pages: 202
9. The fifth metacarpal articulates with the pollex.
Answer: False

Pages: 199
10. The coronoid fossa is a depression on the anterior surface of the distal end of the humerus.
Answer: True

Pages: 199
11. The lesser tubercle projects posteriorly from the humerus.
Answer: False

Pages: 199
12. The scapular notch articulates with the head of the humerus.
Answer: False

Pages: 197
13. During a fall on an outstretched arm, force is transmitted from the upper limb to the trunk by the clavicle.
Answer: True

Pages: 202
14. The distal end of the radius articulates with the capitate and hamate.
Answer: False

Pages: 205
15. The pelvic brim is more heart-shaped in women than in men.
Answer: False

Short Answer

Write the word or phrase that best completes each statement or answers the question.

Pages: 197
1. The scapula articulates with the clavicle and the _____.
Answer: humerus

Pages: 199
2. The former site of the epiphyseal plate on the proximal end of the humerus is the _____.
Answer: anatomical neck

Pages: 203
3. The fossa that is formed by the union of the ilium, ischium, and pubis, and that receives the head of the femur is the _____.
Answer: acetabulum

Pages: 200
4. The prominence of the elbow is formed by the part of the ulna known as the _____.
Answer: olecranon process

Pages: 202
5. The largest foramen in the skeleton is the _____.
Answer: obturator foramen

Pages: 209
6. The medial prominence of the ankle is the medial malleolus of the _____.
Answer: tibia

Pages: 209
7. The only foot bone that articulates with the tibia and fibula is the _____.
Answer: talus

Pages: 211
8. The tarsal bone that is part of the lateral part of the longitudinal arch, but is **not** part of the medial part of the longitudinal arch is the _____.
Answer: cuboid

Pages: 208
9. The superior end of the patella is called the _____.
Answer: base

Pages: 209
10. The proximal end of the fibula is called the _____.
Answer: head

Pages: 208
11. The area between the condyles on the anterior surface of the femur is called the _____.
Answer: patellar surface

Pages: 205
12. The imaginary curved line describing the course taken by the baby's head as it passes through the pelvis is the pelvic _____.
Answer: axis

Pages: 202
13. The indentation just below the posterior inferior iliac spine is the _____.
Answer: greater sciatic notch

Pages: 202
14. The iliacus muscle attaches to a concavity known as the _____.
Answer: iliac fossa

Pages: 202
15. The nerve that passes through the carpal tunnel is the _____.
Answer: median nerve

Pages: 202
16. The proximal row of carpals consists of the scaphoid, lunate, triquetrum, and _____.
Answer: pisiform

Pages: 202
17. The scaphoid and the lunate articulate proximally with the _____.
Answer: radius

Pages: 202
18. The most medial of the distal row of carpals is the _____.
Answer: hamate

Pages: 202
19. The small projection on the lateral side of the distal end of the radius is the _____.
Answer: styloid process

Pages: 199
20. The head of the radius articulates with a rounded knob on the humerus called the _____.
Answer: capitulum

Pages: 199
21. The most laterally palpable bony landmark of the shoulder region is the _____.
Answer: greater tubercle of the humerus

Pages: 199
22. The roughened, V-shaped area in the middle portion of the shaft of the humerus is the

_____.
Answer: deltoid tuberosity

Pages: 200
23. The nerve that lies on the posterior surface of the medial epicondyle of the humerus is the

_____.
Answer: ulnar nerve

Pages: 199
24. A depression on the posterior surface of the distal end of the humerus is the _____.
Answer: olecranon fossa

Pages: 197
25. The prominent indentation along the superior border of the scapula through which the suprascapular nerve passes is the _____.
Answer: scapular notch

Pages: 197
26. The edge of the scapula near the vertebral column is called the _____.
Answer: medial border

Pages: 197
27. The acromion process of the scapula articulates with the _____.
Answer: clavicle

Pages: 197
28. The projection on the inferior surface of the lateral end of the clavicle is the _____.
Answer: conoid tubercle

Pages: 202
29. The auricular surface of the ilium articulates with the _____.
Answer: sacrum

Pages: 208
30. The larger projection on the proximal end of the femur that is lateral to the head of the femur is the _____.
Answer: greater trochanter

Matching

Choose the item from Column 2 that best matches each item in Column 1.

1. trochlear notch   ulna

2. ulnar notch   radius

3. greater sciatic notch   ilium

4. lesser sciatic notch   ischium

5. coronoid fossa   humerus

6. glenoid fossa   scapula

7. intercondylar fossa   femur

8. fibular notch   tibia

9. lateral malleolus   fibula

10. conoid tubercle   clavicle

Essay

Write your answer in the space provided or on a separate sheet of paper.

Pages: 205
1. Describe the differences between the male pelvis and the female pelvis.
   Answer: Female pelvis specialized for childbirth; false pelvis shallower; pelvic brim larger and more oval (vs. heart-shaped in male); pubic arch greater than 90° (male less than 90°); ilium less vertical; iliac fossa shallower; iliac crest less curved; acetabulum smaller; obturator foramen oval (vs. rounded in male)

Pages: 202
2. Name the carpal bones, and describe their arrangement and articulations.
   Answer: · proximal row, lateral to medial = scaphoid, lunate, triquetrum, pisiform; scaphoid and lunate articulate with radius
   · distal row, lateral to medial = trapezium, trapezoid, capitate, hamate; articulate distally with bases of metacarpals
   · articulations with each other may be included as required

Pages: 209
3. Name the tarsal bones, and describe their arrangements and articulations.
   Answer: · talus, most superior, articulates with tibia and fibula
   · calcaneus, largest, strongest, posterior (heelbone)
   · anterior, medial to lateral = first, second, third cuneiforms, navicular, cuboid; articulate anteriorly with bases of metatarsals

Pages: 200
4. Describe the anatomical features and articulations of the bones of the forearm.
    Answer: · ulna on medial aspect; articulates with humerus via trochlear notch, olecranon process (elbow prominence), and coronoid process; articulates with proximal end of radius at radial notch; other features may be described as required
    · radius on lateral aspect; articulates via head with capitulum of humerus and radial notch of ulna, and with distal end of ulna at ulnar notch; other features may be described as required

Pages: 196
5. Name the bones of the pectoral girdle, and describe their anatomical features and articulations.
    Answer: · Scapula and clavicle, which articulate at acromioclavicular joint.
    · Clavicle articulates on medial end with sternum; S-shaped; two inferior projections - conoid tubercle on lateral end, costal tuberosity on medial end.
    · Scapula articulates at glenoid cavity with humerus; triangular; describe projections, indentations, borders, and angles

Pages: 211
6. Discuss the importance of the arched arrangement of the foot bones.
    Answer: Arches enable the foot to support body weight, provide ideal distribution of weight over hard and soft tissues of foot, act as shock absorbers, and provide leverage.

Pages: 202-206
7. Describe the anatomical features of the pelvic girdle.
    Answer: Two coxal bones consisting of fused ilium, ischium, and pubis; all three fused laterally at acetabulum (indentation receiving head of femur); ilium largest and superior; ischium inferior; pubis anterior and fused at pubic symphysis; obturator foramen surrounded by ischium and pubis; specific spines, notches, etc., may be included in answer

Pages: 207
8. You are confronted with a disarticulated femur. How will you determine whether this is a left femur or a right one?
    Answer: Look for patellar surface (anterior) and intercondylar fossa (posterior); also gluteal tuberosity and linea aspera (posterior). Once anterior and posterior determined, aim head medially.

Pages: 199
9. You are confronted with a disarticulated humerus. How will you determine whether this is a left humerus or a right one?
    Answer: Look for lesser tubercle (anterior), sizes of epicondyles (medial larger), depths of fossae on distal end (olecranon deeper), capitulum and trochlea (anterior). Once anterior and posterior have been determined, aim head medially.

Pages: 204
10. You are confronted with a single coxal bone. How will you determine whether this is a left coxal bone or a right one?
    Answer: Look for ischial tuberosity, ischial spine, and sciatic notches, all of which are posterior, and acetabulum on lateral surface. Also look for auricular surface on posterior, medial surface.

Multiple-Choice

Choose the one alternative that best completes the statement or answers the question.

Pages: 216
1. Which of the following is considered to be an immovable joint?
    A) amphiarthrosis
    B) symphysis
    C) diarthrosis
    D) synovial
    E) synarthrosis
    Answer: E

Pages: 216
2. Which of the following is a freely movable joint?
    A) amphiarthrosis
    B) synostosis
    C) diarthrosis
    D) synarthrosis
    E) symphysis
    Answer: C

Pages: 226
3. A movement that increases the angle between articulating bones is:
    A) abduction
    B) adduction
    C) flexion
    D) extension
    E) rotation
    Answer: D

Pages: 226
4. Turning the soles of the feet inward such that they face each other is termed:
    A) dorsiflexion
    B) plantar flexion
    C) eversion
    D) inversion
    E) supination
    Answer: D

Pages: 231
5. The ligament stretched or torn in about 70% of all serious knee injuries is the:
    A) anterior cruciate
    B) arcuate popliteal
    C) lateral collateral
    D) medial collateral
    E) posterior cruciate
    Answer: A

Pages: 217
6. In which of the following joints does fibrocartilage connect the bones?
   A) synovial joint only
   B) symphysis only
   C) suture only
   D) syndesmosis only
   E) both symphysis and suture
   Answer: B

Pages: 216
7. Which of the following joints normally becomes a synostosis?
   A) synovial joint only
   B) symphysis only
   C) suture only
   D) synchondrosis only
   E) both suture and synchondrosis
   Answer: E

Pages: 217
8. When referring to a joint, a **meniscus** is:
   A) a sac of synovial fluid between bones and overlying tissues
   B) a fibrocartilage disc that extends into the joint cavity
   C) connective tissue connecting one bone to another bone
   D) the level of synovial fluid in the joint cavity
   E) the layer of cells lining the articular capsule
   Answer: B

Pages: 219
9. When referring to a joint, a **bursa** is:
   A) a sac of synovial fluid between bones and overlying tissues
   B) a fibrocartilage disc that extends into the joint cavity
   C) connective tissue connecting one bone to another bone
   D) the synovial fluid within the joint cavity
   E) the layer of cells lining the articular capsule
   Answer: A

Pages: 217
10. A **ligament** is:
    A) a sac of synovial fluid between the bones and overlying tissues at a joint
    B) a fibrocartilage disc extending into the joint cavity
    C) connective tissue connecting one bone to another bone
    D) connective tissue connecting a skeletal muscle to a bone
    E) the cartilage covering the ends of long bones at joints
    Answer: C

Pages: 217
11. The articular cartilage covering the ends of long bones at synovial joints is:
    A) hyaline cartilage
    B) elastic cartilage
    C) fibrocartilage
    D) dense fibrous connective tissue
    E) loose connective tissue
    Answer: A

Pages: 217
12. Synovial fluid is produced by:
    A) chondrocytes in the articular cartilage
    B) osteocytes in the articulating bones
    C) cells in the inner layer of the articular capsule
    D) fibroblasts in ligaments surrounding the joint
    E) chondrocytes in menisci
    Answer: C

Pages: 216
13. There is fibrous connective tissue connecting the bones in a:
    A) suture
    B) hinge joint
    C) pivot joint
    D) symphysis
    E) all of these are correct
    Answer: A

Pages: 224
14. Flexion and extension are the principal movements performed at:
    A) gliding joints
    B) pivot joints
    C) syndesmoses
    D) symphyses
    E) hinge joints
    Answer: E

Pages: 224
15. The greatest range of motion occurs at:
    A) hinge joints
    B) ellipsoidal joints
    C) pivot joints
    D) ball-and-socket joints
    E) synchondroses
    Answer: D

Pages: 241
16. The fibrous connective tissue that connects a muscle to a bone is called a:
    A) tendon
    B) meniscus
    C) ligament
    D) bursa
    E) labrum
    Answer: A

Pages: 224
17. The type of movements normally seen at pivot joints is:
    A) abduction and adduction
    B) rotation
    C) flexion and extension
    D) protraction and retraction
    E) inversion and eversion
    Answer: B

Pages: 217
18. The joints formed by the bodies of the vertebrae are:
    A) synovial joints
    B) sutures
    C) symphyses
    D) hinge joints
    E) pivot joints
    Answer: C

Pages: 221
19. The joints between the carpal bones are examples of:
    A) gliding joints
    B) hinge joints
    C) pivot joints
    D) ball-and-socket joints
    E) sutures
    Answer: A

Pages: 226
20. Tipping the head backward is an example of:
    A) flexion
    B) hyperflexion
    C) hyperextension
    D) retraction
    E) abduction
    Answer: C

21. The articulation of the head of the humerus in the glenoid cavity of the scapula is an example of a:
   A) saddle joint
   B) ball-and-socket joint
   C) condyloid joint
   D) pivot joint
   E) hinge joint
   Answer: B

22. A gomphosis is a joint that connects:
   A) the cranial bones of the skull
   B) the bodies of vertebrae
   C) carpals to each other
   D) roots of teeth to their alveoli
   E) the diaphysis to the epiphysis of a long bone
   Answer: D

23. A synchondrosis connects:
   A) cranial bones in the skull
   B) roots of the teeth in their alveoli
   C) the bodies of vertebrae
   D) carpal bones to each other
   E) the ribs to the sternum
   Answer: E

24. The epiphyseal plate is an example of a:
   A) gliding joint
   B) synchondrosis
   C) symphysis
   D) suture
   E) syndesmosis
   Answer: B

25. The mouth is opened when:
   A) depression occurs at the temporomandibular joint
   B) elevation occurs at the temporomandibular joint
   C) flexion occurs at the temporomandibular joint
   D) extension occurs at the temporomandibular joint
   E) protraction occurs at the temporomandibular joint
   Answer: A

Pages: 226
26. Turning of the palm anteriorly or superiorly is referred to as:
    A) abduction of the hand
    B) adduction of the hand
    C) supination of the forearm
    D) pronation of the forearm
    E) rotation of the hand
    Answer: C

Pages: 226
27. Standing on the toes, such as ballet dancers do, is an example of:
    A) inversion
    B) eversion
    C) plantar flexion
    D) dorsiflexion
    E) hyperextension
    Answer: C

Pages: 226
28. Making circles with your arms is best described as:
    A) abduction
    B) hyperextension
    C) circumduction
    D) elevation
    E) flexion
    Answer: C

Pages: 226
29. The transverse humeral ligament extends from the greater tubercle of the humerus to the:
    A) anatomical neck of the humerus
    B) coracoid process of the scapula
    C) glenoid labrum
    D) lesser tubercle of the humerus
    E) deltoid tuberosity of the humerus
    Answer: D

Pages: 226
30. Fibrocartilage that stabilizes the shoulder joint is the:
    A) articular capsule
    B) coracohumeral ligament
    C) glenoid labrum
    D) subscapular bursa
    E) both A and C are fibrocartilage structures
    Answer: C

Pages: 229

31. The most common rotator cuff injury is to the:
    A) anatomical neck of the humerus
    B) supraspinatus muscle tendon
    C) transverse humeral ligaments
    D) articular capsule
    E) articular cartilage
    Answer: B

Pages: 229

32. The fused tendons of the quadriceps femoris muscle and the fascia lata make up the:
    A) collateral ligaments of the knee
    B) cruciate ligaments of the knee
    C) patellar retinacula
    D) articular discs of the knee
    E) rotator cuff
    Answer: C

Pages: 231

33. The anterior cruciate ligament extends from the lateral condyle of the femur to the:
    A) medial condyle of the femur
    B) lateral epicondyle of the femur
    C) medial condyle of the tibia
    D) head of the fibula
    E) area anterior to the intercondylar eminence of the tibia
    Answer: E

Pages: 230

34. The patellar ligament extends from the patella to the:
    A) linea aspera of the femur
    B) tibial tuberosity of the tibia
    C) head of the fibula
    D) medial and lateral condyles of the femur
    E) lesser trochanter of the femur
    Answer: B

Pages: 217

35. The joint between the two parietal bones is an example of a:
    A) suture
    B) synchondrosis
    C) syndesmosis
    D) symphysis
    E) gomphosis
    Answer: A

Pages: 234
36. The damage of rheumatoid arthritis is thought to be caused by:
    A) the body's own immune system
    B) excessive use of joints
    C) deposition of urate crystals in the soft tissues
    D) repeated dislocations of joints
    E) depletion of synovial fluid
    Answer: A

Pages: 235
37. The accumulation of the waste products of nucleic acid metabolism can cause:
    A) rheumatoid arthritis
    B) osteoarthritis
    C) gouty arthritis
    D) Lyme disease
    E) ankylosing spondylitis
    Answer: C

Pages: 226
38. Spreading the legs apart, as in doing jumping jacks, is an example of:
    A) abduction
    B) adduction
    C) flexion
    D) extension
    E) protraction
    Answer: A

Pages: 216
39. Which of the following is an example of a synarthrosis?
    A) pubic symphysis
    B) lambdoidal suture
    C) distal articulation of tibia and fibula
    D) temporomandibular joint
    E) radiocarpal joint
    Answer: B

Pages: 217
40. The outer layer of the articular capsule of a synovial joint is made of:
    A) skeletal muscle
    B) hyaline cartilage
    C) osseous tissue
    D) fibrocartilage
    E) dense irregular connective tissue
    Answer: E

Pages: 217
41. The inner layer of the articular capsule is composed of:
    A) dense irregular connective tissue
    B) areolar connective tissue
    C) hyaline cartilage
    D) elastic cartilage
    E) osseous tissue
    Answer: B

Pages: 219
42. Bursae are filled with:
    A) venous blood
    B) air
    C) synovial fluid
    D) dense irregular connective tissue
    E) spongy bone
    Answer: C

Pages: 224
43. The articulation of the dens of the axis with the atlas is an example of a:
    A) hinge joint
    B) pivot joint
    C) gliding joint
    D) symphysis
    E) saddle joint
    Answer: B

Pages: 226
44. Turning your right shoulder toward your sternum would be an example of:
    A) flexion
    B) extension
    C) lateral rotation
    D) medial rotation
    E) protraction
    Answer: D

Pages: 224
45. Biaxial motion is possible at:
    A) condyloid joints
    B) pivot joints
    C) hinge joints
    D) sutures
    E) all synovial joints
    Answer: A

Pages: 226
46. Curling the fingers into the palm would be an example of:
    A) flexion
    B) extension
    C) abduction
    D) adduction
    E) rotation
    Answer: A

Pages: 224
47. Triaxial movement can occur at the:
    A) radiocarpal joint
    B) glenohumeral joint
    C) atlanto-occipital joint
    D) temporomandibular joint
    E) sacroiliac joint
    Answer: B

Pages: 226
48. Arching the back is an example of:
    A) hyperflexion
    B) dorsiflexion
    C) hyperextension
    D) lateral flexion
    E) supination
    Answer: C

Pages: 224
49. A spool-like surface fits into a concave surface in a:
    A) ball-and-socket joint
    B) pivot joint
    C) hinge joint
    D) gliding joint
    E) symphysis
    Answer: C

Pages: 226
50. You can elevate and depress your:
    A) forearms
    B) legs
    C) mandible
    D) feet at the ankles
    E) all of these are correct
    Answer: C

51. When your palms are lying flat on a table, your forearms are:
    A) inverted
    B) everted
    C) pronated
    D) supinated
    E) dorsiflexed
    Answer: C

52. The articular capsule of the glenohumeral joint extends from the glenoid cavity to the:
    A) glenoid labrum
    B) acromion process of the scapula
    C) coracoid process of the scapula
    D) anatomical neck of the humerus
    E) greater tubercle of the humerus
    Answer: D

53. The arcuate popliteal ligament extends from the lateral condyle of the femur to the:
    A) tibial tuberosity
    B) medial condyle of the femur
    C) medial condyle of the tibia
    D) head of the fibula
    E) patella
    Answer: D

54. The tendon of the biceps femoris muscle is associated with the:
    A) patellar ligament
    B) oblique popliteal ligament
    C) fibular collateral ligament
    D) anterior cruciate ligament
    E) posterior cruciate ligament
    Answer: C

55. The tendons of the sartorius, gracilis, and semitendinosus muscles all cross and help strengthen the:
    A) patellar ligament
    B) anterior cruciate ligament
    C) posterior cruciate ligament
    D) tibial collateral ligament
    E) fibular collateral ligament
    Answer: D

Pages: 231
56. The tendon of the semimembranosus muscle is associated with and helps strengthen the:
    A) patellar ligament
    B) oblique popliteal ligament
    C) anterior cruciate ligament
    D) posterior cruciate ligament
    E) medial meniscus
    Answer: B

Pages: 231
57. An intracapsular ligament that helps connect the tibia and femur within the joint capsule is the:
    A) patellar ligament
    B) medial patellar retinaculum
    C) anterior cruciate ligament
    D) fibular collateral ligament
    E) lateral meniscus
    Answer: C

Pages: 231
58. The coronary ligaments attach the:
    A) ribs to the sternum
    B) temporal bone to the mandible
    C) head of the femur to the acetabulum
    D) medial and lateral menisci to the tibia
    E) muscles of the rotator cuff to the humerus
    Answer: D

Pages: 226
59. Dorsiflexion and plantar flexion occur at the:
    A) coxal joint
    B) talocrural joint
    C) glenohumeral joint
    D) tibiofemoral joint
    E) radiocarpal joint
    Answer: B

Pages: 226
60. You are bent over at the waist touching your toes. **ALL** of the following body parts are **extended EXCEPT** your:
    A) fingers
    B) legs
    C) arms
    D) vertebral column
    E) elbows
    Answer: D

Pages: 226
61. You are sitting up straight in a chair, with your fingers curled into your palm on the table in front of you. **ALL** of the following body parts are **flexed EXCEPT** your:
    A) fingers
    B) forearms
    C) legs
    D) knees
    E) vertebral column
    Answer: E

Pages: 235
62. Microorganisms transmitted via the bite of an infected tick are the cause of:
    A) rheumatoid arthritis
    B) osteoarthritis
    C) gouty arthritis
    D) Lyme disease
    E) ankylosing spondylitis
    Answer: D

Pages: 224
63. The joint between the head of the radius and the radial notch of the ulna is an example of a:
    A) hinge joint
    B) pivot joint
    C) gliding joint
    D) ball-and-socket joint
    E) saddle joint
    Answer: B

Pages: 216
64. A slightly movable joint is referred to as a(n):
    A) amphiarthrosis
    B) synarthrosis
    C) diarthrosis
    D) synostosis
    E) spondylosis
    Answer: A

Pages: 217
65. Synovial fluid is made up of interstitial fluid and:
    A) collagen
    B) elastin
    C) hydroxyapatites
    D) hyaluronic acid
    E) uric acid
    Answer: D

True-False

Write T if the statement is true and F if the statement is false.

Pages: 216
1. A cartilaginous joint has no joint cavity.
    Answer: True

Pages: 226
2. Flexion increases the angle between articulating bones, while extension decreases the angle between articulating bones.
    Answer: False

Pages: 224
3. The primary movement permitted at pivot joints is rotation.
    Answer: True

Pages: 235
4. A sprain is an injury without dislocation.
    Answer: True

Pages: 224
5. Biaxial movement is usually possible at hinge joints.
    Answer: False

Pages: 217
6. The articular cartilage covering the ends of bones at synovial joints is fibrocartilage.
    Answer: False

Pages: 216
7. The epiphyseal plate is a temporary synchondrosis.
    Answer: True

Pages: 219
8. A bursa is a fat pad located at high friction points in large joints.
    Answer: False

Pages: 231
9. Hyperextension of the knee may result in an anterior dislocation.
    Answer: True

Pages: 219
10. Someone who has a "torn cartilage" in the knee has pulled the articular cartilage from the surface of the tibia or femur.
    Answer: False

Pages: 217
11. Syndesmoses and symphyses are both amphiarthrotic joints.
    Answer: True

12. The surfaces of bones articulating at gliding joints are usually flat.
Answer: True

13. Articular discs help modify the surfaces of articulating bones to make them fit together more closely.
Answer: True

14. Synovial fluid is secreted by the chondrocytes of the articular cartilage.
Answer: False

15. Inversion and eversion are special movements that occur only at the wrists.
Answer: False

Short Answer

Write the word or phrase that best completes each statement or answers the question.

1. The functional classification of joints defined as slightly movable is the _____.
Answer: amphiarthrosis

2. A fibrous joint uniting the bones of the skull is a _____.
Answer: suture

3. The angle between articulating bones is decreased by a movement called _____.
Answer: flexion

4. A bone moves away from the body's midline during _____.
Answer: abduction

5. Any painful state of the body's supporting structures is known as _____.
Answer: rheumatism

6. A degenerative joint disease that results from the combined effects of aging, irritation of joints, wear and abrasion, is _____.
Answer: osteoarthritis

7. The forcible wrenching or twisting of a joint without dislocation is called a _____.
Answer: sprain

Pages: 229
8. A partial or incomplete dislocation is called a _____.
Answer: subluxation

Pages: 216
9. The study of motion of the human body is called _____.
Answer: kinesiology

Pages: 216
10. When an epiphyseal plate closes, it is transformed from a synchondrosis into a(n) _____.
Answer: synostosis

Pages: 217
11. In a symphysis, the articulating bones are joined by _____.
Answer: fibrocartilage

Pages: 217
12. A syndesmosis would be classified structurally as a _____ joint.
Answer: fibrous

Pages: 217
13. The function of the inner lining of the articular capsule is to produce _____.
Answer: synovial fluid

Pages: 217
14. Fibrous connective tissue that connects one bone to another in a joint capsule is called a _____.
Answer: ligament

Pages: 217
15. Nutrients are supplied to the chondrocytes of the articular cartilage by _____.
Answer: synovial fluid

Pages: 219
16. Fluid-filled sacs located between bones and overlying tissues that help alleviate pressure are called _____.
Answer: bursae

Pages: 217
17. Fibrocartilage pads that extend from the fibrous capsule into the joint cavity between articulating bones in large synovial joints are called _____.
Answer: articular discs (menisci)

Pages: 219
18. Someone who has a "torn cartilage" in the knee has damaged a(n) _____.
Answer: articular disc (meniscus)

Pages: 224
19. The elbow is an example of a(n) _____ joint.
Answer: hinge

Pages: 224

20. The articulation of the clavicle and the acromion process of the scapula is an example of a(n) _____ joint.
Answer: gliding

Pages: 226

21. The combined movements of flexion, extension, abduction, and adduction is called _____.
Answer: circumduction

Pages: 222

22. The movement of the thumb so that the tip of the thumb can meet the tip of any other digits on the same hand is referred to as _____.
Answer: opposition

Pages: 224

23. The joint between the trapezium and the metacarpal of the thumb is an example of a _____ joint.
Answer: saddle (sellaris)

Pages: 224

24. The glenohumeral joint and the coxal joint are the only examples of the _____ joint.
Answer: ball-and-socket

Pages: 224

25. The periodontal membrane separates the articulating bones of a _____.
Answer: gomphosis

Pages: 216

26. A freely movable joint is called a _____.
Answer: diarthrosis

Pages: 226

27. Bending the soles of the feet outward (laterally) so that the soles face away from each other is called _____.
Answer: eversion

Pages: 226

28. The narrow rim of fibrocartilage around the glenoid cavity is called the _____.
Answer: glenoid labrum

Pages: 229

29. The anterior surface of the knee is strengthend by the fused tendons of the quadriceps femoris muscle and the fascia lata, known as the _____.
Answer: patellar retinacula

Pages: 231

30. The ligament stretched or torn in 70% of all serious knee injuries is the _____.
Answer: anterior cruciate ligament

Matching

Choose the item from Column 2 that best matches each item in Column 1.

1. movement of the sole of the foot inward so that the soles face toward each other

    inversion

2. movement of the sole of the foot outward so that the soles face away from each other

    eversion

3. bending of the foot in the direction of the upper surface

    dorsiflexion

4. bending the foot in the direction of the sole

    plantar flexion

5. movement of the mandible forward on a plane parallel to the ground

    protraction

6. movement of the mandible backward on a plane parallel to the ground

    retraction

7. movement of the forearm in which the palm is turned anteriorly or superiorly

    supination

8. movement of the forearm in which the palm is turned posteriorly or inferiorly

    pronation

9. movement of the mandible upward

    elevation

10. movement of the mandible downward

    depression

Essay

Write your answer in the space provided or on a separate sheet of paper.

Pages: 216
1. Discuss the general factors that affect the degree of movement at a joint.
   Answer: 1) tightness of fit of articulating bones
   2) manner in which articulating bones fit together
   3) flexibility of tissues binding bones together
   4) position of ligaments, muscles, and tendons

Pages: 219
2. Describe the factors that affect the degree of movement at diarthroses.
   Answer: 1) structure of shape of articulating bones
   2) strength and tension of joint ligaments
   3) arrangement and tension of surrounding muscles
   4) apposition of soft parts
   5) hormones [note: examples may be given to amplify answer]

Pages: 217
3. Describe the functions of synovial fluid.
   Answer: 1) lubricates joint
   2) shock absorber
   3) provides nutrients to articular cartilage
   4) contains phagocytes for removal of debris and microbes
   5) removes metabolic wastes from articular cartilage

Pages: 217
4. Describe the structure of a synovial joint.
   Answer: Sleevelike articular capsule surrounds synovial cavity, unites articulating bones; articular capsule has outer fibrous layer of dense irregular connective tissue, possibly arranged into ligaments; inner layer is synovial membrane that secretes synovial fluid; synovial fluid fills joint cavity; may see accessory ligaments, articular discs, bursae

Pages: 226
5. Explain why the shoulder has more freedom of movement than other joints.
   Answer: loose articular capsule; shallow glenoid cavity receives barely more than one third of head of humerus; most of strength relies on muscle strength of rotator cuff

Pages: 216
6. Name and describe the three types of synarthroses.
   Answer: 1) suture - between skull bones; fibrous connective tissue between bones; may become synostosis
   2) gomphosis - between roots of teeth and alveoli; bones connected by periodontal ligaments
   3) synchondrosis - unclosed epiphyseal plates and between first rib and sternum; hyaline cartilage between bones; epiphyseal plates become synostoses

Pages: 220
7. Name the six types of diarthroses, and give an example of each.
   Answer: 1) gliding (arthrodial) - between carpals; between tarsals; between ribs and
             vertebrae; between clavicle and scapula or sternum
             2) hinge (ginglymus) - knee, elbow, ankle, interphalangeal
             3) pivot (trochoid) - atlas around dens of axis; proximal ends of radius and ulna
             4) condyloid (ellipsoidal) - radius to carpals
             5) saddle (sellaris) - trapezium and metacarpal of thumb
             6) ball-and-socket (spheroid) - shoulder, hip [note: other answers may be acceptable]

Pages: 226
8. Name and define the types of angular movements possible at synovial joints.
   Answer: · flexion - decrease angle between articulating bones
             · extension - increase angle between articulating bones
             · hyperextension - continuation of extension beyond anatomical position
             · abduction - move bone away from midline
             · adduction - move bone toward midline
             · circumduction - combination of flexion, extension, abduction, and adduction in
             succession

Pages: 216
9. Name and briefly describe the structural classifications of joints.
   Answer: · fibrous joints - no joint cavity; bones held together by fibrous connective tissue
             · cartilaginous joints - no joint cavity; bones held together by hyaline cartilage or
             fibrocartilage
             · synovial joints - have a joint cavity; bones held together by articular capsule and
             possibly accessory ligaments

Chapter 10   Muscle Tissue

Multiple-Choice

Choose the one alternative that best completes the statement or answers the question.

Pages: 239
1. Striations are seen in:
    A) osseous tissue
    B) smooth muscle cells
    C) muscle cells that contain sarcomeres
    D) deep fascia
    E) tendons and ligaments
    Answer: C

Pages: 239
2. A type of muscle tissue that is both striated and involuntary is:
    A) skeletal muscle tissue
    B) smooth muscle tissue
    C) cardiac muscle tissue
    D) both cardiac and smooth muscle tissues
    E) no type of muscle tissue is both striated and involuntary
    Answer: C

Pages: 244
3. Which of the following is the **smallest** unit?
    A) myofilament
    B) myofibril
    C) myofiber
    D) fascicle
    E) motor unit
    Answer: A

Pages: 245
4. Cylindrical muscle cells that contain multiple nuclei located peripherally within the cell
    would be:
    A) skeletal muscle cells only
    B) single unit smooth muscle cells
    C) multiunit smooth muscle cells
    D) cardiac muscle cells only
    E) both skeletal and cardiac muscle cells
    Answer: A

5. Branching muscle cells with single nuclei and intercalated discs at cellular junctions would be:
   A) skeletal muscle cells
   B) single unit smooth muscle cells only
   C) multiunit smooth muscle cells only
   D) cardiac muscle cells
   E) both single unit and multiunit smooth muscle cells
   Answer: D

6. Smooth muscle tissue is found:
   A) attached to bones
   B) lining hollow organs and body tubes (only)
   C) in the wall of the heart (only)
   D) lining long bones
   E) both B and C are correct
   Answer: B

7. Dense irregular (fibrous) connective tissue that separates fascicles from each other in skeletal muscle tissue is called:
   A) deep fascia
   B) a tendon
   C) epimysium
   D) endomysium
   E) perimysium
   Answer: E

8. The fibrous connective tissue that connects muscles to bones is a(n):
   A) tendon
   B) ligament
   C) epimysium
   D) perimysium
   E) endomysium
   Answer: A

9. The sarcolemma is the:
   A) storage site for calcium ions in myofibers
   B) cell membrane of a myofiber
   C) compound that binds oxygen for use in slow, oxidative muscle cells
   D) separation between sarcomeres in a myofiber
   E) structure that produces acetylcholine
   Answer: B

Pages: 245
10. Thick myofilaments are made mostly of:
    A) actin
    B) troponin
    C) myosin
    D) tropomyosin
    E) myoglobin
    Answer: C

Pages: 239
11. The property of muscle tissue that allows it to return to its original shape after stretching or contracting is called:
    A) excitability
    B) elasticity
    C) contractility
    D) extensibility
    E) conductivity
    Answer: B

Pages: 239
12. The property of muscle tissue that describes its ability to receive and respond to stimuli is:
    A) excitability
    B) elasticity
    C) contractility
    D) extensibility
    E) conductivity
    Answer: A

Pages: 250
13. The function of calcium ions in skeletal muscle contraction is to:
    A) bind to receptors on the sarcolemma at the neuromuscular junction to stimulate muscle contraction
    B) cause a pH change in the sarcoplasm to trigger muscle contraction
    C) bind to the myosin binding sites on actin so that myosin will have something to attach to
    D) bind to the troponin on the thin myofilaments so that the myosin binding sites on actin can be exposed
    E) bind oxygen to fuel cellular respiration in oxidative myofibers
    Answer: D

Pages: 242
14. Skeletal muscles are stimulated to contract when:
    A) calcium ions bind to the sarcolemma, causing an electrical disturbance
    B) acetylcholine diffuses into the sarcoplasm
    C) acetylcholine binds to receptors on the sarcolemma, causing an electrical disturbance
    D) ATP is released from the sarcoplasmic reticulum
    E) oxygen binds to myoglobin
    Answer: C

15. Energy released during the complete oxidation of glucose is either used for ATP production or:
    A) given off as heat
    B) turned into water
    C) given off as light
    D) given off as carbon dioxide
    E) converted into glycogen
    Answer: A

16. What structures meet at the neuromuscular junction?
    A) T tubules and sarcoplasmic reticulum
    B) the sarcolemma and T tubules
    C) an axon and the sarcoplasmic reticulum
    D) an axon and the sarcolemma
    E) an axon and thick myofilaments
    Answer: D

17. The dense connective tissue surrounding individual muscle fibers is the:
    A) deep fascia
    B) epimysium
    C) perimysium
    D) endomysium
    E) sarcolemma
    Answer: D

18. The function of myoglobin is to:
    A) bind oxygen for aerobic respiration
    B) bind actin to shorten myofibrils
    C) block the myosin binding sites on thin myofilaments
    D) store ATP
    E) separate one sarcomere from another
    Answer: A

19. The purpose of the phosphagen system is to:
    A) bind oxygen for aerobic respiration
    B) bind actin to shorten myofibrils
    C) block the myosin binding sites on thin myofilaments
    D) store high energy phosphate groups for ATP production
    E) actively transport acetylcholine into skeletal muscle cells
    Answer: D

Pages: 247
20. The Sliding Filament Theory of muscle contraction says that myofibers shorten when:
    A) actin filaments become shorter when they combine with myosin heads
    B) thin myofilaments are pulled toward the center of the sarcomere by swiveling of the myosin heads
    C) myosin heads rotate when they attach to actin, causing the myosin filaments to fold in the middle
    D) acetylcholine reduces the friction between thin and thick myofilaments, so they slide over each other more easily
    E) a neurotransmitter alters the arrangement of collagen fibers in the endomysium, causing it to shrink more tightly around the myofiber
    Answer: B

Pages: 252
21. The time following a stimulus during which a muscle cell is unable to respond to another stimulus is called the:
    A) treppe period
    B) refractory period
    C) relaxation period
    D) tonus period
    E) latent period
    Answer: B

Pages: 254
22. The sustained partial contraction of a portion of skeletal muscle is called:
    A) treppe
    B) incomplete tetany
    C) a simple twitch
    D) tonus
    E) a spasm
    Answer: D

Pages: 257
23. Slow oxidative skeletal muscle fibers are:
    A) specialized for aerobic muscle activity
    B) lighter in color than other types of skeletal muscle fibers
    C) used primarily for short-term strenuous activity
    D) rich in glycogen
    E) both A and C are correct
    Answer: A

Pages: 241
24. A motor unit is:
    A) all the muscles that act as prime movers for a particular action
    B) the sarcomeres of an individual myofibril
    C) all of the neurons that stimulate a particular muscle
    D) a motor neuron plus all the skeletal muscle fibers it stimulates
    E) the quantity of neurotransmitter that is sufficient to stimulate muscle contraction
    Answer: D

Pages: 242
25. The synaptic vesicles seen in an axon terminal at a neuromuscular junction contain:
    A) acetylcholine
    B) calcium ions
    C) myoglobin
    D) actin
    E) myosin
    Answer: A

Pages: 245
26. Sarcomeres are separated from each other by the:
    A) A band
    B) H zone
    C) M line
    D) Z disc
    E) T tubules
    Answer: D

Pages: 246
27. Titin is a protein found in:
    A) thin myofilaments
    B) thick myofilaments
    C) elastic filaments
    D) dense bodies
    E) synaptic vesicles
    Answer: C

Pages: 245
28. The muscle protein whose function is related to its golf club-like shape is:
    A) actin
    B) troponin
    C) tropomyosin
    D) myosin
    E) titin
    Answer: D

Pages: 246
29. The function of elastic filaments in skeletal muscle contraction is to:
    A) block the myosin binding sites on the thin myofilaments in resting muscle cells
    B) anchor thick myofilaments to Z discs for stability during contraction
    C) anchor thin myofilaments to Z discs to prevent their dislocation during contraction
    D) allow the connective tissues associated with the myofiber to return to their original shapes after contraction
    E) connect thin myofilaments across the H zone so that sarcomeres do not rupture during muscle stretching
    Answer: B

Pages: 247
30. Which of the following conditions exists in a resting myofiber?
    A) levels of calcium ions are relatively high in the sarcoplasm
    B) thin myofilaments extend across the H zone
    C) ATP is attached to myosin cross bridges
    D) levels of acetylcholine are high inside the sarcoplasmic reticulum
    E) both A and C are correct
    Answer: C

Pages: 248
31. The term "power stroke," when used relative to skeletal muscle contraction, refers to the:
    A) flooding of the sarcoplasm with calcium ions
    B) release of acetylcholine from the motor neuron via exocytosis
    C) sudden increase in the sarcolemma's permeability to sodium ions following the binding of acetylcholine
    D) pulling of the tendon on the bone as a whole muscle contracts
    E) swiveling of the myosin heads as they combine with actin
    Answer: E

Pages: 249
32. The function of acetylcholinesterase is to:
    A) produce acetylcholine
    B) stimulate release of acetylcholine from the axon terminals
    C) provide an anchor between acetylcholine and the sarcolemma
    D) break down acetylcholine
    E) bind calcium ions to each acetylcholine molecule
    Answer: D

Pages: 249
33. The function of calsequestrin is to:
    A) bind calcium ions within the sarcoplasmic reticulum
    B) provide a means by which calcium ions can attach to myosin heads
    C) actively transport calcium ions into the T tubules
    D) break down the ATP on the myosin cross bridges
    E) open calcium channels in the sarcoplasmic reticulum so that calcium ions can diffuse into the sarcoplasm
    Answer: A

Pages: 252
34. What is happening during the latent period of a muscle contraction?
    A) nothing
    B) calcium ions are being actively transported back into the sarcoplasmic reticulum
    C) new contractile proteins are being synthesized
    D) calcium ions are beginning to enter the sarcoplasm from the sarcoplasmic reticulum
    E) ATP molecules are attaching to myosin heads
    Answer: D

Pages: 252
35. A sustained muscle contraction that lacks even partial relaxation between stimuli is referred to as:
    A) treppe
    B) incomplete tetanus
    C) complete tetanus
    D) spastic paralysis
    E) rigor mortis
    Answer: C

Pages: 254
36. If the overall length of a muscle increases during a contraction, the contraction is referred to as a(n):
    A) concentric isometric contraction
    B) eccentric isometric contraction
    C) concentric isotonic contraction
    D) eccentric isotonic contraction
    E) muscle spasm
    Answer: D

Pages: 255
37. During the process of glycolysis, glucose is:
    A) broken down into two molecules of pyruvic acid
    B) converted into glycogen
    C) stored as fat
    D) converted into creatine phosphate
    E) attached directly to myosin cross bridges
    Answer: A

Pages: 256
38. The aerobic processes of cellular respiration occur in the:
    A) sarcoplasm
    B) mitochondria
    C) T tubules
    D) sarcolemma
    E) nucleus
    Answer: B

Pages: 258
39. The myofibers with the largest diameter are the:
    A) slow oxidative skeletal muscle fibers
    B) fast oxidative skeletal muscle fibers
    C) fast glycolytic skeletal muscle fibers
    D) cardiac muscle fibers
    E) smooth muscle fibers
    Answer: C

40. Muscle cells with relatively few mitochondria that generate most of their ATP via glycolysis and that have low resistance to fatigue are most likely:
A) slow oxidative skeletal muscle fibers
B) fast oxidative skeletal muscle fibers
C) fast glycolytic skeletal muscle fibers
D) cardiac muscle fibers
E) smooth muscle fibers
Answer: C

41. **ALL** of the following statements about training-induced changes in skeletal muscle are **TRUE EXCEPT**:
A) Endurance-type exercises can cause a gradual transformation of fast glycolytic fibers into fast oxidative fibers.
B) Exercises requiring great strength for short periods of time increase the synthesis of thin and thick myofilaments.
C) Endurance-type exercises are thought to greatly increase the total number of skeletal muscle cells.
D) Anabolic steroids can increase muscle size, but moderate doses probably do not increase either strength or endurance.
E) Endurance-type exercises increase the efficiency of oxygen delivery to skeletal muscle cells.
Answer: C

42. The function of intercalated discs in cardiac muscle cells is to:
A) separate the sarcomeres from each other
B) generate the appropriate neurotransmitters to regulate heart rate
C) slow the rate at which calcium ions are transported back into the sarcoplasmic reticulum
D) provide a mechanism by which all the cells in a network can contract as a functional unit
E) store ATP
Answer: D

43. When calmodulin binds calcium ions in smooth muscle, it then:
A) actively transports them into the interstitial fluid
B) causes the appropriate motor nueuron to release its neurotransmitter
C) makes the dense bodies pull on the intermediate filaments
D) stimulates the adjacent muscle cells to contract
E) activates an enzyme that phosphorylates myosin heads
Answer: E

44. The function of satellite cells is to:
A) generate extra ATP during periods of high muscle activity
B) form new skeletal muscle cells
C) produce the material that forms Z discs
D) assist in contraction by pulling on the connective tissues surrounding the muscle fiber
E) store calcium ions that have diffused into the extracellular fluid
Answer: B

45. The purpose of T tubules is to:
    A) generate ATP
    B) store calcium ions
    C) produce additional myofilaments in response to exercise
    D) conduct the muscle action potential toward the sarcoplasmic reticulum
    E) trap and break down acetylcholine molecules
    Answer: D

46. In the absence of oxygen, pyruvic acid is converted into:
    A) creatine
    B) lactic acid
    C) myoglobin
    D) ADP
    E) oxygen
    Answer: B

47. **ALL** of the following are considered functions of the muscular system **EXCEPT**:
    A) production of motion
    B) thermogenesis
    C) maintaining posture
    D) protection
    E) regulation of organ volume
    Answer: D

48. The process by which a motor neuron releases acetylcholine is:
    A) exocytosis
    B) simple diffusion
    C) facilitated diffusion
    D) primary active transport
    E) filtration
    Answer: A

49. The role of acetylcholine in skeletal muscle contraction is to:
    A) open calcium ion channels in the sarcoplasmic reticulum
    B) bind to thick myofilaments when the muscle is at rest
    C) provide an additional energy source during periods of high activity
    D) open sodium ion channels in the sarcolemma
    E) block the attachment of thick myofilaments to thin myofilments while the muscle is resting
    Answer: D

50. The region of the sarcomere that contains thin myofilaments, but not thick myofilaments is the:
    A) A band
    B) I band
    C) H zone
    D) M line
    E) T tubule
    Answer: B

51. The role of tropomyosin in skeletal muscle is to:
    A) provide an additional source of ATP during periods of high intensity exercise
    B) bind to the myosin cross bridges during the power stroke
    C) block the myosin binding sites on actin during periods of rest
    D) actively transport calcium ions back into the sarcoplasmic reticulum following the power stroke
    E) connect the thick myofilaments to the Z discs
    Answer: C

52. Thin myofilaments contain:
    A) actin only
    B) myosin only
    C) titin only
    D) both actin and myosin
    E) actin, troponin, and tropomyosin, but not myosin
    Answer: E

53. During contraction of a sarcomere, thin myofilaments are:
    A) pulled toward the H zone
    B) wrapped around the thick myofilaments in a helix
    C) folded like an accordion
    D) absorbed into the Z discs
    E) separated from the Z discs
    Answer: A

54. One of the roles of the myosin heads is to:
    A) act as an ATPase
    B) hold open calcium channels in the sarcoplasmic reticulum
    C) break down acetylcholine
    D) pull troponin away from the thin myofilaments
    E) trigger the attachment of a phosphate group to creatine
    Answer: A

Pages: 254
55. Contraction of a muscle fiber without shortening of the muscle fiber is called a(n):
    A) isometric contraction
    B) concentric isotonic contraction
    C) eccentric isotonic contraction
    D) spasm
    E) fasciculation
    Answer: A

Pages: 251
56. In the negative feedback loop that describes the role of skeletal muscle in thermal homeostasis, shivering is initiated by:
    A) autorhythmicity
    B) a sudden drop in the level of calcium ions in the extracellular fluid
    C) the hypothalamus of the brain
    D) special receptors in the tendons
    E) reduced blood flow to the muscles
    Answer: C

Pages: 251
57. The all-or-none principle of skeletal muscle contraction states that:
    A) a muscle will contract to its fullest extent or not at all
    B) all of the muscle cells of an individual motor unit will contract to their fullest extent or not at all
    C) all of the myosin heads in a myofibril will combine with actin simultaneously or none will
    D) a motor neuron will release all or none of the neurotransmitter it contains, depending on the strength of the stimulus
    E) a muscle will contract more forcefully if it is stretched first
    Answer: B

Pages: 251
58. The amount of force (tension) developed by a skeletal muscle depends on **ALL** of the following **EXCEPT**:
    A) rate of conduction of the muscle action potential along the sarcolemma
    B) number of motor units recruited
    C) length of muscle fiber prior to contraction
    D) frequency of stimulation
    E) size of motor units recruited
    Answer: A

Pages: 253
59. Multiple motor unit summation is defined as the:
    A) ability of a single motor neuron to stimulate multiple myofibers
    B) stimulation of a single myofiber by multiple motor neurons
    C) continuous release of calcium ions into the sarcoplasm
    D) simultaneous contraction of all muscle cells in a single-unit smooth muscle network
    E) process of increasing the number of active motor units during the contraction of a skeletal muscle
    Answer: E

60. Passive tension is generated during a skeletal muscle contraction by the:
    A) Z discs
    B) troponin and tropomyosin
    C) connective tissues associated with the muscle
    D) dense bodies associated with the sarcolemma
    E) movement of calcium ions in and out of the sarcoplasmic reticulum
    Answer: C

61. The glycogen stored in skeletal muscle is used:
    A) as a neurotransmitter
    B) as one of the components of thin myofilaments
    C) to fill the spaces between fascicles
    D) as a source of glucose for energy production
    E) to bind acetylcholine in the sarcolemma
    Answer: D

62. The complete aerobic oxidation of pyruvic acid yields:
    A) carbon dioxide and lactic acid
    B) carbon dioxide and water
    C) oxygen and water
    D) glycogen
    E) actin and myosin
    Answer: B

63. Postural muscles contain a relatively high proportion of:
    A) slow oxidative skeletal muscle fibers
    B) fast oxidative skeletal muscle fibers
    C) fast glycolytic skeletal muscle fibers
    D) single unit smooth muscle fibers
    E) multiunit smooth muscle fibers
    Answer: A

64. Nerve stimulation of cardiac muscle results in:
    A) no response whatsoever, because cardiac muscle fibers are autorhythmic
    B) failure of cardiac muscle cells to contract, because the stimulation by nerves negates the autorhythmicity
    C) a change in the rate of discharging muscle action potentials
    D) depletion of the neurotransmitters of autorhythmic cells, and thus, heart failure
    E) alteration of the structure of the intercalated discs
    Answer: C

65. When smooth muscle fibers are stretched, they:
    A) cannot function properly because the intermediate filaments separate from the dense bodies
    B) increase the rate of ATP production
    C) maintain relatively constant tension as they change length
    D) lose all their calcium ions to the extracellular fluid
    E) temporarily appear striated
    Answer: C

True-False

Write T if the statement is true and F if the statement is false.

Pages: 240

1. To say that muscle tissue exhibits the property of **elasticity** means that muscle can be stretched without damaging the tissue.
   Answer: False

Pages: 241

2. A motor unit consists of a motor neuron plus all the skeletal muscle fibers it stimulates.
   Answer: True

Pages: 242

3. Acetylcholine stimulates skeletal muscle cells by diffusing into the T tubules and binding to the sarcoplasmic reticulum.
   Answer: False

Pages: 245

4. The sarcolemma of a myofiber is the connective tissue layer that surrounds the cell.
   Answer: False

Pages: 245

5. The A band of a sarcomere consists mostly of thick myofilaments, and includes portions of the thin myofilaments where they overlap the thick myofilaments.
   Answer: True

Pages: 245

6. Myosin binding sites are found on troponin.
   Answer: False

Pages: 245

7. Elastic filaments anchor thick filaments to Z discs.
   Answer: True

Pages: 247

8. In mammalian skeletal muscle, a transverse tubule is found at each A-I band junction.
   Answer: True

Pages: 247
9. The function of the sarcoplasmic reticulum is to store acetylcholine.
Answer: False

Pages: 248
10. In resting skeletal muscle tissue, ATP molecules are bound to myosin heads.
Answer: True

Pages: 252
11. During the relaxation period of a twitch contraction, calcium ions are being actively transported into the sarcoplasm from the sarcoplasmic reticulum.
Answer: False

Pages: 258
12. Fast glycolytic skeletal muscle fibers contain relatively large amounts of myoglobin.
Answer: False

Pages: 261
13. Cardiac muscle tissue remains contracted 10 to 15 times longer than skeletal muscle tissue because cardiac muscle breaks down acetylcholine more slowly.
Answer: False

Pages: 261
14. When calmodulin binds calcium ions in smooth muscle tissue, it causes troponin to move away from the myosin binding sites on the actin filaments.
Answer: False

Pages: 261
15. The more common type of smooth muscle tissue is visceral (single-unit) smooth muscle tissue.
Answer: True

Short Answer

Write the word or phrase that best completes each statement or answers the question.

Pages: 239
1. Muscle tissue that is both non-striated and involuntary is _____.
Answer: smooth muscle tissue

Pages: 239
2. The ability of muscle tissue to respond to certain stimuli by producing electrical signals (action potentials) is called _____.
Answer: excitability (irritability)

Pages: 240
3. The ability of muscle tissue to return to its original shape after contracting or stretching is called _____.
Answer: elasticity

Pages: 240
4. The dense irregular connective tissue that carries nerves and blood vessels, fills the spaces between muscles, and separates muscles into functional groups is the _____.
Answer: deep fascia

Pages: 241
5. A broad, flat tendon that connects muscle to bone, muscle, or skin is called a(n) _____.
Answer: aponeurosis

Pages: 242
6. The region of a sarcolemma adjacent to the axon terminals at a neuromuscular junction is called the _____.
Answer: motor end plate

Pages: 245
7. Acetylcholine receptors are located on the _____.
Answer: sarcolemma (motor end plate)

Pages: 245
8. The region of the sarcomere that contains only thick myofilaments is the _____.
Answer: H zone

Pages: 245
9. The region of the sarcomere that contains only thin myofilaments is the _____.
Answer: I band

Pages: 245
10. The thick myofilaments are anchored to the Z discs and stabilized during contraction and relaxation by _____.
Answer: elastic filaments (titin)

Pages: 247
11. A triad consists of two terminal cisterns and a(n) _____.
Answer: transverse tubule

Pages: 248
12. Myosin binding sites on actin are exposed when troponin changes shape as a result of binding _____.
Answer: calcium ions

Pages: 248
13. The shape change (swiveling) that occurs as myosin heads bind to actin produces the _____ of contraction.
Answer: power stroke

Pages: 254
14. Muscle contraction without muscle shortening is called a(n) _____ contraction.
Answer: isometric

Pages: 252
15. The time between the application of a stimulus and the beginning of contraction, when calcium ions are being released from the sarcoplasmic reticulum, is called the _____ period.
Answer: latent

Pages: 252
16. The time following a stimulus during which a muscle cell is unable to respond to another stimulus is called the _____ period.
Answer: refractory

Pages: 252
17. A sustained contraction with no relaxation between stimuli is called _____.
Answer: complete (fused) tetanus

Pages: 252
18. In a well-relaxed muscle, several identical stimuli administered in quick succession, but allowing complete relaxation between stimuli, will result in a phenomenon known as _____.
Answer: treppe (the staircase effect)

Pages: 254
19. Involuntary activation of a small number of motor units causes sustained, small contractions that give relaxed skeletal muscle a firmness known as _____.
Answer: muscle tone

Pages: 254
20. Muscles that exhibit hypotonia are said to be _____.
Answer: flaccid

Pages: 254
21. Tension generated by tendons, elastic filaments, and connective tissues surrounding muscle fibers is called _____ tension.
Answer: passive

Pages: 254
22. An isotonic contraction in which the muscle shortens to produce movement and to reduce the angle at a joint is called a(n) _____ contraction.
Answer: concentric

Pages: 254
23. An isotonic contraction in which the muscle lengthens to produce movement and to increase the angle at a joint is called a(n) _____ contraction.
Answer: eccentric

Pages: 254
24. The wasting away of muscles due to progressive loss of myofibrils is called _____.
Answer: atrophy

Pages: 255
25. In the phosphagen system, high-energy phosphate groups can be stored for future ATP production by combining with _____ .
Answer: creatine

Pages: 255
26. During anaerobic glycolysis, glucose is broken down into two molecules of _____, resulting in a net gain of _____ molecules of ATP.
    Answer: pyruvic acid; two

Pages: 256
27. The reactions of cellular respiration occurring in the mitochondria are said to be aerobic because they require _____.
    Answer: oxygen

Pages: 256
28. During long-term exercise, most ATP is produced as a result of _____.
    Answer: aerobic cellular respiration

Pages: 259
29. Each cardiac muscle fiber in a network is connected to its neighbors by an irregular transverse thickening of the sarcolemma known as a(n) _____.
    Answer: intercalated disc

Pages: 261
30. In smooth muscle, myosin binds to actin only after the myosin head has been phosphorylated via the action of the enzyme _____.
    Answer: myosin light chain kinase

Matching

Choose the item from Column 2 that best matches each item in Column 1.

1. protein in thin myofilaments that binds to myosin heads

   actin

2. acts as an ATPase in skeletal and cardiac muscle

   myosin

3. protein that anchors thick myofilaments to the Z discs

   titin

4. regulatory protein that binds calcium ions in smooth muscle tissue

   calmodulin

5. regulatory protein that binds calcium ions in skeletal and cardiac muscle

   troponin

6. binds oxygen for use in aerobic cellular respiration

   myoglobin

7. released by motor neurons to stimulate skeletal muscle contraction

acetylcholine

8. binds calcium ions inside the sarcoplasmic reticulum

calsequestrin

9. binds a high-energy phosphate group for use in future ATP production

creatine

10. regulatory protein in thin myofilaments that binds neither calcium ions nor thick myofilaments

tropomyosin

Essay

Write your answer in the space provided or on a separate sheet of paper.

Pages: 250
1. Explain how/why rigor mortis develops.
   Answer: After death, calcium ions leak out of the SR, then bind to troponin, thus triggering the attachment of myosin to actin. However, ATP cannot be produced, so myosin cannot separate from actin, and muscle stays contracted.

Pages: 250
2. Outline the steps of skeletal muscle contraction.
   Answer: 1) Stimulus provided by binding of ACh to the sarcolemma.
   2) Resulting action potential travels along sarcolemma and into T tubules, triggering release of calcium ions from SR.
   3) Calcium ions bind to troponin, and resulting shape change causes myosin binding sites on actin to be exposed.
   4) Myosin heads bind to actin, and swivel (power stroke), pulling Z discs closer together, shortening myofiber. [add details of ATP use, as desired]

Pages: 263
3. Compare and contrast the structural features of the two types of striated muscle tissue.
   Answer: Skeletal muscle cells - long, cylindrical, multinucleate, peripheral nuclei, unbranched; cells generally parallel to each other; T tubules aligned with A-I band junctions.
   Cardiac muscle cells - contain same contractile and regulatory proteins; cells are shorter with branches and single central nuclei; cells joined via intercalated discs; T tubules aligned with each Z disc; mitochondria larger and more numerous

Pages: 255-256
4. Describe the mechanisms by which skeletal muscle tissue obtains ATP to fuel contraction.
    Answer: 1) ATP attached to resting myosin heads.
        2) Phosphagen system - creatine stores high-energy phosphate group that can be added to ADP as needed
        3) Glycolysis - anaerobic process for breaking down glucose to pyruvic acid, releasing enough energy to net 2 ATP molecules per glucose
        4) Aerobic cellular respiration - mitochondria oxidize pyruvic acid to carbon dioxide and water; energy released nets about 36 ATP per glucose molecule plus heat; requires oxygen

Pages: 239
5. Describe the role of muscle tissue in thermal homeostasis.
    Answer: Smooth muscle controls the diameter of blood vessels in the skin, allowing regulation of blood flow, and thus, regulation of heat transfer from blood to environment. Also, energy released by catabolic reactions that is not used to fuel contraction is released as heat. Shivering can increase the amount of heat generated.

Pages: 251
6. State the All-or-None Principle as it applies to skeletal muscle tissue. Describe the factors that affect the amount of tension/force generated during a muscle contraction.
    Answer: Individual muscle fibers contract to their fullest extent, or not at all (no partial contraction of individual fibers). Tension depends on frequency of stimulation, initial length of muscle fiber, total number of motor units recruited and size of the motor units recruited, and the structural components of fibers themselves.

Pages: 261-262
7. Compare and contrast the processes by which striated and non-striated muscle tissues contract.
    Answer: Smooth muscle: contraction begins more slowly, lasts longer due to slow influx of calcium ions; regulator protein (calmodulin) binds calcium ions and activates myosin light chain kinase to phosphorylate myosin and trigger attachment to actin; wide variety of stimuli
        Striated muscle: regulator protein (troponin) binds calcium ions, changes shape, and exposes myosin binding sites on actin, causing spontaneous attachment of myosin to actin; acetylcholine stimulates contraction of skeletal muscle

Pages: 261
8. Compare the two types of smooth muscle with regard to structure and function.
    Answer: Visceral smooth muscle is found in wrap-around sheets in the walls of small arteries and veins and hollow organs. The appropriate stimulus causes all cells in network to contract as one, due to gap junctions. Multiunit smooth muscle has a one-to-one relationship between neuron and myofiber, and is seen in walls of large arteries, bronchioles, and the eye.

Pages: 257
9. What is recovery oxygen consumption, and for what physiological activities is this oxygen used?
    Answer: Recovery oxygen consumption refers to elevated oxygen use after exercise. Extra oxygen is used to convert lactic acid into glycogen in the liver, to resynthesize creatine phosphate and ATP, to replace oxygen removed from myoglobin, to fuel faster chemical reactions resulting from elevated temperature, to fuel increased heart and respiratory activity, and to fuel tissue repair processes.

Pages: 254

10. You need to lift a pail of water and carry it across the room. Describe the types of whole muscle contractions that occur during this activity.

    Answer: Concentric isotonic contractions to lift pail; isometric contractions to maintain pail in position; eccentric isotonic contractions to set pail down [include complete tetanus, etc., as desired]

Multiple-Choice

Choose the one alternative that best completes the statement or answers the question.

Pages: 271
1. In a muscle group, the muscle that relaxes during a particular action is the:
   A) prime mover
   B) antagonist
   C) fixator
   D) synergist
   E) both B and D are correct
   Answer: B

Pages: 271
2. The insertion of a skeletal muscle is the:
   A) connection to the bone that remains stationary while the muscle contracts
   B) connection to the bone that moves while the muscle contracts
   C) point at which effort is applied in an anatomical lever system
   D) point at which the tendon attaches to the muscle itself
   E) both B and C are correct
   Answer: E

Pages: 273
3. Which of the following would allow a greater range of motion around a joint?
   A) having the insertion point of the muscle be very close to the joint
   B) having the insertion point of the muscle be as far as possible from the joint
   C) having very short fascicles
   D) both B and C are correct
   E) range of motion is not related to the point of muscle insertion
   Answer: A

Pages: 280
4. The mandible is elevated by the:
   A) platysma
   B) buccinator
   C) masseter
   D) sternocleidomastoid
   E) orbicularis oris
   Answer: C

Pages: 292
5. Thoracic volume is increased during normal breathing by the:
   A) internal intercostals
   B) external oblique
   C) diaphragm
   D) trapezius
   E) both B and C are correct
   Answer: C

Pages: 312
6. **ALL** of the following extend the vertebral column **EXCEPT** the:
   A) splenius muscles
   B) iliocostalis muscles
   C) intercostal muscles
   D) longissimus muscles
   E) semispinalis muscles
   Answer: C

Pages: 297
7. The trapezius is the prime mover for most movements of the:
   A) cervical vertebrae
   B) thighs
   C) forearms
   D) scapula
   E) face
   Answer: D

Pages: 300
8. The arm is extended, medially rotated, and adducted by the:
   A) latissimus dorsi
   B) triceps brachii
   C) biceps brachii
   D) deltoid
   E) sternocleidomastoid
   Answer: A

Pages: 300
9. The arm is abducted and rotated by the:
   A) latissimus dorsi
   B) triceps brachii
   C) biceps brachii
   D) deltoid
   E) brachialis
   Answer: D

Pages: 303
10. The forearm is flexed by the:
   A) latissimus dorsi
   B) triceps brachii
   C) biceps brachii
   D) deltoid
   E) pectoralis major
   Answer: C

Pages: 315
11. The thigh is extended and laterally rotated by the:
   A) tensor fasciae latae
   B) adductor magnus
   C) gluteus maximus
   D) gastrocnemius
   E) biceps femoris
   Answer: C

Pages: 321
12. The lower leg is extended by the:
   A) gastrocnemius
   B) quadriceps femoris
   C) gluteus maximus
   D) hamstrings
   E) tibialis anterior
   Answer: B

Pages: 323
13. The lower leg is flexed by the:
   A) vastus lateralis
   B) vastus medialis
   C) gluteus maximus
   D) hamstrings
   E) tibialis anterior
   Answer: D

Pages: 321
14. To cross one's legs, the crossing thigh is flexed and laterally rotated by the:
   A) rectus femoris
   B) gluteus medius
   C) sartorius
   D) semitendinosus
   E) biceps femoris
   Answer: C

Pages: 321
15. The group of muscles known as the hamstrings includes the:
   A) rectus femoris, vastus lateralis, and vastus medialis
   B) sartorius and gracilis
   C) gluteus maximus, medius, and minimus
   D) external oblique, internal oblique, and rectus abdominis
   E) biceps femoris, semitendinosus, and semimembranosus
   Answer: E

Pages: 321
16. The primary action of the hamstrings is to:
    A) extend the lower leg
    B) flex the lower leg
    C) flex the thigh
    D) rotate the pelvis
    E) plantar flex the foot
    Answer: B

Pages: 323
17. The foot is plantar flexed by the:
    A) gastrocnemius only
    B) tibialis anterior
    C) soleus only
    D) biceps femoris
    E) both the gastrocnemius and the soleus
    Answer: E

Pages: 278
18. The eye is closed by the:
    A) orbicularis oris
    B) buccinator
    C) orbicularis oculi
    D) platysma
    E) frontalis
    Answer: C

Pages: 289
19. The origin of the sternocleidomastoid muscle is on the sternum and:
    A) temporal
    B) zygomatic
    C) clavicle
    D) occipital
    E) atlas
    Answer: C

Pages: 290
20. The aponeurosis of the abdominal muscles that extends from the xiphoid process to the pubic symphysis is the:
    A) inguinal ligament
    B) linea alba
    C) galea aponeurotica
    D) fascia lata
    E) flexor retinaculum
    Answer: B

Pages: 290
21. A hole in the linea alba that allows passage of the spermatic cord in males is the:
    A) rotator cuff
    B) inguinal canal
    C) extensor retinaculum
    D) levator ani
    E) carpal tunnel
    Answer: B

Pages: 300
22. **ALL** of the following muscles are part of the rotator cuff **EXCEPT** the:
    A) teres minor
    B) infraspinatus
    C) supraspinatus
    D) subscapularis
    E) pectoralis major
    Answer: E

Pages: 303
23. The biceps brachii inserts onto the:
    A) coracoid process
    B) olecranon process
    C) deltoid tuberosity
    D) radial tuberosity
    E) acromion process
    Answer: D

Pages: 315
24. The gluteus maximus inserts, in part, onto the:
    A) ilium
    B) scapula
    C) femur
    D) lumbar vertebrae
    E) tibia
    Answer: C

Pages: 321
25. The hamstrings have their origin on the:
    A) ischial tuberosity
    B) iliac fossa
    C) iliac crest
    D) tibial tuberosity
    E) head of the fibula
    Answer: A

Pages: 321
26. The tibial tuberosity is the insertion point for the:
    A) hamstrings
    B) quadriceps femoris
    C) quadratus lumborum
    D) gastrocnemius
    E) gluteus medius
    Answer: B

Pages: 323
27. The calcaneal (Achilles) tendon attaches which muscle to the calcaneus?
    A) gastrocnemius
    B) quadriceps femoris
    C) biceps femoris
    D) tibialis anterior
    E) peroneus longus
    Answer: A

Pages: 315
28. A muscle that originates on the lumbar vertebrae and inserts onto the lesser trochanter, and that flexes both the thigh and the vertebral column is the:
    A) gluteus medius
    B) psoas major
    C) transverse abdominis
    D) gracilis
    E) serratus anterior
    Answer: B

Pages: 271
29. The **origin** of a muscle refers to the:
    A) embryonic derivation from a particular germ layer
    B) attachment to the moving bone
    C) attachment to the stationary bone
    D) point at which the tendon meets the muscle
    E) point at which blood vessels and nerves enter the muscle
    Answer: C

Pages: 270
30. The word "**brevis**" in a muscle name means:
    A) long
    B) short
    C) large
    D) small
    E) straight
    Answer: B

Pages: 270

31. A word that refers to the relative shape of a muscle is:
    A) rectus
    B) minimus
    C) oblique
    D) brevis
    E) rhomboideus
    Answer: E

Pages: 270

32. A word in a muscle name that indicates that the muscle decreases the size of a opening is:
    A) sphincter
    B) extensor
    C) levator
    D) tensor
    E) rotator
    Answer: A

Pages: 273

33. In a third class lever system, the arrangement of the system's components is such that the:
    A) fulcrum is between the effort and the resistance
    B) effort is between the fulcrum and the resistance
    C) resistance is between the fulcrum and the effort
    D) effort and resistance are equidistant from the fulcrum
    E) there is no fulcrum
    Answer: B

Pages: 272

34. In an anatomical lever system, the lever is the:
    A) bone
    B) muscle that is contracting
    C) weight of the part to be moved
    D) joint
    E) tendon that is pulling on the bone
    Answer: A

Pages: 326

35. An intramuscular injection into the buttocks is usually administered into the:
    A) gluteus maximus
    B) gluteus medius
    C) vastus lateralis
    D) vastus medialis
    E) tensor fasciae latae
    Answer: B

Pages: 326
36. Intramuscular injections in the thigh are usually administered into the:
    A) rectus femoris
    B) biceps femoris
    C) vastus lateralis
    D) vastus medialis
    E) sartorius
    Answer: C

Pages: 326
37. Intramuscular injections into the arm are usually administered into the:
    A) biceps brachii
    B) triceps brachii
    C) brachialis
    D) deltoid
    E) pectoralis major
    Answer: D

Pages: 277
38. The aponeurosis covering the superior and lateral portions of the skull is the:
    A) linea alba
    B) galea aponeurotica
    C) fascia lata
    D) Achilles tendon
    E) flexor retinaculum
    Answer: B

Pages: 282
39. The superior and inferior rectus muscles are prime movers for:
    A) flexion and extension of the lower leg
    B) flexion and extension of the thigh
    C) elevation and depression of the eyelid
    D) flexion and extension of the vertebral column
    E) movements of the eyeball
    Answer: E

Pages: 284
40. Muscles with the combining form "-glossus" in the name cause movements of the:
    A) eyeball
    B) eyelid
    C) tongue
    D) vocal cords
    E) ossicles in the ear
    Answer: C

Pages: 290
41. Which of the following muscles is deepest?
    A) rectus abdominis
    B) external oblique
    C) internal oblique
    D) transverse abdominis
    E) pectoralis major
    Answer: D

Pages: 292
42. **ALL** of the following pass through the diaphragm **EXCEPT** the:
    A) aorta
    B) spinal cord
    C) esophagus
    D) vagus nerve
    E) inferior vena cava
    Answer: B

Pages: 292
43. The phrenic nerve provides innervation to the:
    A) biceps brachii
    B) trapezius
    C) rectus abdominis
    D) diaphragm
    E) quadriceps femoris
    Answer: D

Pages: 292
44. The principal muscles of both inspiration and expiration during normal breathing are the diaphragm and the:
    A) external oblique
    B) rectus abdominis
    C) external intercostals
    D) internal intercostals
    E) pectoralis major
    Answer: C

Pages: 294
45. Which of the following passes through the pelvic diaphragm?
    A) aorta
    B) inferior vena cava
    C) esophagus
    D) anal canal
    E) all of these
    Answer: D

Pages: 295
46. The central tendon of the perineum is the insertion for the:
    A) external anal sphincter
    B) levator ani
    C) gluteus maximus
    D) external oblique
    E) rectus abdominis
    Answer: A

Pages: 297
47. The serratus anterior is a prime mover for rotation and abduction of the:
    A) humerus
    B) scapula
    C) cervical vertebrae
    D) thigh
    E) thoracic vertebrae
    Answer: B

Pages: 300
48. Which of the following muscles that move the humerus is designated as an **axial** muscle?
    A) latissimus dorsi
    B) deltoid
    C) teres major
    D) subscapularis
    E) supraspinatus
    Answer: A

Pages: 303
49. The radial nerve provides innervation to the:
    A) latissmus dorsi
    B) biceps brachii
    C) triceps brachii
    D) deltoid
    E) trapezius
    Answer: C

Pages: 305
50. The superficial muscles that flex the wrist originate on the:
    A) radius
    B) ulna
    C) humerus
    D) flexor retinaculum
    E) metacarpals
    Answer: C

51. The thenar muscles act on the:
    A) tongue
    B) eyeball
    C) perineum
    D) ribs
    E) thumb
    Answer: E

52. The linea aspera of the femur is the insertion point for the:
    A) adductors (longus, brevis, magnus)
    B) iliopsoas
    C) hamstrings
    D) sartorius and gracilis
    E) gastrocnemius and soleus
    Answer: A

53. The femoral nerve provides innervation to the:
    A) hamstrings
    B) quadriceps femoris
    C) gastrocnemius and soleus
    D) gluteus muscles
    E) internal and external oblique
    Answer: B

54. Which of the following would act as an antagonist to the tibialis anterior?
    A) rectus femoris
    B) biceps femoris
    C) tensor fasciae latae
    D) peroneus longus
    E) gracilis
    Answer: D

55. Which of the following acts as an antagonist to the rectus femoris?
    A) vastus lateralis
    B) semimembranosus
    C) gastrocnemius
    D) tibialis anterior
    E) adductor magnus
    Answer: B

Pages: 300
56. Which of the following acts as an antagonist to the deltoid?
    A) biceps brachii
    B) triceps brachii
    C) pectoralis major
    D) sternocleidomastoid
    E) external intercostals
    Answer: C

Pages: 280
57. Which of the following acts as an antagonist to the platysma?
    A) occipitalis
    B) trapezius
    C) sternocleidomastoid
    D) masseter
    E) orbicularis oculi
    Answer: D

Pages: 273
58. In a first class lever system, the components are arranged such that the:
    A) fulcrum is between the effort and the resistance
    B) effort is between the fulcrum and the resistance
    C) resistance is between the fulcrum and the effort
    D) lever rotates around the effort rather than the fulcrum
    E) fulcrum is not necessary
    Answer: A

Pages: 272
59. In an anatomical lever system, the fulcrum is the:
    A) bone on which the muscle originates
    B) joint
    C) force of muscle contraction
    D) weight of the part to be moved
    E) point at which the tendon attaches to the muscle
    Answer: B

Pages: 272
60. In an anatomical lever system, the effort is the:
    A) bone on which the muscle originates
    B) joint
    C) muscular contraction pulling on the insertion point
    D) weight of the body part to be moved
    E) brain activity that stimulates movement
    Answer: C

Pages: 273
61. In an anatomical lever system, the resistance is the:
    A) bone on which the muscle originates
    B) connective tissues surrounding the joint
    C) muscle contraction pulling on the insertion point
    D) weight of the body part to be moved
    E) amount of ATP required to perform the action
    Answer: D

Pages: 273
62. Two structurally identical muscles cross a joint. Muscle X inserts one inch from the joint. Muscle Y inserts three inches from the joint. Which of the following statements is most likely to be **TRUE**?
    A) Muscle X will produce a stronger movement due to greater leverage.
    B) Muscle Y will produce a stronger movement due to greater leverage.
    C) Muscle Y will produce a stronger movement because the tendon must be longer.
    D) Muscle Y will produce a movement with a greater range of motion because the lever is longer.
    E) Muscles X and Y must produce movements of equal strength, regardless of point of insertion, because they are structurally identical.
    Answer: B

Pages: 270
63. The word "**biceps**" in a muscle name means that the muscle has two:
    A) origins
    B) insertions
    C) tendons
    D) primary actions
    E) types of myofibers
    Answer: A

Pages: 287
64. The muscles with the word "arytenoid" in their names are involved in:
    A) movements of the eyeball
    B) movements of the tongue
    C) changing the amount of space between the vocal folds
    D) flexing and extending the fingers
    E) changing facial expression
    Answer: C

Pages: 295
65. A muscle that acts to maintain the erection of the penis/clitoris is the:
    A) levator ani
    B) external anal sphincter
    C) urethral sphincter
    D) iliacus
    E) ischiocavernosus
    Answer: E

True-False

Write T if the statement is true and F if the statement is false.

Pages: 271
1. The attachment of a muscle tendon to the movable bone is called the origin.
   Answer: False

Pages: 270
2. The word "**rectus**" in a muscle name means that the muscle fibers run parallel to the midline.
   Answer: True

Pages: 273
3. A greater range of motion is achieved by placing the insertion further away from the joint.
   Answer: False

Pages: 273
4. Most lever systems in the body are first class levers.
   Answer: False

Pages: 273
5. In a third class lever, the fulcrum is located between the effort and the resistance.
   Answer: False

Pages: 273
6. The longer the fibers in a muscle, the greater the range of motion it can produce.
   Answer: True

Pages: 277
7. The zygomaticus major muscle is a prime mover in the act of smiling.
   Answer: True

Pages: 282
8. The eyeball is rolled by the orbicularis oculi.
   Answer: False

Pages: 290
9. The rectus abdominis has its origin on the pubic crest and pubic symphysis.
   Answer: True

Pages: 321
10. The hamstrings flex the thigh.
    Answer: False

Pages: 292
11. During inspiration, the diaphragm flattens as it contracts.
    Answer: True

Pages: 326
12. Intramuscular injections are usually administered into the biceps brachii.
Answer: False

Pages: 323
13. The tibialis anterior and the gastrocnemius are antagonists in the movement of the foot.
Answer: True

Pages: 323
14. The Achilles tendon attaches the quadriceps femoris to the tibial tuberosity.
Answer: False

Pages: 290
15. The transverse abdominis is the most superficial of the muscles that compress the abdomen.
Answer: False

Short Answer

Write the word or phrase that best completes each statement or answers the question.

Pages: 271
1. The attachment of a muscle tendon to the movable bone is called the _____.
Answer: insertion

Pages: 271
2. The fleshy portion of a muscle between the tendons of the origin and insertion is called the

_____.
Answer: belly

Pages: 270
3. The word "**maximus**" in a muscle name means _____.
Answer: largest

Pages: 270
4. The word "**rectus**" in a muscle name means that the muscle fibers run _____ to the midline.
Answer: parallel

Pages: 272
5. In an anatomical lever system, bones act as _____.
Answer: levers

Pages: 272
6. In an anatomical lever system, the fulcrum is the _____.
Answer: joint

Pages: 273
7. A muscle that causes a particular action is called the _____.
Answer: prime mover

Pages: 273
8. When a prime mover contracts, the _____ relaxes.
   Answer: antagonist

Pages: 273
9. Muscles that help stabilize movements and help prime movers work more efficiently are called
   _____ and _____.
   Answer: synergists; fixators

Pages: 326
10. The thigh muscle that is the site of most intramuscular injections into the leg is the
    _____.
    Answer: vastus lateralis

Pages: 277
11. The galea aponeurotica is an aponeurosis that unites the _____ and the _____ muscles.
    Answer: frontalis; occipitalis

Pages: 278
12. The eye is closed by the _____.
    Answer: orbicularis oculi

Pages: 278
13. The cheeks are compressed into a sucking action by the _____.
    Answer: buccinator

Pages: 283
14. The superior and inferior oblique muscles cause rotation of the _____.
    Answer: eyeball

Pages: 289
15. The sternocleidomastoid has its origin on the sternum and the _____.
    Answer: clavicle

Pages: 290
16. The spermatic cord passes through an opening in the aponeurosis of the abdominal wall called
    the _____.
    Answer: inguinal ring

Pages: 292
17. The floor of the thoracic cavity is formed by the _____.
    Answer: diaphragm

Pages: 292
18. In a hiatus hernia, the stomach protrudes through the opening in the diaphragm through which
    the _____ passes.
    Answer: esophagus (and vagus nerve)

Pages: 300

19. The muscles that move the humerus that originate on the axial skeleton are the latissimus dorsi and the _____.
Answer: pectoralis major

Pages: 303

20. The forearm is extended by the anconeus and the _____.
Answer: triceps brachii

Pages: 303

21. The triceps brachii inserts on the _____ of the _____.
Answer: olecranon process; ulna

Pages: 310

22. The _____ muscles act on the little finger.
Answer: hypothenar

Pages: 321

23. The hamstrings are the biceps femoris, semimembranosus, and the _____.
Answer: semitendinosus

Pages: 323

24. The longest muscle in the body is the _____.
Answer: sartorius

Pages: 321

25. The antagonist to the hamstrings regarding movements of the lower leg is the _____.
Answer: quadriceps femoris

Pages: 326

26. In RICE therapy, the "E" stands for _____.
Answer: elevation

Pages: 326

27. In RICE therapy, the "C" stands for _____.
Answer: compression

Pages: 274

28. The fascicular arrangement in which the fasciculi are short in relation to muscle length is called _____.
Answer: pennate

Pages: 321

29. The common tendon for the four muscles of the quadriceps femoris is the _____.
Answer: patellar ligament

Pages: 305

30. At the wrist, the deep fascia is thickened into fibrous bands known as the flexor and extensor _____.
Answer: retinacula

Matching

Choose the item from Column 2 that best matches each item in Column 1.

1. semitendinosus   flexes leg

2. vastus lateralis   extends leg

3. tibialis anterior   dorsiflexes foot

4. soleus   plantar flexes foot

5. gluteus medius   abducts thigh

6. gracilis   adducts thigh

7. deltoid   abducts arm

8. latissimus dorsi   adducts arm

9. platysma   depresses mandible

10. masseter   elevates mandible

Essay

Write your answer in the space provided or on a separate sheet of paper.

Pages: 271-273

1. Identify the anatomical parts corresponding to the generic components of a lever system. Describe the arrangement of these parts in first, second, and third class lever systems.
    Answer: Bone = lever; joint = fulcrum; muscle contraction pulling on insertion point = effort; weight of part to be moved = resistance
    - First class lever has fulcrum between effort and resistance
    - Second class lever has the resistance between the fulcrum and effort
    - Third class lever has effort between the fulcrum and resistance
    - In all cases, lever moves around fulcrum

Pages: 273

2. Discuss the role of prime movers, antagonists, synergists and fixators in movement.
    Answer: The prime mover contracts to cause a particular action. The antagonist causes the opposite action, and so, must relax while the prime mover contracts. Synergists prevent unwanted movements during an action, while fixators stabilize the origin of the prime mover. Both allow the prime mover to work more efficiently.

3. Discuss the relationships among fascicle length, point of insertion, strength of movement, and range of movement.
    Answer: Stronger movement if fascicles short and insertion point further from joint; greater range of motion if fascicles long and insertion point nearer to joint.

4. What is RICE therapy? What do the letters stand for?
    Answer: RICE therapy is a treatment for sports injuries involving immediate **Rest**, application of **Ice**, **Compression** with an elastic bandage, and **Elevation** of injured part.

5. Chuck has a rotator cuff injury. What muscles and associated structures might be involved, and what sorts of activities might have led up to this injury? What movements might be inhibited by this injury?
    Answer: The tendons of the four deep muscles of the shoulder make up rotator cuff - subscapularis, supraspinatus, infraspinatus, teres minor. Any activity involving these muscles could be the problem - from throwing baseballs to shoveling coal. Inhibited movements depend on specific muscle involved - medial and lateral rotation, adduction, abduction, or extension of arm.

6. Roger has "pulled a hamstring," and is out of the line-up for two weeks. What muscles might be involved in this injury, and what movements are inhibited because of this injury?
    Answer: Possibly biceps femoris, semitendinosus, or semimembranosus; flexion of leg and extension of thigh inhibited

7. Identify the muscles most commonly used for the administration of intramuscular injections. Describe the landmarks used for each site.
    Answer: · Gluteus medius - use upper outer quadrant using iliac crest as landmark
    · Vastus lateralis - use knee and greater trochanter as landmarks
    · Deltoid - two to three fingerbreadths below acromion of scapula, lateral to axilla

8. Pat is doing jumping jacks. What muscles and actions are involved in the movements of the upper extremities?
    Answer: Deltoid for abduction of arm; latissimus dorsi and pectoralis major for adduction of arm [include other muscles of rotator cuff as desired]; triceps brachii and anconeus for keeping forearm extended

9. Carolyn is doing jumping jacks. What muscles and actions are involved in the movements of the lower extremities?
    Answer: Gluteus medius and minimus, tensor fasciae latae for abduction of thigh; adductor magnus, longus, and brevis, pectineus, and gracilis for adduction of thigh; quadriceps femoris for keeping leg extended; gluteus maximus for keeping thigh extended [include other muscles as desired]

Pages: 321

10. Joe is doing deep knee bends. What muscles and actions are involved in the movements of the lower extremities?

Answer: Quadriceps femoris for extension of leg; hamstrings for flexion of leg; gluteus maximus for extension of thigh; iliopsoas for flexion of thigh [include other muscles as desired; [include foot movements as desired]

Multiple-Choice

Choose the one alternative that best completes the statement or answers the question.

Pages: 339
1. The sympathetic nervous system contains:
   A) special somatic afferent neurons
   B) general somatic afferent neurons
   C) general somatic efferent neurons
   D) general visceral efferent neurons
   E) parasympathetic neurons
   Answer: D

Pages: 332
2. Effectors in the autonomic nervous system would include the:
   A) brain
   B) spinal cord
   C) cardiac muscle
   D) skeletal muscles
   E) both C and D are correct
   Answer: C

Pages: 333
3. Neuroglia in the central nervous system that are phagocytes are the:
   A) astrocytes
   B) oligodendrocytes
   C) microglia
   D) ependymal cells
   E) neurolemmocytes
   Answer: C

Pages: 333
4. Neuroglia in the central nervous system that produce the myelin sheath are the:
   A) astrocytes
   B) oligodendrocytes
   C) microglia
   D) ependymal cells
   E) neurolemmocytes
   Answer: B

5. Neuroglia that are positioned between neurons and capillaries to form part of the blood-brain barrier are the:
   A) astrocytes
   B) oligodendrocytes
   C) microglia
   D) ependymal cells
   E) neurolemmocytes
   Answer: A

6. Neuroglia that have cilia to facilitate the flow of cerebrospinal fluid through the spinal cord are the:
   A) astrocytes
   B) oligodendrocytes
   C) microglia
   D) ependymal cells
   E) neurolemmocytes
   Answer: D

7. The nucleus of a neuron is located in:
   A) the cell body
   B) the axon
   C) the myelin sheath
   D) the dendrite
   E) any part of the cell
   Answer: A

8. A function of the myelin sheath is to:
   A) synthesize neurotransmitters
   B) generate ATP for use during conduction of action potentials
   C) store glycogen for energy production during impulse conduction
   D) connect several axons together to form a nerve
   E) provide electrical insulation to help increase the rate of impulse conduction
   Answer: E

9. Gaps in the myelin sheath are:
   A) not normal
   B) spaces between individual neurolemmocytes
   C) the parts of the membrane that do not become depolarized during conduction of an action potential
   D) seen only along the myelin sheath of dendrites
   E) the channels through which neurotransmitters are released
   Answer: B

Pages: 338
10. A **tract** is:
    A) the connective tissue surrounding a bundle of axons
    B) a bundle of fascicles of axons in the peripheral nervous system
    C) a bundle of fascicles of axons in the central nervous system
    D) the grooves along bones in which nerves run
    E) the connective tissue that surrounds an entire nerve
    Answer: C

Pages: 334
11. The part of a neuron that might be myelinated is the:
    A) axon
    B) neurofibral node
    C) cell body
    D) nucleus
    E) all of these except the nucleus
    Answer: A

Pages: 338
12. The axon terminals of a neuron are found:
    A) at the end of axons
    B) in the axolemma
    C) where dendrites meet axons of the same neuron
    D) where the axon leaves the cell body
    E) within the myelin sheath
    Answer: A

Pages: 336
13. The branch of a neuron that carries a nerve impulse away from the cell body is the:
    A) axon
    B) dendrite
    C) perikaryon
    D) neurolemmocyte
    E) neurofibril node
    Answer: A

Pages: 332
14. Most cells within nervous tissue are:
    A) neurons
    B) neuroglia
    C) fibroblasts
    D) axons
    E) myofibers
    Answer: B

Pages: 335
15. The neurofibril nodes (of Ranvier) are:
   A) sites of neurotransmitter storage
   B) where the nucleus of a neuron is located
   C) gaps in the myelin sheath
   D) sites of myelin production
   E) sites of neurotransmitter production
   Answer: C

Pages: 336
16. Chromatophilic substance (Nissl bodies) is:
   A) the membrane of a neurolemmocyte
   B) a phagocytic type of neuroglia
   C) the DNA of a neuron
   D) the endoplasmic reticulum of a neuron
   E) a type of neurotransmitter
   Answer: D

Pages: 332
17. The central nervous system contains the brain and the:
   A) spinal nerves
   B) ganglia
   C) spinal cord
   D) motor nerves
   E) all of these
   Answer: C

Pages: 332
18. Afferent nerves conduct nerve impulses from:
   A) the central nervous system to effectors
   B) effectors to the central nervous system
   C) receptors to the central nervous system
   D) the central nervous system to receptors
   E) one effector to another
   Answer: C

Pages: 336
19. Nerve impulses are conducted toward the cell body by the:
   A) axon
   B) dendrite
   C) cell body
   D) synaptic knobs
   E) neurolemmocyte
   Answer: B

Pages: 344
20. During the depolarization phase of an action potential, which of the following is the primary activity?
    A) Potassium ions are flowing into the cell.
    B) Potassium ions are flowing out of the cell.
    C) Sodium ions are flowing into the cell.
    D) Sodium ions are flowing out of the cell.
    E) Neurotransmitter is diffusing into the cell.
    Answer: C

Pages: 344
21. During the depolarization phase of an action potential, which of the following situations exists?
    A) The inside of the membrane is becoming more negative with respect to the outside.
    B) The inside of the membrane is becoming more positive with respect to the outside.
    C) The membrane is becoming less permeable to all ions.
    D) The membrane potential remains constant.
    E) The threshold is changing.
    Answer: B

Pages: 345
22. During the repolarization phase of an action potential, which of the following is the primary activity?
    A) Potassium ions are flowing into the cell.
    B) Potassium ions are flowing out of the cell.
    C) Sodium ions are flowing into the cell.
    D) Sodium ions are flowing out of the cell.
    E) The membrane is impermeable to all ions.
    Answer: B

Pages: 345
23. During the repolarization phase of an action potential, which of the following situations exists?
    A) The inside of the membrane is becoming more negative with respect to the outside.
    B) The inside of the membrane is becoming more positive with respect to the outside.
    C) The membrane is becoming less permeable to all ions.
    D) The membrane potential remains constant.
    E) The threshold is changing.
    Answer: A

Pages: 344
24. When the ions are moving across the membrane during the depolarization and repolarization phases of an action potential, they are moving by:
    A) primary active transport
    B) secondary active transport
    C) exocytosis
    D) filtration
    E) simple diffusion
    Answer: E

Pages: 346
25. The after-hyperpolarization that follows an action potential occurs due to a rapid:
    A) outflow of sodium ions
    B) outflow of potassium ions
    C) influx of sodium ions
    D) influx of potassium ions
    E) release of neurotransmitter
    Answer: B

Pages: 344
26. Which of the following events occurs first in an action potential?
    A) after-hyperpolarization
    B) depolarization
    C) repolarization
    D) saltatory conduction
    E) refractory period
    Answer: B

Pages: 346
27. The term "saltatory conduction" refers to the:
    A) "leaping" of an action potential across a synapse
    B) movement of sodium ions into the cell during depolarization
    C) one-way conduction of a nerve impulse across a synapse
    D) conduction of a nerve impulse along a myelinated axon
    E) action of the sodium-potassium pump
    Answer: D

Pages: 349
28. Neurotransmitter is released into the synaptic cleft by:
    A) exocytosis
    B) osmosis
    C) facilitated diffusion
    D) filtration
    E) primary active transport
    Answer: A

Pages: 349
29. The effect of a neurotransmitter on the postsynaptic cell occurs when the neurotransmitter:
    A) diffuses into the postsynaptic cell
    B) flows along the postsynaptic cell membrane into transverse tubules
    C) is broken down by enzymes in the synaptic cleft
    D) actively transports sodium ions into the postsynaptic cell
    E) binds to specific receptors on the postsynaptic cell membrane
    Answer: E

Pages: 334
30. The myelin sheath is composed mostly of:
    A) glycogen
    B) sodium and potassium ions
    C) various neurotransmitters
    D) phospholipids
    E) DNA
    Answer: D

Pages: 338
31. Nerve cell bodies in the peripheral nervous system generally cluster together to form:
    A) tracts
    B) nerves
    C) ganglia
    D) nuclei
    E) white matter
    Answer: C

Pages: 340
32. Collections of nerve cell bodies within the central nervous system are referred to as:
    A) tracts
    B) nerves
    C) ganglia
    D) nuclei
    E) white matter
    Answer: D

Pages: 349
33. A neurotransmitter that allows sodium ions to leak into a postsynaptic neuron causes:
    A) excitatory postsynaptic potentials
    B) inhibitory postsynaptic potentials
    C) no changes in the resting potential
    D) an alteration of the membrane threshold
    E) damage to the myelin sheath
    Answer: A

Pages: 349
34. Excitatory postsynaptic potentials triggered by release of neurotransmitters from several presynaptic end bulbs are added together in an effect known as:
    A) convergence
    B) divergence
    C) reverberation
    D) temporal summation
    E) spatial summation
    Answer: E

35. When depolarization reaches the axon terminal of a presynaptic neuron, the next event is:
   A) immediate release of neurotransmitter
   B) uptake of neurotransmitter from the synaptic cleft
   C) diffusion of calcium ions out of the cell
   D) diffusion of calcium ions into the cell
   E) active transport of calcium ions out of the cell
   Answer: D

36. Synaptic delay occurs because:
   A) there is no myelin in the synaptic cleft
   B) it takes time for the neurotransmitter to diffuse across the synaptic cleft
   C) neurotransmitter must be manufactured once depolarization reaches the axon terminal
   D) the postsynaptic cell may have run out of the carrier molecules necessary to transport the neurotransmitter into the cell
   E) the presynaptic cell tries to transport the neurotransmitter back in as soon as it is released
   Answer: B

37. A neurotransmitter that allows negatively charged chloride ions to enter the postsynaptic cell causes:
   A) excitatory postsynaptic potentials, because the membrane would begin to depolarize as the inside becomes more negative
   B) inhibitory postsynaptic potentials, because the membrane would become hyperpolarized as the inside becomes more negative
   C) no change in resting potential, because only sodium and potassium ions are involved in maintaining resting potential
   D) immediate cell death as chloride blocks the action of ATP
   E) inactivation of the enzymes that break down the neurotransmitter
   Answer: B

38. An amino acid that acts as an inhibitory neurotransmitter in the brain is:
   A) acetylcholine
   B) epinephrine
   C) GABA
   D) glutamate
   E) nitric oxide
   Answer: C

39. The function of neurotropins is to:
   A) act as precursor molecules to neurotransmitters
   B) regulate growth and development of neurons
   C) act as carrier molecules during the initiation of an action potential
   D) break down neurotransmitters
   E) facilitate circulation of cerebrospinal fluid
   Answer: B

Pages: 341
40. The plasma membrane of a neuron is more permeable to potassium ions than to sodium ions because the membrane has:
    A) more voltage-gated sodium channels
    B) more chemically-gated potassium channels
    C) more potassium leakage channels
    D) fewer voltage-gated sodium channels
    E) more carrier molecules for potassium ions
    Answer: C

Pages: 341
41. Neurotransmitters control the passage of ions through:
    A) leakage channels
    B) voltage-gated channels
    C) chemically-gated channels
    D) mechanically-gated channels
    E) light-gated channels
    Answer: C

Pages: 342
42. Regarding a neuron at rest, **ALL** of the following conditions exist **EXCEPT**:
    A) there are relatively more sodium ions in the extracellular fluid than in the intracellular fluid
    B) there are relatively more potassium ions in the intracellular fluid than in the extracellular fluid
    C) most intracellular anions cannot leave the intracellular fluid
    D) the sodium/potassium active transport pump is operating
    E) the membrane is relatively more permeable to sodium ions than to potassium ions
    Answer: E

Pages: 344
43. The threshold of a neuron is the:
    A) time between binding of the neurotransmitter and firing of an action potential
    B) voltage at which the inflow of sodium ions causes reversal of the resting potential
    C) total number of sodium ions that enter the cell before sodium inactivation gates close
    D) total amount of neurotransmitter it takes to cause an action potential
    E) voltage across the resting cell membrane
    Answer: B

Pages: 342
44. In the resting membrane state, which of the following situations exists for voltage-gated sodium ion channels?
    A) The activation gate is open and the inactivation gate is closed.
    B) The inactivation gate is open and the activation gate is closed
    C) The activation and inactivation gates are both open.
    D) The activation and inactivation gates are both closed.
    E) The activation and inactivation gates alternate being closed and open.
    Answer: B

Pages: 346
45. The absolute refractory period for a neuron is the time:
   A) between application of a stimulus and the achievement of threshold
   B) it takes to restore sodium and potassium ions to their resting positions
   C) necessary to break down the neurotransmitter molecules that have bound to the membrane
   D) during which the membrane is hyperpolarized after repolarization
   E) during which a second action potential cannot be initiated, no matter how strong the stimulus
   Answer: E

Pages: 347
46. The most important factors that determine the propagation speed of a nerve impulse are:
   A) amount of blood flow to the neuron and the thickness of the membrane
   B) the number of dendrites and their total length
   C) amount of neurotransmitter required to reach threshold and the rate at which it is broken down
   D) stimulus strength and the length of the axon
   E) the diameter of the nerve fiber and presence or absence of a myelin sheath
   Answer: E

Pages: 346
47. Local anesthetics, such as procaine and lidocaine, block pain sensations by:
   A) opening sodium ion leakage channels
   B) opening potassium ion leakage channels
   C) blocking blood flow to the region
   D) preventing opening of voltage-gated sodium ion channels
   E) preventing the opening of chemically-gated sodium ion channels
   Answer: D

Pages: 347
48. Which of the following statements is **TRUE** regarding Types A, B and C nerve fibers?
   A) Type A fibers conduct nerve impulses more slowly than Types B and C fibers.
   B) Type A fibers have the longest absolute refractory period.
   C) Type A fibers conduct nerve impulses related principally to the activities of the autonomic nervous system.
   D) Type A and B fibers have a larger diameter and are myelinated, while Type C fibers have the smallest diameter and are unmyelinated.
   E) Type C fibers conduct impulses from the central nervous system to skeletal muscles.
   Answer: D

Pages: 348
49. At an electrical synapse, conduction of current occurs by means of:
   A) saltatory conduction
   B) diffusion of excitatory neurotransmitters
   C) direct passage of ions through tunnels called connexons between cells
   D) passage of sodium ions along neurofibrils across the synaptic cleft
   E) active transport of neurotransmitters across adjoining cell membranes
   Answer: C

50. One way a neurotransmitter can cause an excitatory postsynaptic potential is by:
    A) opening chemically-gated calcium ion channels
    B) closing chemically-gated sodium ion channels
    C) opening chemically-gated potassium ion channels
    D) opening chemically-gated chloride ion channels
    E) binding to anions trapped inside the cell
    Answer: A

51. One way a neurotransmitter can cause an inhibitory postsynaptic potential is by:
    A) opening chemically-gated calcium ion channels
    B) opening chemically-gated sodium ion channels
    C) opening chemically-gated potassium ion channels
    D) closing chemically-gated chloride ion channels
    E) binding to anions trapped inside the cell
    Answer: C

52. Presynaptic facilitation occurs when:
    A) a neuron produces neurotransmitter at a more rapid rate than normal
    B) a neurotransmitter released by the postsynaptic neuron diffuses back across the synaptic cleft to the presynaptic neuron
    C) several presynaptic end bulbs release neurotransmitter simultaneously
    D) a second presynaptic neuron is stimulated to release more neurotransmitter by the action of neurotransmitter from a first presynaptic neuron
    E) one presynaptic neuron inhibits the release of neurotransmitter from a second presynaptic neuron
    Answer: D

53. Caffeine and nicotine have excitatory effects because they:
    A) reduce the threshold for excitation
    B) increase the threshold for excitation
    C) open potassium ion leakage channels
    D) open chemically-gated chloride ion channels
    E) close chemically-gated calcium ion channels
    Answer: A

54. Which of the following best describes a diverging neuronal circuit?
    A) Multiple presynaptic neurons stimulate a single postsynaptic neuron.
    B) A single presynaptic neuron stimulates several postsynaptic neurons.
    C) A single presynaptic neuron releases several different neurotransmitters to a single postsynaptic neuron.
    D) A single postsynaptic neuron has receptors for several different neurotransmitters.
    E) A single postsynaptic neuron has multiple responses to a single type of neurotransmitter.
    Answer: B

Pages: 354

55. A single presynaptic neuron stimulates a group of neurons, each of which synapses with a common postsynaptic cell. This best describes a:
    A) simple circuit
    B) diverging circuit
    C) converging circuit
    D) reverberating circuit
    E) parallel after-discharge circuit
    Answer: E

Pages: 355

56. Virtually all developing neurons lose their ability to undergo mitosis by the age of:
    A) six months
    B) five years
    C) puberty
    D) 25 years
    E) 65 years
    Answer: A

Pages: 355

57. In the PNS, damage to axons and dendrites may be repaired if the cell body is intact, and if:
    A) neurotransmitters are still being produced
    B) the neurolemmocytes are still active
    C) astrocytes multiply rapidly to form a protective scar
    D) drugs are administered to make the remaining parts of the neuron less excitable so it can rest
    E) connexons can be formed between the severed ends
    Answer: B

Pages: 344

58. Action potentials occur only when:
    A) leakage ion channels open
    B) chemically-gated ion channels open
    C) light-gated ion channels open
    D) voltage-gated ion channels open
    E) leakage ion channels close
    Answer: D

Pages: 346

59. The all-or-none principle of neuron function states that:
    A) once an action potential is generated, it travels at a constant speed and maximum strength for the existing conditions
    B) all ion channels in a nerve cell membrane are open, or none of them are
    C) a neuron releases all its stored neurotransmitter, or none of it
    D) a postsynaptic neuron is capable of responding to a particular neurotransmitter, or it is not
    E) a neuron may exhibit graded potentials only if the myelin sheath is continuous along the entire neuron
    Answer: A

60. The voltage across a neuron's membrane has changed from its resting potential, which is -70mV, to -60mV. Which of the following is most likely to be **TRUE** about this situation?
    A) The membrane has bound an excitatory neurotransmitter.
    B) The membrane has bound an inhibitory neurotransmitter.
    C) An anesthetic drug has been applied to the neuron.
    D) Voltage-gated potassium ion channels have been opened.
    E) The metabolic activities of the cell have caused production of more anions that are now trapped inside the cell.
    Answer: A

61. The voltage across a neuron's membrane has changed from its resting potential of -70mV to -80mV. Which of the following is most likely to be **TRUE** regarding this situation?
    A) The neuron has bound a neurotransmitter that opens chemically-gated sodium ion channels.
    B) The neuron has bound a neurotransmitter that opens chemically-gated chloride ion channels.
    C) Large pores have opened in the membrane to allow release of large intracellular anions.
    D) The sodium/potassium pump has ceased functioning.
    E) Additional myelin has been added to the membrane to provide greater electrical insulation.
    Answer: B

62. Buildup of neurotransmitter released by a single presynaptic end-bulb firing two or more times in rapid succession results in:
    A) convergence
    B) divergence
    C) parallel-after discharge
    D) spatial summation
    E) temporal summation
    Answer: E

63. Association neurons carry nerve impulses from:
    A) a receptor to the central nervous system
    B) one neuron to another
    C) a receptor to an effector
    D) the central nervous system to an effector
    E) an effector to a receptor
    Answer: B

64. The white color of white matter in the brain and spinal cord is due to the:
    A) types of neurotransmitters present
    B) large amount of DNA in the neurons of the CNS
    C) presence of myelin
    D) collagen in the connective tissues surrounding the tracts
    E) pigment in the neuroglia of the CNS
    Answer: C

Pages: 349
65. What would happen to a postsynaptic neuron that binds a neurotransmitter that closes potassium ion channels?
    A) An EPSP would result because potassium ions are positively charged, and as they accumulate inside the cell, the membrane moves closer to threshold.
    B) An EPSP would result because the sodium/potassium pump would work to force potassium ions out of the cell, and sodium ions would be brought in as a result.
    C) An IPSP would result because potassium ions are negatively charged, and as they accumulate inside the cell, the membrane becomes hyperpolarized.
    D) There would be no change in membrane potential because the membrane potential is dependent only on the opening and closing of sodium ion channels.
    E) There would be no change in membrane potential because there would be an increase in production of intracellular anions to compensate for the added potassium ions.
    Answer: A

True-False

Write T if the statement is true and F if the statement is false.

Pages: 355
1. Mature neurons normally do not undergo mitosis.
   Answer: True

Pages: 339
2. General somatic efferent neurons conduct impulses from the central nervous system to smooth muscle, cardiac muscle, and glands.
   Answer: False

Pages: 342
3. At rest, a neuron's cell membrane is slightly more positively charged inside than outside.
   Answer: False

Pages: 333
4. Astrocytes form part of the blood-brain barrier.
   Answer: True

Pages: 338
5. Bipolar neurons are so-called because they have two axons.
   Answer: False

Pages: 342
6. A resting neuron cell membrane is more permeable to sodium ions than to potassium ions.
   Answer: False

Pages: 346
7. During the relative refractory period, a neuron can initiate a second action potential in response to a suprathreshold stimulus.
   Answer: True

Pages: 343
8. A neuron cell membrane can be hyperpolarized either by an influx of negatively charged ions or by an outflow of positively charged ions.
Answer: True

Pages: 346
9. An action potential becomes weaker and slower as it nears the end of an axon.
Answer: False

Pages: 346
10. Saltatory conduction occurs on myelinated axons, but not on unmyelinated axons.
Answer: True

Pages: 348
11. At an electrical synapse, ions flow directly from one cell to another with no need for a neurotransmitter.
Answer: True

Pages: 349
12. Some neurotransmitters can be either excitatory or inhibitory, depending on the characteristics of the responding cell.
Answer: True

Pages: 349
13. Conduction is faster across a chemical synapse than across an electrical synapse.
Answer: False

Pages: 349
14. A graded potential that causes the membrane potential to move closer to threshold is called an EPSP.
Answer: True

Pages: 347
15. Nerve fibers with smaller diameters conduct action potentials more rapidly than larger ones because of the smaller total membrane surface area.
Answer: False

Short Answer

Write the word or phrase that best completes each statement or answers the question.

Pages: 332
1. The two principal divisions of the nervous system are the central nervous system and the _____ nervous system.
Answer: peripheral

Pages: 346
2. The period of time in which an excitable cell cannot generate another action potential is called the _____ period.
Answer: refractory

Pages: 332

3. The central nervous system contains the _____ and the _____.
    Answer: brain; spinal cord

Pages: 332

4. _____ neurons carry nerve impulses from receptors to the central nervous system.
    Answer: Afferent (sensory)

Pages: 332

5. The peripheral nervous system is subdivided into the somatic nervous system and the _____ nervous system.
    Answer: autonomic

Pages: 333

6. The neuroglia that produce the myelin sheath in the central nervous system are the _____.
    Answer: oligodendrocytes

Pages: 336

7. The nucleus of a neuron is located in the _____ of the neuron.
    Answer: cell body

Pages: 336

8. Nerve impulses are conducted toward the cell body by a neuronal process called a(n) _____.
    Answer: dendrite

Pages: 336

9. Nerve impulses arise in an area at the junction of the axon hillock and the initial segment called the _____.
    Answer: trigger zone

Pages: 338

10. Synaptic vesicles store _____.
    Answer: neurotransmitter

Pages: 338

11. The general term for any neuronal process is a nerve _____.
    Answer: fiber

Pages: 338

12. A neuron with several dendrites and one axon is classified as a(n) _____ neuron.
    Answer: multipolar

Pages: 339

13. The effectors for general somatic efferent neurons are _____.
    Answer: skeletal muscles

Pages: 338

14. Nerve cell bodies in the peripheral nervous system generally are clustered together to form _____.
    Answer: ganglia

Pages: 341
15. An ion channel that opens in response to direct changes in the membrane potential is called a(n) _____ channel.
Answer: voltage-gated

Pages: 342
16. A cell that exhibits a membrane potential is said to be _____.
Answer: polarized

Pages: 343
17. A membrane whose polarization is more negative than the resting level is said to be

_____.

Answer: hyperpolarized

Pages: 343
18. Rapid opening of voltage-gated sodium ion channels brings about _____.
Answer: depolarization

Pages: 345
19. Recovery of the resting potential due to opening of voltage-gated potassium ion channels and closing of voltage-gated sodium ion channels is called _____.
Answer: repolarization

Pages: 346
20. Impulse conduction that appears to jump from one neurofibral node to the next is called _____ conduction.
Answer: saltatory

Pages: 347
21. The largest, fastest conducting, myelinated nerve fibers are called Type _____ fibers.
Answer: A

Pages: 349
22. When a nerve impulse arrives at a synaptic end-bulb or varicosity, the depolarization phase opens voltage-gated _____ channels in addition to opening voltage-gated sodium ion channels.
Answer: calcium ion

Pages: 349
23. A neurotransmitter that causes hyperpolarization of the membrane is said to cause a(n) _____ postsynaptic potential.
Answer: inhibitory

Pages: 351
24. Integration of the effects of the neurotransmitters from several presynaptic neurons by a postsynaptic neuron is referred to as _____.
Answer: spatial summation

Pages: 353
25. GABA and glycine act as inhibitory neurotransmitters because they open chemically-gated _____ channels.
Answer: chloride ion

Pages: 353
26. Alkalosis results in _____ excitability of neurons.
Answer: increased

Pages: 354
27. A neuronal circuit in which a postsynaptic neuron receives input from several different sources is called a _____ circuit.
Answer: converging

Pages: 355
28. Degeneration of the distal portion of a damaged neuronal process is called _____ degeneration.
Answer: Wallerian

Pages: 335
29. Gaps in the myelin sheath of an axon are called _____.
Answer: neurofibral nodes (Nodes of Ranvier)

Pages: 348
30. The site of functional contact between two neurons or between a neuron and an effector is called a(n) _____.
Answer: synapse

Matching

Choose the item from Column 2 that best matches each item in Column 1.

1. neuronal process that conducts impulses away from the cell body

   axon

2. neuronal process that conducts impulses toward the cell body

   dendrite

3. location of the nucleus of a neuron

   cell body

4. storage sites for neurotransmitter

   synaptic vesicles

5. provide framework along which fast axonal transport occurs

   microtubules   plexuses

6. phospholipid substance that provides electrical insulation and increases propagation speed — myelin

7. unmyelinated areas of a myelinated axon — neurofibral nodes

8. encloses the myelin sheath of axons in the PNS — neurolemma   sarcolemma

9. site of protein synthesis in neurons — chromatophilic substance

10. site at which nerve impulses are initiated in most neurons — trigger zone   synaptic end bulb

Essay

Write your answer in the space provided or on a separate sheet of paper.

Pages: 332

1. State the functions of the nervous system.
   Answer: The nervous system works with the endocrine system to maintain homeostasis via:
   · Sensory function - senses changes in internal and external environments; communicates information to integrating centers
   · Integrative function - analyzes sensory input; stores some information; decides appropriate responses; communicates decisions to effectors
   · Motor function - responds to stimuli by initiating appropriate muscular contractions or glandular secretions

Pages: 336

2. Describe the structural features of a typical neuron.
   Answer: · Cell body - contains nucleus and typical organelles · Chromatophilic substance - rough ER
   · Neurofibrils - form cytoskeleton
   · Dendrites - short, tapering, branched processes extending from cell body
   · Axon - single, long, thin process arising from cone-shaped axon hillock from cell body; may have collaterals; ends in axon terminals containing synaptic end-bulbs with synaptic vesicles containing neurotransmitter; may be myelinated

Pages: 342

3. Describe the characteristics of the neuron cell membrane and its environment that contribute to the existence of a resting potential.
   Answer: Sodium and chloride ions predominate in the ECF, while potassium, organic phosphates, and amino acids predominate in the ICF. The membrane is moderately permeable to potassium, but only slightly permeable to sodium. Any sodium that leaks in is removed via active transport pumps, which pump out three sodium ions for each two potassium ions imported. Anions in the ICF are generally too large to escape. The net effect is an accumulation of positive charges outside, while the inner surface of the membrane becomes more negatively charged.

Pages: 344

4. Describe the phases of an action potential, including all appropriate ion movements and the mechanisms by which such movements occur.
   Answer: · Depolarization - graded potential brings the membrane to threshold, voltage-gated sodium ion channels open, sodium rushes in by diffusion and creates positive feedback situation; sodium inactivation gates close just after activation gates open
   · Repolarization - voltage-gated potassium ion channels open as sodium ion channels are closing; potassium diffuses out to restore resting potential; voltage-gated sodium ion channels revert to resting state; after-hyperpolarization may occur as large outflow of potassium passes normal resting potential

Pages: 349

5. Describe the events of transmission of an impulse across a chemical synapse.
   Answer: Voltage-gated calcium ion channels open when the synaptic end-bulb is depolarized. Calcium ions flow into the presynaptic cell, causing exocytosis of synaptic vesicles. Neurotransmitter is released into the synaptic cleft, and diffuses across to the postsynaptic cell, where it binds to specific receptors, causing a change in membrane permeability. The neurotransmitter is then removed, either by diffusion, enzyme action, or uptake into the presynaptic cell or local neuroglia.

Pages: 346

6. How do saltatory conduction and continuous conduction of an action potential differ? What are the advantages of saltatory conduction over continuous conduction?
   Answer: Saltatory conduction involves depolarization only at neurofibril nodes, while the entire membrane is depolarized in continuous conduction. Saltatory conduction is faster and more energy efficient, since less ATP is needed to actively transport ions to resting positions.

Pages: 343-344

7. Outline the differences between graded potentials and action potentials.
   Answer: · Amplitude - graded potentials have variable strength; action potential are all-or-none
   · Duration - graded potentials longer
   · Channels - action potentials involve voltage-gated channels; graded potentials involve chemically-, mechanically-, or light-gated channels
   · Location - graded potentials arise mainly on dendrites and cell bodies; action potentials arise mainly on axons
   · Propagation - graded potentials localized; action potentials travel over longer distances
   · Refractory period - action potentials have one, graded potentials do not

Pages: 333-334

8. Name and describe the functions of the different types of neuroglia.

    Answer: · Astrocytes - involved in metabolism of neurotransmitters, maintain potassium ion balance of neurons, participate in brain development, help form blood-brain barrier
    · Oligodendrocytes - provide support and produce myelin sheath for CNS neurons
    · Microglia - provide protection in CNS via phagocytosis
    · Ependymal cells - line brain ventricles and central canal of spinal cord; assist in circulation of CSF
    · Neurolemmocytes - produce myelin around PNS neurons
    · satellite cells - support neurons in ganglia of PNS

Pages: 342-344

9. Predict the effects on levels of excitability of A) increased extracellular concentrations of sodium ions, and B) increased extracellular concentrations of potassium ions. Explain your answers.

    Answer: Many answers could be acceptable, depending on what you expect your students to know at this point - i.e., the answers do not necessarily have to be absolutely correct, as long as the students are thinking about the right things. Students should mention increases in concentration gradients and possible changes in resting potential. You might also expect comments about ion movements through leakage channels.

Pages: 349

10. You need to develop a drug that will block transmission of impulses across a chemical synapse. By what possible mechanisms of action could this drug work?

    Answer: The drug could block release of the neurotransmitter, block the binding of the neurotransmitter to the postsynaptic cell, or promote removal of the neurotransmitter from the synaptic cleft.

Chapter 13   The Spinal Cord and Spinal Nerves

Multiple-Choice

Choose the one alternative that best completes the statement or answers the question.

Pages: 365

1. What would normally be found within the central canal of the spinal cord?
   A) blood
   B) myelin
   C) cerebrospinal fluid
   D) air
   E) gray matter
   Answer: C

Pages: 362

2. The extension of the meninges inferior to the termination of the spinal cord is called the:
   A) cauda equina
   B) posterior median sulcus
   C) conus medullaris
   D) filum terminale
   E) gray commissure
   Answer: D

Pages: 361

3. The outermost layer of the meninges is the:
   A) dura mater
   B) arachnoid
   C) pia mater
   D) cauda equina
   E) rami communicantes
   Answer: A

Pages: 365

4. What would normally be found immediately surrounding the central canal of the spinal cord?
   A) white matter
   B) gray matter
   C) cerebrospinal fluid
   D) the pia mater
   E) the dura mater
   Answer: B

Pages: 361

5. Cerebrospinal fluid normally circulates in the:
   A) epidural space
   B) subdural space
   C) subarachnoid space
   D) ascending spinal tracts
   E) descending spinal tracts
   Answer: C

Pages: 362
6. In the adult, the spinal cord extends from the medulla to the:
    A) coccyx
    B) sacral promontory
    C) point of attachment of the most inferior pair of ribs
    D) sacral hiatus
    E) upper border of vertebra L2
    Answer: E

Pages: 377
7. Spinal nerves are considered mixed, which means that:
    A) they contain both nerves and tracts
    B) they contain both gray and white matter
    C) they contain both afferent and efferent neurons
    D) they use multiple types of neurotransmitters
    E) a single nerve arises from multiple segments of the spinal cord
    Answer: C

Pages: 365
8. The part of a spinal nerve that contains only sensory neurons is the:
    A) dorsal root
    B) ventral root
    C) dorsal ramus
    D) ventral ramus
    E) rami communicantes
    Answer: A

Pages: 365
9. The part of a spinal nerve that contains only efferent fibers is the:
    A) dorsal root
    B) ventral root
    C) dorsal ramus
    D) ventral ramus
    E) plexus
    Answer: B

Pages: 365
10. To do a lumbar puncture, the needle is inserted into the:
    A) central canal
    B) sacral plexus
    C) nucleus pulposus
    D) subarachnoid space
    E) gray commissure
    Answer: D

Pages: 384
11. The largest nerve in the body arises from the:
    A) cervical plexus
    B) lumbar plexus
    C) brachial plexus
    D) sacral plexus
    E) solar plexus
    Answer: D

Pages: 379
12. The nerve that stimulates the diaphragm to contract arises from the:
    A) cervical plexus
    B) lumbar plexus
    C) brachial plexus
    D) sacral plexus
    E) intercostal nerves
    Answer: A

Pages: 379
13. The nerve that stimulates the diaphragm to contract is the:
    A) median nerve
    B) phrenic nerve
    C) sciatic nerve
    D) radial nerve
    E) second intercostal nerve
    Answer: B

Pages: 380
14. The nerve that is usually damaged in carpal tunnel syndrome is the:
    A) median nerve
    B) phrenic nerve
    C) sciatic nerve
    D) femoral nerve
    E) axillary nerve
    Answer: A

Pages: 377
15. The connective tissue enclosing an individual nerve fiber is the:
    A) endoneurium
    B) perineurium
    C) epineurium
    D) neurilemma
    E) axolemma
    Answer: A

16. Cell bodies within the central nervous system are grouped into:
    A) white matter
    B) ganglia
    C) the subdural space
    D) nuclei
    E) plexuses
    Answer: D

17. The cauda equina is formed by:
    A) parallel bundles of collagen in connective tissue
    B) myelin
    C) blood vessels extending downward from the end of the spinal cord
    D) spinal nerves extending downward from the end of the spinal cord
    E) damage from herpes viruses
    Answer: D

18. The spinal cord passes through the:
    A) vertebral bodies
    B) intervertebral discs
    C) intervertebral foramina
    D) vertebral foramina
    E) all of the above except the intervertebral foramina
    Answer: D

19. The dura mater is composed of:
    A) myelin
    B) neuroglia
    C) dense irregular connective tissue
    D) cerebrospinal fluid
    E) neurons
    Answer: C

20. The epidural space contains:
    A) fat
    B) cerebrospinal fluid
    C) blood
    D) air
    E) myelin
    Answer: A

Pages: 361
21. The subdural space normally contains:
    A) fat
    B) interstitial fluid
    C) cerebrospinal fluid
    D) blood
    E) air
    Answer: B

Pages: 361
22. Denticulate ligaments are extensions of the:
    A) dura mater
    B) arachnoid
    C) pia mater
    D) cauda equina
    E) gray commissure
    Answer: C

Pages: 361
23. The highly vascular, innermost layer of the meninges is the:
    A) dura mater
    B) arachnoid
    C) pia mater
    D) gray commissure
    E) conus medullaris
    Answer: C

Pages: 361
24. The avascular, middle layer of the meninges is the:
    A) dura mater
    B) arachnoid
    C) pia mater
    D) gray commissure
    E) conus medullaris
    Answer: B

Pages: 362
25. The spinal cord generally stops growing longer around the age of:
    A) six months
    B) eighteen months
    C) five years
    D) puberty
    E) the spinal cord never stops growing until death
    Answer: C

Pages: 362
26. Nerves to and from the upper limbs arise from the:
    A) cervical plexus
    B) cervical enlargement
    C) conus medullaris
    D) cauda equina
    E) filum terminale
    Answer: B

Pages: 362
27. The conus medullaris is the:
    A) site of production of cerebrospinal fluid in the spinal cord
    B) array of spinal nerves extending downward from the end of the spinal cord
    C) extension of the meninges beyond the end of the spinal cord
    D) the tapered end of the spinal cord below the lumbar enlargement
    E) the upper part of the spinal cord where it is attached to the brain
    Answer: D

Pages: 365
28. The dorsal root ganglia contain:
    A) cerebrospinal fluid
    B) the cell bodies of peripheral sensory nerves
    C) the cell bodies of peripheral motor nerves
    D) the cell bodies of nerves in descending spinal tracts
    E) the dorsal rami of several spinal nerves
    Answer: B

Pages: 361
29. Which of the following lists the anatomical features in the correct order from outermost to innermost?
    A) dura mater, epidural space, arachnoid, subdural space, pia mater, subarachnoid space
    B) pia mater, epidural space, dura mater, subdural space, arachnoid, subarachnoid space
    C) arachnoid, subarachnoid space, pia mater, epidural space, dura mater, subdural space
    D) epidural space, dura mater, subdural space, arachnoid, subarachnoid space, pia mater
    E) dura mater, epidural space, subdural space, arachnoid, subarachnoid space, pia mater
    Answer: D

Pages: 362
30. The spinal cord is continuous with the:
    A) occipital bone
    B) cerebral cortex
    C) medulla oblongata
    D) thalamus
    E) coccyx
    Answer: C

Pages: 361
31. The function of denticulate ligaments is to:
    A) connect sensory and motor neurons in a reflex arc
    B) produce cerebrospinal fluid in the central canal
    C) transmit sensations from the face to the spinal cord
    D) bind the ventral rami of spinal nerves into plexuses
    E) protect the spinal cord against shock and displacement
    Answer: E

Pages: 361
32. The dura mater of the spinal cord extends from the foramen magnum to the:
    A) second sacral vertebra
    B) coccyx
    C) second lumbar vertebra
    D) conus medullaris
    E) upper border of the lumbar enlargement
    Answer: A

Pages: 362
33. The cervical enlargement extends from vertebrae:
    A) C1-C7
    B) C4-T1
    C) C1 to the brachial plexus
    D) C7-T12
    E) T9-T12
    Answer: B

Pages: 362
34. The lumbar enlargement extends from vertebrae:
    A) L1-L4
    B) T1 to the lumbar plexus
    C) T1 to the sacral plexus
    D) T9-T12
    E) L1-S2
    Answer: D

Pages: 365
35. A spinal tap is safely done on an adult:
    A) where the brain and spinal cord meet
    B) just below the cervical enlargement
    C) between vertebrae T12 and L1
    D) between vertebrae L4 and L5
    E) anywhere along the length of the spinal cord
    Answer: D

Pages: 365
36. At the superior end, the central canal of the spinal cord is continuous with the:
    A) foramen magnum
    B) brain sinuses
    C) internal jugular veins
    D) cerebral cortex
    E) fourth ventricle
    Answer: E

Pages: 365
37. Cell bodies of motor neurons to skeletal muscles are located in the:
    A) anterior gray horns
    B) lateral gray horns
    C) anterior white columns
    D) posterior white columns
    E) central canal
    Answer: A

Pages: 365
38. Cell bodies for motor neurons to cardiac muscle are located in the:
    A) anterior gray horns
    B) lateral gray horns
    C) anterior white columns
    D) posterior white columns
    E) central canal
    Answer: B

Pages: 367
39. Sensations of pain and temperature are conveyed by the:
    A) spinothalamic tracts
    B) posterior column tracts
    C) rubrospinal tract
    D) corticospinal tracts
    E) tectospinal tract
    Answer: A

Pages: 367
40. Sensations of proprioception and discriminative touch are conveyed by the:
    A) spinothalamic tracts
    B) posterior column tracts
    C) corticospinal tracts
    D) rubrospinal tract
    E) vestibulospinal tract
    Answer: B

Pages: 367
41. Motor impulses causing precise, voluntary movements of skeletal muscles are conveyed by the:
   A) spinothalamic tracts
   B) posterior column tracts
   C) corticospinal tracts
   D) rubrospinal tract
   E) vestibulospinal tract
   Answer: C

Pages: 367
42. Extrapyramidal tracts convey:
   A) sensory information about pain and temperature
   B) sensory information about body position
   C) sensory information about touch and pressure
   D) motor impulses coordinating movements and maintaining muscle tone
   E) motor impulses triggering precise, voluntary skeletal movements
   Answer: D

Pages: 370
43. **ALL** of the following are **TRUE** for the stretch reflex arc **EXCEPT**:
   A) it is a contralateral reflex arc
   B) the receptors in the arc are called muscle spindles
   C) it is a monosynaptic reflex arc
   D) the sensory neuron synapses with a motor neuron in the anterior gray horn of the spinal cord
   E) the effectors in the arc are skeletal muscles
   Answer: A

Pages: 369
44. Which of the following lists the components of a reflex arc in the correct order of functioning?
   A) receptor, motor neuron, integrating center, sensory neuron, effector
   B) motor neuron, receptor, integrating center, sensory neuron, effector
   C) receptor, sensory neuron, effector, motor neuron, integrating center
   D) receptor, sensory neuron, integrating center, motor neuron, effector
   E) effector, sensory neuron, integrating center, motor neuron, receptor
   Answer: D

Pages: 373
45. The flexor reflex is considered to be an intersegmental reflex because it:
   A) involves reciprocal innervation
   B) is a contralateral reflex
   C) can occur in any region of the spinal cord
   D) always occurs following the same sequence of events
   E) involves one sensory neuron activating association neurons in different segments of the
      spinal cord
   Answer: E

Pages: 374

46. The contralateral reflex that helps you maintain balance when the flexor reflex is initiated is the:
    A) stretch reflex
    B) tendon reflex
    C) crossed extensor reflex
    D) abdominal reflex
    E) patellar reflex
    Answer: C

Pages: 370

47. Tapping the knee during a physical exam tests for the patellar reflex, which is a:
    A) stretch reflex
    B) tendon reflex
    C) flexor reflex
    D) crossed extensor reflex
    E) Babinski reflex
    Answer: A

Pages: 374

48. Tapping the Achilles tendon normally elicits:
    A) contraction of the quadriceps femoris
    B) a positive Babinski sign
    C) contraction of the gastrocnemius and soleus
    D) compression of the abdominal wall
    E) contraction of the tibialis anterior
    Answer: C

Pages: 375

49. All spinal nerves except the first cervical pair emerge from the spinal cord through the:
    A) vertebral bodies
    B) intervertebral foramina
    C) intervertebral discs
    D) cauda equina
    E) filum terminale
    Answer: B

Pages: 385

50. A dermatome is:
    A) the segment of the spinal cord from which a particular spinal nerve arises
    B) all the muscles innervated by the motor neurons in a single spinal segment
    C) the area of skin that provides sensory input to one pair of spinal nerves
    D) severing of the sensory portion of a spinal nerve
    E) the union of the ventral rami of spinal nerves from several spinal cord segments
    Answer: C

Pages: 378
51. The ventral rami of spinal nerves T2-T12 are those that:
    A) form the thoracic plexus
    B) form the cervical plexus
    C) provide innervation to the diaphragm
    D) contain only sensory neurons coming from thoracic structures
    E) are known as the intercostal nerves
    Answer: E

Pages: 379
52. Clark has fallen off his horse, and landed on his head. He now can breathe only with the aid of
    a respirator. This is most likely because he has:
    A) damaged the spinal cord above the level of C3
    B) punctured his lungs via broken vertebrae and ribs
    C) severed the intercostal nerves innervating the thoracic structures
    D) crushed the centers in the brain that regulate breathing
    E) ruptured the upper air passages as his head compressed onto his neck
    Answer: A

Pages: 379
53. The ventral rami of spinal nerves C5-C8 and T1 form the:
    A) cervical plexus
    B) brachial plexus
    C) intercostal nerves
    D) cervical enlargement
    E) coccygeal plexus
    Answer: B

Pages: 380
54. Someone who is unable to contract the biceps brachii most likely has damaged the:
    A) axillary nerve
    B) phrenic nerve
    C) radial nerve
    D) median nerve
    E) musculocutaneous nerve
    Answer: E

Pages: 380
55. A misplaced intramuscular injection has rendered a patient unable to abduct his arm. This
    patient most likely has damage to the:
    A) axillary nerve
    B) median nerve
    C) radial nerve
    D) ulnar nerve
    E) phrenic nerve
    Answer: A

Pages: 380
56. A misplaced intramuscular injection has rendered a patient unable to extend the wrist. This patient most likely has damage to the:
A) sciatic nerve
B) ulnar nerve
C) radial nerve
D) phrenic nerve
E) third intercostal nerve
Answer: C

Pages: 382
57. The plexus located between the psoas major muscle and the quadratus lumborum muscle is the:
A) brachial plexus
B) cervical plexus
C) sacral plexus
D) lumbar plexus
E) coccygeal plexus
Answer: D

Pages: 382
58. A friend has pain in his back, and wants to know why, when he presses on a certain spot, his knee gives out from under him. The most likely reason is that he is pressing on the:
A) tibial portion of the sciatic nerve
B) common peroneal portion of the sciatic nerve
C) femoral nerve
D) phrenic nerve
E) blood vessels supplying the lower extremity
Answer: C

Pages: 385
59. The two nerves that make up the sciatic nerve are the:
A) femoral and pudendal
B) tibial and common peroneal
C) obturator and pudendal
D) femoral and genitofemoral
E) tibial and pudendal
Answer: B

Pages: 385
60. The sciatic nerve usually splits into its two divisions at the:
A) knee
B) pubic symphysis
C) obturator foramen
D) ankle
E) sacral promontory
Answer: A

Pages: 385
61. In order for a positive Achilles reflex to occur, the effectors must receive stimulation from the:
    A) femoral nerve
    B) tibial nerve
    C) pudendal nerve
    D) common peroneal nerve
    E) obturator nerve
    Answer: B

Pages: 387
62. The virus that causes shingles stays latent in the:
    A) plexuses
    B) anterior gray horns of the spinal cord
    C) ascending spinal tracts
    D) central canal of the spinal cord
    E) dorsal root ganglia
    Answer: E

Pages: 387
63. The polio virus produces paralysis by destroying:
    A) the cell bodies of motor neurons in the anterior gray horns of the spinal cord
    B) the contractile proteins in the affected muscles
    C) neuromuscular junctions
    D) cell bodies of sensory neurons in dorsal root ganglia
    E) supporting neuroglia in the spinal cord
    Answer: A

Pages: 369
64. The part of the spinal cord that acts as the integrating center in reflex arcs is the:
    A) white matter
    B) gray matter
    C) meninges
    D) central canal
    E) cervical and lumbar plexuses
    Answer: B

Pages: 370
65. An ipsilateral monosynaptic reflex that is important in maintaining muscle tone and muscle coordination during exercise is the:
    A) stretch reflex
    B) plantar flexion reflex
    C) crossed extensor reflex
    D) tendon reflex
    E) withdrawal reflex
    Answer: A

True-False

Write T if the statement is true and F if the statement is false.

Pages: 365
1. Descending spinal tracts have a sensory function.
Answer: False

Pages: 375
2. Spinal nerves are part of the central nervous system.
Answer: False

Pages: 374
3. After the age of 18 months a gentle stroking of the outer margin of the sole of the foot should elicit a positive Babinski sign.
Answer: False

Pages: 384
4. The femoral nerve is the largest nerve in the body.
Answer: False

Pages: 361
5. The arachnoid layer of the meninges is avascular.
Answer: True

Pages: 361
6. The subdural space normally contains cerebrospinal fluid.
Answer: False

Pages: 362
7. The spinal cord extends from the medulla to the coccyx.
Answer: False

Pages: 365
8. The dorsal root of a spinal nerve contains only sensory nerve fibers.
Answer: True

Pages: 365
9. A spinal tap is done to remove cerebrospinal fluid from the central canal of the spinal cord.
Answer: False

Pages: 365
10. The nervous tissue surrounding the central canal of the spinal cord is gray matter.
Answer: True

Pages: 370
11. The stretch reflex is an example of an ipsilateral monosynaptic reflex arc.
Answer: True

12. The ventral rami of spinal nerves, other than T2-T12, form plexuses before going to their effectors.
    Answer: True

13. The phrenic nerve provides innervation to the diaphragm.
    Answer: True

14. Carpal tunnel syndrome usually results from damage to the axillary nerve.
    Answer: False

15. The sciatic nerve arises from the lumbar plexus.
    Answer: False

Short Answer

Write the word or phrase that best completes each statement or answers the question.

1. The outermost layer of the spinal meninges is the _____.
   Answer: dura mater

2. Cerebrospinal fluid circulates in the _____ space.
   Answer: subarachnoid

3. In the adult, the spinal cord extends from the medulla to the _____ vertebra.
   Answer: second lumbar

4. The tapering end of the spinal cord below the lumbar enlargement is called the _____.
   Answer: conus medullaris

5. The roots of the spinal nerves that angle inferiorly in the vertebral canal from the end of the spinal cord form the _____.
   Answer: cauda equina

6. The cell bodies of the peripheral sensory neurons are located in swellings known as _____.
   Answer: dorsal root ganglia

7. Motor neuron axons are contained in the _____ root of a spinal nerve.
   Answer: ventral

Pages: 365

8. The cross-bar of the gray matter "H" surrounding the central canal of the spinal cord is called the gray _____ , while the arms of the "H" are called gray _____ .
Answer: commissure; horns

Pages: 365

9. Clusters of neuron cell bodies within the spinal cord are called _____ .
Answer: nuclei

Pages: 367

10. Sensory information regarding pain, temperature, touch, and deep pressure is transmitted via the _____ tracts.
Answer: spinothalamic

Pages: 367

11. Motor impulses stimulating precise, voluntary movements of skeletal muscles are transmitted via the _____ tracts.
Answer: pyramidal

Pages: 369

12. Changes in the internal or external environment are sensed by the component of a reflex arc known as the _____ .
Answer: receptor

Pages: 369

13. If the effector in a reflex arc is the heart, a gland, or smooth muscle, the reflex is called a(n) _____ reflex.
Answer: autonomic (visceral)

Pages: 370

14. The receptors in the stretch reflex arc are the _____ .
Answer: muscle spindles

Pages: 370

15. When a sensory nerve impulse enters the spinal cord on the same side that the motor impulse leaves it, the arrangement is called a(n) _____ reflex arc.
Answer: ipsilateral

Pages: 374

16. When a single sensory neuron activates several muscles via the action of association neurons in other parts of the spinal cord, the arrangement is called a(n) _____ reflex arc.
Answer: intersegmental

Pages: 374

17. When sensory impulses enter one side of the spinal cord and motor impulses leave on the opposite side, as in the crossed extensor reflex, the arrangement is called a(n) _____ reflex arc.
Answer: contralateral

Pages: 374
18. Extension of the great toe in response to gentle stroking of the outer margin of the sole of the foot is called a positive _____.
Answer: Babinski sign

Pages: 369
19. If the effector in a reflex arc is a skeletal muscle, the reflex is called a(n) _____ reflex.
Answer: somatic

Pages: 375
20. There are _____ pairs of spinal nerves.
Answer: 31

Pages: 377
21. The connective tissue wrapping each fascicle of nerve fibers is called the _____.
Answer: perineurium

Pages: 377
22. The branch of a spinal nerve containing the autonomic components is the _____.
Answer: rami communicantes

Pages: 377
23. The ventral rami of spinal nerves T2-T12 are known as _____ nerves.
Answer: intercostal

Pages: 379
24. The phrenic nerve arises from the _____ plexus.
Answer: cervical

Pages: 380
25. Carpal tunnel syndrome usually results from compression of the _____ nerve.
Answer: median

Pages: 380
26. The entire nerve supply of the shoulder and upper limb is provided by the _____ plexus.
Answer: brachial

Pages: 382
27. Damage to the _____ nerve results in the inability to extend the leg.
Answer: femoral

Pages: 384
28. The largest nerve in the body is the _____ nerve.
Answer: sciatic

Pages: 385
29. The area of skin providing sensory input to one pair of spinal nerves is called a(n) _____.
Answer: dermatome

Pages: 387
30. Shingles is caused by the same herpes virus that causes _____.
    Answer: chickenpox

Matching

Choose the item from Column 2 that best matches each item in Column 1.

1. tough, outermost layer of meninges

   dura mater

2. avascular, middle layer of meninges

   arachnoid

3. highly vascular, innermost layer of meninges

   pia mater

4. contains fat and connective tissue

   epidural space

5. contains interstitial fluid

   subdural space

6. contains cerebrospinal fluid

   subarachnoid space

7. tapered end of the spinal cord below the lumbar enlargement

   conus medullaris

8. extension of the pia mater attaching spinal cord to coccyx

   filum terminale

9. axons conveying sensory information to brain

   ascending tracts

10. axons conveying motor impulses from brain to effectors

    descending tracts

Essay

Write your answer in the space provided or on a separate sheet of paper.

Pages: 365
1. Discuss the role of the spinal cord in maintaining homeostasis.
   Answer: Spinal cord has two main functions: 1) conduction of nerve impulses between periphery and brain via white matter tracts, and 2) integration of sensory and motor information to trigger reflex actions for rapid changes necessary to maintain homeostasis.

Pages: 365
2. Describe the cross-sectional anatomy of the spinal cord.
    Answer: Spinal cord is slightly flattened in anterior-posterior dimension. It has two grooves
    - anterior median fissure and shallower posterior median sulcus. Central canal in
    center contains CSF. Gray commissure surrounds central canal. Gray commissure has gray
    horns (anterior, posterior, lateral), together forming "H" arrangement. White matter
    surrounds "H" and is subdivided into columns (anterior, posterior, lateral). Anterior
    white commissure is anterior to gray commissure.

Pages: 376-378
3. Describe the structure and functioning of a spinal nerve.
    Answer: Spinal nerves attach to spinal cord at dorsal (sensory) root and ventral (motor) root.
    Dorsal root has ganglion where cell bodies are located. After passing through
    intervertebral foramen, nerve divides into dorsal ramus (to deep muscles and skin of
    dorsum of trunk), ventral ramus (to structures of limbs and lateral and ventral
    trunk), meningeal branch (to vertebrae and associated structures), and rami
    communicantes (autonomic structures).

Pages: 369
4. Identify the components of a spinal reflex arc, and describe the function of each.
    Answer: · Receptor - responds to specific change in environment (stimulus) by producing graded
    potential
    · Sensory neuron - conducts impulse from receptor to integrating center in gray matter
    of spinal cord
    · Integrating center - site of synapse between sensory neuron and other neurons;
    decision-making area in gray matter of spinal cord
    · Motor neuron - conducts impulse from integrating center to effector
    · Effector - responds to motor nerve impulse; either muscle or gland

Pages: 369-374
5. Define the terms monosynaptic reflex, polysynaptic reflex, ipsilateral reflex, and
    contralateral reflex. Give an example of each.
    Answer: · Monosynaptic reflex - sensory nerve synapses directly with motor neuron in the
    spinal cord; e.g., stretch reflex
    · Polysynaptic reflex - involves more than two types of neurons and more than one CNS
    synapse; e.g., flexor (withdrawal) reflex
    · Ipsilateral reflex - sensory impulse enters spinal cord on same side motor impulse
    leaves; e.g., stretch reflex
    · Contralateral reflex - sensory impulse enters spinal cord on opposite side motor
    impulse leaves; e.g., crossed extensor reflex
    [Other examples are acceptable.]

Pages: 382-385
6. A lesion has formed in the anterior gray horns in the segment of the spinal cord from which
    spinal nerves L1-L4 arise. Predict the possible effects such a lesion might have on body
    function, and explain your answer.
    Answer: Anterior gray horns contain motor neurons to skeletal muscles, so paralysis of
    whatever muscles are served by the affected neurons is likely. Most likely lower limbs
    served by nerves from the lumbar and/or sacral plexuses would be affected.

Pages: 379

7. During a motorcycle accident, a patient's spinal cord was severed at the level of C2. Predict the effects on body functioning, and explain your answer.

Answer: Death most likely, from severing of phrenic nerve. Diaphragm would be unable to contract, thus breathing stops.

Pages: 361

8. Describe the protective coverings of the spinal cord.

Answer: · Dura mater - outer layer of dense, irregular connective tissue extending from foramen magnum to S2; separated from bone by epidural space containing fat and connective tissue
· Arachnoid - middle, avascular layer; web of collagen and elastic fibers; separated from dura mater by subdural space containing interstitial fluid
· Pia mater - inner, highly vascular layer of transparent connective tissue adhering to surface of spinal cord; separated from arachnoid by subarachnoid space containing CSF; denticulate ligaments suspend spinal cord in middle

Pages: 384-385

9. What is sciatica, and how does it develop?

Answer: Sciatica is neuritis characterized by severe pain along the path of the sciatic nerve, usually caused by a herniated disc, but can be caused by any injury or physiological condition that puts prolonged pressure on the sciatic nerve.

Pages: 384-385

10. In observing a person walking along the sidewalk, you notice that the person's right foot seems to go limp every time he takes a step. Offer a plausible neurological reason to explain this phenomenon.

Answer: The person possibly has foot drop, caused by damage to the common peroneal branch of the sciatic nerve, which causes inability to dorsiflex the foot.
[Other explanations are acceptable.]

Multiple-Choice

Choose the one alternative that best completes the statement or answers the question.

Pages: 393
1. The cavities within the brain are called:
   A) sulci
   B) choroid plexuses
   C) nuclei
   D) ventricles
   E) commissures
   Answer: D

Pages: 401
2. Fine control of body coordination and balance is a function of the:
   A) cerebellum
   B) hypothalamus
   C) thalamus
   D) midbrain
   E) reticular activating system
   Answer: A

Pages: 405
3. Body temperature is regulated by the:
   A) cerebellum
   B) hypothalamus
   C) thalamus
   D) midbrain
   E) pons
   Answer: B

Pages: 406
4. The cerebrum is divided into right and left hemispheres by the:
   A) longitudinal fissure
   B) transverse fissure
   C) corpus callosum
   D) tentorium cerebelli
   E) hippocampus
   Answer: A

Pages: 409
5. Paired masses of gray matter within the white matter of the cerebrum that are rich in dopamine and are involved in maintenance of muscle tone are the:
   A) corpora quadrigemina
   B) basal ganglia
   C) mammillary bodies
   D) substantia nigra
   E) supraoptic nuclei
   Answer: B

Pages: 393
6. Cerebrospinal fluid is produced by the:
   A) neurofibral nodes
   B) falx cerebri
   C) arachnoid villi
   D) substantia nigra
   E) choroid plexuses
   Answer: E

Pages: 403
7. The main relay center for conducting information between the spinal cord and the cerebrum is the:
   A) thalamus
   B) insula
   C) corpus callosum
   D) cerebellar peduncles
   E) tentorium cerebelli
   Answer: A

Pages: 394
8. Abnormally rapid production or blockage of flow of cerebrospinal fluid may result in:
   A) prolonged REM sleep
   B) Alzheimer's Disease
   C) Parkinson's Disease
   D) hydrocephalus
   E) dehydration of brain tissue
   Answer: D

Pages: 397
9. The brain stem is made up of the:
   A) cerebellum, pons, and hypothalamus
   B) medulla oblongata, thalamus, and midbrain
   C) medulla oblongata, hypothalamus, and pons
   D) medulla oblongata, pons, and midbrain
   E) midbrain, hypothalamus, and thalamus
   Answer: D

10. A part of the brain involved in emotions and including the insula, hippocampus, and parts of the diencephalon is the:
    A) corpus callosum
    B) basal ganglia
    C) limbic system
    D) cerebellar peduncles
    E) cerebral cortex
    Answer: C

11. The function of a choroid plexus is to:
    A) receive sensations from the viscera
    B) send motor impulses to the diaphragm
    C) produce cerebrospinal fluid
    D) reabsorb cerebrospinal fluid
    E) transmit impulses from one cerebral hemisphere to the other
    Answer: C

12. The function of arachnoid villi is to:
    A) reabsorb cerebrospinal fluid
    B) produce cerebrospinal fluid
    C) hold the meninges onto the brain
    D) provide nourishment for neurons in the CNS
    E) conduct impulses from one cerebral hemisphere to the other
    Answer: A

13. What is the significance of the decussation of pyramids?
    A) It causes paralysis.
    B) It produces cerebrospinal fluid.
    C) It makes the left side of the brain control the right side of the body, and vice versa.
    D) It reabsorbs cerebrospinal fluid.
    E) It transmits nerve impulses from one cerebral hemisphere to the other.
    Answer: C

14. The centers that coordinate swallowing, vomiting, coughing, sneezing, and hiccuping are located in the:
    A) cerebral cortex
    B) cerebellum
    C) hypothalamus
    D) medulla oblongata
    E) insula
    Answer: D

Pages: 400
15. The substantia nigra of the midbrain is rich in the neurotransmitter:
    A) acetylcholine
    B) serotonin
    C) epinephrine
    D) dopamine
    E) glycine
    Answer: D

Pages: 403
16. The main function of the thalamus is to:
    A) control body position
    B) control breathing patterns
    C) interpret sensory input
    D) produce hormones
    E) control body temperature
    Answer: C

Pages: 400
17. The pneumotaxic and apneustic areas are located in the:
    A) medulla oblongata
    B) pons
    C) cerebellum
    D) hypothalamus
    E) cerebral cortex
    Answer: B

Pages: 400
18. The function of the corpora quadrigemina is to:
    A) control reflex movements of the body in response to visual and auditory stimuli
    B) produce cerebrospinal fluid
    C) set the basic pattern of breathing
    D) produce hormones that control the anterior pituitary gland
    E) control blood pressure
    Answer: A

Pages: 409
19. White fibers that transmit impulses between corresponding gyri in opposite cerebral hemispheres
    are called:
    A) association fibers
    B) projection fibers
    C) commissural fibers
    D) ganglia
    E) choroid plexuses
    Answer: C

20. Damage to the cerebellum would result in:
    A) loss of memory
    B) uncoordinated movement
    C) inability to dream
    D) altered pituitary function
    E) uncontrollable body temperature
    Answer: B

21. The hippocampus is most important for:
    A) conversion of short-term to long-term memory
    B) production of cerebrospinal fluid
    C) maintenance of posture
    D) setting the basic pattern of breathing
    E) controlling blood pressure
    Answer: A

22. Cranial nerve I is the:
    A) optic
    B) olfactory
    C) oculomotor
    D) abducens
    E) accessory
    Answer: B

23. Which of the following has only a sensory function?
    A) optic
    B) oculomotor
    C) trochlear
    D) abducens
    E) trigeminal
    Answer: A

24. Which of the following combinations of cranial nerves cause movements of the eyeball?
    A) optic, trochlear, and abducens
    B) optic, oculomotor, and trochlear
    C) optic, oculomotor, and abducens
    D) trochlear, oculomotor, and abducens
    E) trochlear, oculomotor, and trigeminal
    Answer: D

25. Motor impulses for chewing and sensory impulses from most of the face and tongue are carried by cranial nerve number:
    A) XII
    B) VII
    C) IX
    D) V
    E) IV
    Answer: D

26. Cranial nerve number X is the:
    A) vagus
    B) facial
    C) hypoglossal
    D) glossopharyngeal
    E) trigeminal
    Answer: A

27. Cranial nerve number VIII conveys sensory information about:
    A) visual stimuli
    B) facial expression
    C) taste
    D) equilibrium
    E) smell
    Answer: D

28. The embryonic hindbrain is the:
    A) prosencephalon
    B) mesencephalon
    C) rhombencephalon
    D) diencephalon
    E) telencephalon
    Answer: C

29. The metencephalon develops into the:
    A) cerebrum
    B) cerebellum
    C) thalamus
    D) medulla oblongata
    E) midbrain
    Answer: B

Pages: 393
30. The third ventricle is located:
    A) between the layers of the cranial dura mater
    B) between the brain stem and the cerebellum
    C) within the left cerebral hemisphere
    D) above the hypothalamus and between the right and left halves of the thalamus
    E) between the medulla oblongata and the spinal cord
    Answer: D

Pages: 393
31. The interventricular foramina of the brain are the:
    A) holes through which the first pair of spinal nerves emerges
    B) route of cerebrospinal fluid flow from the lateral to the third ventricle
    C) vessels through which cerebrospinal fluid passes to re-enter the blood
    D) holes through which cranial nerves leave the brain stem
    E) route of cerebrospinal fluid flow from the third to the fourth ventricles
    Answer: B

Pages: 424
32. The marginal layer of the neural tube develops into:
    A) white matter
    B) gray matter
    C) cerebrospinal fluid
    D) meninges
    E) blood vessels
    Answer: A

Pages: 424
33. Spina bifida and anencephaly are neural tube defects associated with deficiencies of:
    A) vitamin A
    B) folic acid
    C) vitamin C
    D) calcium
    E) riboflavin
    Answer: B

Pages: 397
34. The most inferior part of the brain stem is the:
    A) cerebral cortex
    B) cerebellum
    C) hypothalamus
    D) pons
    E) medulla oblongata
    Answer: E

Pages: 421
35. Sensory impulses related to equilibrium are carried by a branch of the same nerve that conveys sensory impulses associated with:
A) vision
B) hearing
C) smell
D) taste
E) visceral pain
Answer: B

Pages: 394
36. The superior sagittal sinus is normally filled with:
A) air
B) blood
C) cerebrospinal fluid
D) gray matter
E) interstitial fluid
Answer: B

Pages: 406
37. Releasing hormones that control the anterior pituitary gland are produced by the:
A) pineal gland
B) thalamus
C) hypothalamus
D) medulla oblongata
E) corpus callosum
Answer: C

Pages: 394
38. Brain sinuses are located:
A) inside each hemisphere of the cerebrum
B) within each lobe of the cerebellum
C) between the two layers of the cranial dura mater
D) between the arachnoid and the pia mater
E) within the insula of the cerebrum
Answer: C

Pages: 397
39. The function of circumventricular organs is to:
A) produce cerebrospinal fluid
B) reabsorb cerebrospinal fluid
C) transmit nerve impulses from one side of a ventricle to the other
D) monitor chemical changes in the blood
E) protect the ventricles from collapsing following blows to the head
Answer: D

Pages: 397
40. The diameter of blood vessels is controlled primarily by centers located in the:
A) thalamus
B) midbrain
C) cerebellum
D) medulla oblongata
E) occipital lobe of the cerebrum
Answer: D

Pages: 399
41. An oval nucleus called the olive is involved in transmission of impulses between the:
A) medulla oblongata and the pons
B) pons and the cerebellum
C) medulla oblongata and the cerebellum
D) two halves of the thalamus
E) thalamus and the hypothalamus
Answer: C

Pages: 403
42. The portion of the diencephalon that is involved in emotional responses to smells is the:
A) pineal gland
B) subthalamus
C) hypothalamus
D) habenular nucleus
E) intermediate mass
Answer: D

Pages: 405
43. The portion of the hypothalamus that acts as a relay station in reflexes related to the sense of smell is the:
A) mammillary bodies
B) supraoptic region
C) preoptic region
D) infundibulum
E) median eminence
Answer: A

Pages: 403
44. The hormone produced by the pineal gland that is thought to promote sleepiness is:
A) oxytocin
B) antidiuretic hormone
C) melatonin
D) acetylcholine
E) monoamine oxidase
Answer: C

Pages: 406
45. The neurosecretory cells that produce the hormones stored and released by the posterior pituitary gland are located in the:
A) hypothalamus
B) pineal gland
C) intermediate mass
D) midbrain
E) cerebellum
Answer: A

Pages: 405
46. The function of the superchiasmatic nucleus of the hypothalamus is to:
A) stimulate the sensation of thirst
B) inhibit feeding behavior
C) produce hormones released by the posterior pituitary gland
D) control muscle tone
E) establish diurnal sleep patterns
Answer: E

Pages: 412
47. The primary motor area of the cerebral cortex is located in the:
A) precentral gyrus
B) postcentral gyrus
C) temporal lobe
D) occipital lobe
E) insula
Answer: A

Pages: 410
48. Large automatic movements of skeletal muscles, such as swinging the arms while walking, are controlled by the portions of the basal ganglia known as the caudate nucleus and the:
A) globus pallidus
B) arbor vitae
C) putamen
D) intermediate mass
E) infundibulum
Answer: C

Pages: 409
49. The internal capsule of the basal ganglia is an example of a group of:
A) association fibers
B) projection fibers
C) commissural fibers
D) reticular activating fibers
E) gyri and sulci
Answer: B

Pages: 410
50. The limbic system is thought to be involved primarily in control of:
    A) muscle tone
    B) precise, voluntary motor activities
    C) involuntary aspects of behavior
    D) autonomic nervous system activities
    E) heart rate and blood pressure
    Answer: C

Pages: 411
51. Localization of sensations of touch and proprioception is the major function of the:
    A) primary motor area
    B) primary gustatory area
    C) primary somatosensory area
    D) frontal eye field area
    E) gnostic area
    Answer: C

Pages: 411,413
52. The primary visual area and visual association area of the cerebral cortex are both located in the:
    A) frontal lobe
    B) temporal lobe
    C) parietal lobe
    D) insula
    E) occipital lobe
    Answer: E

Pages: 413
53. A patient has had a cerebrovascular accident involving the middle region of the temporal cerebral cortex. Which of the following would most likely be a problem for this patient?
    A) recognizing friends and family
    B) identifying foods by taste
    C) determining the relationships between body parts
    D) determining whether a sound is speech, music, or noise
    E) translating thoughts into speech
    Answer: D

Pages: 413
54. Brain waves on an EEG that are seen normally in children and in adults experiencing emotional stress are:
    A) alpha waves
    B) beta waves
    C) theta waves
    D) delta waves
    E) gamma waves
    Answer: C

Pages: 416
55. Antianxiety drugs, such as Valium, enhance the action of:
    A) acetylcholine
    B) glutamate
    C) GABA
    D) norepinephrine
    E) dopamine
    Answer: C

Pages: 416
56. **ALL** of the following are in the biogenic amine family of neurotransmitters **EXCEPT**:
    A) acetylcholine
    B) epinephrine
    C) norepinephrine
    D) dopamine
    E) serotonin
    Answer: A

Pages: 417
57. The function of MAO and COMT is to:
    A) enhance the effects of GABA
    B) break down the catecholamines
    C) break down acetylcholine
    D) reduce sensations of pain
    E) induce sleep
    Answer: B

Pages: 417
58. The most frequently prescribed antidepressant drug, Prozac, is a selective inhibitor of the re-uptake of the neurotransmitter:
    A) acetylcholine
    B) GABA
    C) dopamine
    D) nitric oxide
    E) serotonin
    Answer: E

Pages: 418
59. Damage to the cribriform plate of the ethmoid bone would most likely result in loss of:
    A) vision
    B) sensations to the side of the face
    C) the sense of smell
    D) the ability to speak
    E) equilibrium
    Answer: C

Pages: 403
60. The cerebellar peduncles transmit nerve impulses between the cerebellum and the:
    A) cerebral cortex
    B) hypothalamus
    C) thalamus
    D) brain stem
    E) lateral ventricles
    Answer: D

Pages: 401
61. Damage to the reticular activating system would most likely result in:
    A) coma
    B) ataxia
    C) inability to fall into REM sleep
    D) inability to form new memories
    E) excessive production of cerebrospinal fluid
    Answer: A

Pages: 421
62. The origins for cranial nerves VIII-XII are in nuclei located in the:
    A) cerebral cortex
    B) hypothalamus
    C) cerebellum
    D) midbrain
    E) medulla oblongata
    Answer: E

Pages: 403
63. The blood-cerebrospinal fluid barrier consists primarily of:
    A) white matter
    B) gray matter
    C) astrocytes
    D) ependymal cells
    E) the pia mater
    Answer: C

Pages: 403
64. The subthalamus connects primarily to the:
    A) motor areas of the cerebrum
    B) pituitary gland
    C) basal ganglia
    D) cerebellum
    E) medulla oblongata
    Answer: C

65. A person who tells you, "You're so left-brained," is most likely implying that you are:
A) very artistic and imaginative
B) good with words, reasoning, and numbers
C) very athletic
D) uncontrollably emotional
E) tone-deaf
Answer: B

True-False

Write T if the statement is true and F if the statement is false.

Pages: 397
1. The brain stem consists of the thalamus and the hypothalamus.
Answer: False

Pages: 423
2. The prosencephalon develops into the telencephalon and the diencephalon.
Answer: True

Pages: 394
3. Brain sinuses are normally filled with air.
Answer: False

Pages: 393
4. The fourth ventricle is located between the brain stem and the cerebellum.
Answer: True

Pages: 396
5. The adult brain consumes about 20% of the oxygen used at rest.
Answer: True

Pages: 398
6. The rate and force of heartbeat are regulated by the cardiovascular center in the pons.
Answer: False

Pages: 400
7. Reflex centers for movements of the eyes, head, and neck in response to visual stimuli are located in the cerebellum.
Answer: False

Pages: 401
8. The reticular activating system is responsible for maintaining consciousness.
Answer: True

Pages: 418
9. Nitric oxide is a neurotransmitter that is stored in the synaptic vesicles of the corpus callosum.
Answer: False

Pages: 403
10. The pineal gland is located in the sella turcica of the sphenoid bone.
    Answer: False

Pages: 406
11. Oxytocin and antidiuretic hormone are produced by cell bodies of neurons in the hypothalamus.
    Answer: True

Pages: 409
12. The corpus callosum is a group of commissural fibers.
    Answer: True

Pages: 411
13. A contusion is a temporary loss of consciousness without obvious bruising of the brain.
    Answer: False

Pages: 416
14. GABA is an important excitatory neurotransmitter in the brain.
    Answer: False

Pages: 417
15. Serotonin is concentrated primarily in the raphe nucleus of the brain stem.
    Answer: True

Short Answer

Write the word or phrase that best completes each statement or answers the question.

Pages: 417
1. COMT and MAO are enzymes that break down _____.
   Answer: catecholamines (NE, epinephrine, DA)

Pages: 417
2. _____ is a neuropeptide that transmits pain-related input from peripheral pain receptors to the central nervous system.
   Answer: Substance P

Pages: 406
3. The neuropeptides produced by the hypothalamus that are released as hormones from the posterio pituitary gland are _____ and _____.
   Answer: oxytocin; antidiuretic hormone

Pages: 420
4. Cranial nerve V is the _____ nerve.
   Answer: trigeminal

Pages: 421
5. The _____ nerve conducts sensory information concerning hearing and equilibrium.
   Answer: vestibulocochlear

Pages: 423
6. The mantle layer of the neural tube develops into _____.
   Answer: gray matter

Pages: 423
7. The cranial and spinal nerves develop from the embryonic structure known as the _____.
   Answer: neural crest

Pages: 423
8. The cerebrum develops from the primary brain vesicle known as the _____.
   Answer: prosencephalon

Pages: 392
9. The extension of the dura mater that separates the cerebral hemispheres is the _____.
   Answer: falx cerebri

Pages: 393
10. The ventricles of the brain normally are filled with _____.
    Answer: cerebrospinal fluid

Pages: 393
11. Networks of capillaries involved in the production of cerebrospinal fluid are called

    _____.
    Answer: choroid plexuses

Pages: 393
12. Cerebrospinal fluid passes from the third to the fourth ventricle through the _____ in
    the midbrain.
    Answer: cerebral aqueduct

Pages: 394
13. Cerebrospinal fluid is reabsorbed into the blood through structures called _____.
    Answer: arachnoid villi

Pages: 397
14. Small brain regions in the walls of the third and fourth ventricles that can monitor chemical
    changes in blood because they lack a blood-brain barrier are called _____.
    Answer: circumventricular organs

Pages: 397
15. The brain stem consists of the _____, _____, and _____.
    Answer: medulla oblongata; pons; midbrain

Pages: 397
16. The nucleus gracilis and nucleus cuneatus are prominent nuclei on the dorsal side of the

    _____.
    Answer: medulla oblongata

17. Rising osmotic pressure of the extracellular fluid stimulates the thirst center located in the

    _____.
    Answer: hypothalamus

18. Upfolds of the cerebral cortex are called _____ or convolutions.
    Answer: gyri

19. Regions in the pons that help control respiration are the _____ and the _____ areas.
    Answer: pneumotaxic; apneustic

20. The red nucleus is a region of the _____ involved in coordinating muscular movements.
    Answer: midbrain

21. The _____ produces the hormone melatonin.
    Answer: pineal gland

22. The middle cerebellar peduncles conduct afferent impulses from the _____ to the

    _____.
    Answer: pons; cerebellum

23. The largest portion of the diencephalon is the _____.
    Answer: thalamus

24. The infundibulum is a stalklike structure that attaches the _____ to the _____.
    Answer: pituitary gland; hypothalamus

25. The groove that separates the frontal and parietal lobes of the cerebrum is the _____.
    Answer: central sulcus

26. The region/lobe of the cerebrum that cannot be seen from the exterior surface of the brain is
    the _____.
    Answer: insula

27. White fibers that connect and transmit nerve impulses between gyri in the same cerebral
    hemisphere are called _____ fibers.
    Answer: association

28. Amino acids that act as excitatory neurotransmitters in the brain are _____ and _____; the amino acid, _____, acts as an inhibitory neurotransmitter in the brain.
Answer: glutamate; aspartate; GABA

29. The substantia nigra is particularly rich in the neurotransmitter _____.
Answer: dopamine

30. The two principal types of cerebrovascular accident are 1) _____, due to decreased blood supply, and 2) _____, due to rupture of a blood vessel in the brain.
Answer: ischemic; hemorrhagic

Matching

Choose the item from Column 2 that best matches each item in Column 1.

1.	olive	medulla oblongata
2.	apneustic area	pons
3.	substantia nigra	midbrain
4.	arbor vitae	cerebellum
5.	intermediate mass	thalamus
6.	supraoptic region	hypothalamus
7.	putamen	basal ganglia
8.	hippocampus	limbic system
9.	primary visual area	occipital lobe of cerebrum
10.	primary motor area	frontal lobe of cerebrum

Essay

Write your answer in the space provided or on a separate sheet of paper.

Pages: 392-393
1. Describe the contributions of cerebrospinal fluid to homeostasis.
   Answer: 1) Mechanical protection - shock absorber
   2) Chemical protection - provides optimal environment for impulse transmission
   3) Circulation - medium for exchange of nutrients, wastes, and gases between blood and nervous tissue

Pages: 397
2. Describe the structure and function of the blood-brain barrier.
   Answer: The barrier is formed by capillaries whose endothelial cells have tight junctions and a continuous basement membrane. Astrocytes also press against the capillaries to help control what can leave the blood and enter the brain. Small molecules and lipid-soluble substances pass easily; water-soluble substances may pass via carrier. Large molecules typically do not pass at all.

Pages: 406
3. List and briefly describe the functions of the hypothalamus.
   Answer: 1) Control of ANS - main regulator of visceral activities
   2) Control of pituitary gland - hypothalamic regulating hormones control anterior pituitary; neurosecretory cells produce OT and ADH released by posterior pituitary
   3) Regulation of emotional and behavioral patterns - works with limbic system
   4) Controls eating behavior via feeding and satiety centers, and drinking behavior via thirst center
   5) Control of body temperature via ANS
   6) Regulation of diurnal rhythms and states of consciousness

Pages: 417
4. Predict the effect on brain biochemistry of a drug known as an MAO inhibitor. Explain your answer.
   Answer: An MAO inhibitor would inhibit the breakdown of catecholamines, thus these neurotransmitters should increase in concentration. Therefore, one would expect higher levels of norepinephrine and dopamine, particularly, as well as increased levels of epinephrine.

Pages: 412
5. A patient is experiencing smell sensations that are not real. She is subsequently found to have a brain tumor. Where would you predict the tumor to be located? Why was she experiencing false sensations of smell? Explain your answers.
   Answer: The tumor is likely in any of the olfactory regions of the brain - e.g., olfactory bulbs or primary olfactory area on medial aspect of temporal lobe of cerebrum. False smell sensations are likely forms of seizures that result from pressure created by the tumor on sensory olfactory neurons or neurons in the association areas related to smell.

Pages: 392-394

6. Describe the formation and route of circulation of cerebrospinal fluid.

    Answer: CSF is formed from blood plasma via filtration and secretion by ependymal cells covering capillaries in choroid plexuses. CSF formed in the lateral ventricles flows through interventricular foramina into the third ventricle, through the cerebral aqueduct to the fourth ventricle, through the median and lateral apertures to the subarachnoid space. It is reabsorbed via arachnoid villi into the dural venous sinuses. CSF also circulates in the central canal of the spinal cord.

Pages: 413

7. What is the function of association areas of the cerebral cortex? Give a specific example and describe its function.

    Answer: Association areas connect motor and sensory areas, and integrate and interpret sensations. They store memories of past experiences, compare new imput to past input, and interpret new input. Possible examples include the somatosensory area, visual association area, auditory association area, gnostic area, premotor area, and frontal eye field.

Pages: 400

8. Several patients were admitted to the emergency rooms of several hospitals in a large urban area exhibiting the effects of damage to the substantia nigra by an illegal drug. What would you expect these effects to be? Explain your answer.

    Answer: Symptoms similar to those of Parkinson's disease (inability to control skeletal movements, etc.) would result, due to loss of dopamine-producing cells. The substantia nigra would be unable to contribute to the activities of the basal ganglia, cerebellum, and cerebrum.

Pages: 396

9. Why does the brain need such a large and constant blood supply?

    Answer: The brain consumes 20% of the oxygen used at rest. Low oxygen levels trigger damage from enzymes released from lysosomes. Oxygen is also needed for ATP production, as is glucose. Little carbohydrate can be stored in the brain, so a constant supply for ATP production is necessary. Blood transports both oxygen and glucose to the brain, and removes wastes.

Pages: 397-401

10. Explain why a crushing injury to the occipital bone is often fatal.

    Answer: The crushing of the bone also crushes the brain stem, particularly the medulla oblongata. Damage to important nuclei regulating vital functions, such as respiration, heart rate and force of contraction, and diameter of blood vessels, may result in death.

Chapter 15   Sensory, Motor, and Integrative Systems

Multiple-Choice

Choose the one alternative that best completes the statement or answers the question.

Pages: 435
1. The degree of muscle stretch is monitored by:
   A) tendon organs
   B) Merkel discs
   C) joint kinesthetic receptors
   D) muscle spindles
   E) lamellated corpuscles
   Answer: D

Pages: 436
2. The function of tendon organs is to monitor the:
   A) change in angles at joints
   B) degree of muscle stretch
   C) force of muscle contraction
   D) change in angle between tendon and muscle
   E) heat generated by muscle contraction
   Answer: C

Pages: 430
3. A sensory neuron's receptive field is the:
   A) range of stimuli within a modality to which the neuron can respond
   B) area of skin that is served by a particular sensory neuron
   C) part of the neuron that is able to respond to the sensation
   D) range of strengths of possible generator potentials for a particular neuron
   E) all the different modalities to which a sensory neuron can respond
   Answer: C

Pages: 432
4. Photoreceptors respond to:
   A) changes in temperature
   B) pressure
   C) chemical changes
   D) light
   E) sound waves
   Answer: D

Pages: 430
5. The conscious awareness and interpretation of sensations is called:
   A) modality
   B) transduction
   C) reception
   D) perception
   E) conduction
   Answer: D

Pages: 430
6. Most conscious sensations or perceptions occur in the:
    A) brain stem
    B) spinal cord
    C) skin
    D) cerebral cortex
    E) dorsal root ganglia
    Answer: D

Pages: 431
7. **ALL** of the following are considered "special" senses **EXCEPT**:
    A) smell
    B) taste
    C) touch
    D) vision
    E) hearing
    Answer: C

Pages: 432
8. Proprioceptors are located in **ALL** of the following **EXCEPT** the:
    A) walls of blood vessels
    B) muscles
    C) tendons
    D) joints
    E) internal ear
    Answer: A

Pages: 432
9. Receptors stimulated by changes in the concentration of ions in blood plasma would be classified as:
    A) proprioceptors
    B) chemoreceptors
    C) thermoreceptors
    D) photoreceptors
    E) mechanoreceptors
    Answer: B

Pages: 432
10. A stimulus that elicits a receptor potential causes:
    A) direct release of neurotransmitter via exocytosis from synaptic vesicles
    B) conduction of a nerve impulse along a first-order sensory nerve fiber
    C) initiation of an action potential in the receptor
    D) a subthreshold generator potential
    E) re-uptake of neurotransmitter from the extracellular fluid
    Answer: A

Pages: 432
11. Nerve impulses generated by cutaneous receptors travel to the:
    A) temporal lobe of the cerebral cortex
    B) parietal lobe of the cerebral cortex
    C) occipital lobe of the cerebral cortex
    D) cerebellum
    E) frontal lobe of the cerebral cortex
    Answer: B

Pages: 432
12. Nerve impulses conveying cutaneous sensations are interpreted by the:
    A) premotor area
    B) gnostic area
    C) somatosensory area
    D) cerebellum
    E) brain stem
    Answer: C

Pages: 433
13. Which of the following are considered to be rapidly adapting receptors?
    A) corpuscles of touch
    B) Type I cutaneous mechanoreceptors (Merkel discs)
    C) Type II cutaneous mechanoreceptors (end organs of Ruffini)
    D) nociceptors
    E) all of the above are rapidly adapting receptors
    Answer: A

Pages: 433
14. Receptors for discriminative touch that are located in the dermal papillae are the:
    A) Type II cutaneous mechanoreceptors
    B) corpuscles of touch (Meissner's corpuscles)
    C) lamellated (Pacinian) corpuscles
    D) nociceptors
    E) Golgi tendon organs
    Answer: B

Pages: 433-434
15. **ALL** of the following are **TRUE** regarding pain receptors **EXCEPT**:
    A) pain receptors are called nociceptors
    B) pain receptors are sensitive to all stimuli
    C) pain receptors are free nerve endings
    D) pain receptors are rapidly adapting receptors
    E) pain receptors are found in almost all body tissues
    Answer: D

Pages: 442

16. Output from the basal ganglia comes mainly from the:
    A) red nucleus
    B) substantia nigra
    C) caudate nucleus
    D) putamen
    E) globus pallidus
    Answer: E

Pages: 442

17. The part of the cerebrum most involved in programming habitual or automatic movement sequences, such as walking, is the:
    A) somatosensory area
    B) basal ganglia
    C) thalamus
    D) premotor area
    E) reticular activating system
    Answer: B

Pages: 441

18. Nerve impulses controlling voluntary movements of eyes, tongue, neck, chewing, facial expression, and speech are conveyed by the:
    A) anterior corticospinal tracts
    B) posterior spinocerebellar tracts
    C) lateral spinothalamic tracts
    D) corticobulbar tracts
    E) medial lemniscus
    Answer: D

Pages: 443

19. Impulses coordinating axial skeletal movements are conveyed by the:
    A) anterior spinothalamic tracts
    B) posterior spinocerebellar tracts
    C) lateral spinothalamic tracts
    D) corticobulbar tracts
    E) medial lemniscus
    Answer: B

Pages: 437

20. Cell bodies for first order sensory neurons conveying impulses for proprioception and most tactile sensations are located in the:
    A) dorsal root ganglia
    B) nucleus cuneatus and nucleus gracilis
    C) medial lemniscus
    D) posterior gray horn of the spinal cord
    E) postcentral gyrus
    Answer: A

Pages: 437
21. Cell bodies for second order sensory neurons conveying impulses for proprioception and most tactile sensations are located in the:
A) dorsal root ganglia
B) nucleus cuneatus and nucleus gracilis
C) medial lemniscus
D) posterior gray horn of the spinal cord
E) postcentral gyrus
Answer: B

Pages: 437
22. Axons of first order sensory neurons conveying impulses for proprioception and most tactile sensations form the:
A) anterior spinothalamic tract
B) fasciculus cuneatus and fasciculus gracilis
C) medial lemniscus
D) internal capsule
E) somatosensory area of the cerebral cortex
Answer: B

Pages: 437
23. Axons of second order sensory neurons conveying impulses for proprioception and most tactile sensations form the:
A) anterior spinothalamic tract
B) fasciculus cuneatus and fasciculus gracilis
C) medial lemniscus
D) internal capsule
E) somatosensory area of the cerebral cortex
Answer: C

Pages: 437
24. A projection tract conveying sensory impulses for proprioception and most tactile sensations is the:
A) anterior spinothalamic tract
B) corticobulbar tract
C) medial lemniscus
D) internal capsule
E) precentral gyrus
Answer: C

Pages: 437
25. Stereognosis is the ability to:
A) assess the weight of an object
B) distinguish two separate frequencies of sound simultaneously
C) make two-point touch discriminations
D) visually distinguish two very close objects
E) recognize by "feel" the size, shape, and texture of an object
Answer: E

Pages: 437
26. Kinesthesia is:
    A) inability to recognize exact locations of light touch
    B) awareness of the direction of movement
    C) ability to sense rapidly fluctuating touch
    D) paralysis of the hands and fingers
    E) inability to recognize objects by "feel"
    Answer: B

Pages: 437
27. Lateral spinothalamic pathways convey:
    A) motor impulses controlling skeletal movements of distal extremities
    B) sensory impulses related to stereognosis and kinesthesia
    C) motor impulses controlling movements of facial expressions and speech
    D) sensory impulses related to pain and temperature
    E) motor impulses controlling muscle tone
    Answer: D

Pages: 437
28. The posterior column-medial lemniscus pathway conveys:
    A) motor impulses controlling skeletal movements of the distal extremities
    B) sensory impulses related to stereognosis and kinesthesia
    C) motor impulses controlling movements of facial expression and speech
    D) sensory impulses related to pain and temperature
    E) motor impulses controlling muscle tone
    Answer: B

Pages: 437
29. Third-order sensory neurons in the posterior column-medial lemniscus pathway extend from the:
    A) skin to dorsal root ganglia
    B) dorsal root ganglia to the posterior gray horn of the spinal cord
    C) spinal cord to the medulla oblongata
    D) medulla oblongata to the thalamus
    E) thalamus to the somatosensory area of the cerebral cortex
    Answer: E

Pages: 438
30. The somatosensory area of the cerebral cortex is located in the:
    A) precentral gyrus
    B) postcentral gyrus
    C) basal ganglia
    D) hippocampus
    E) occipital lobe
    Answer: B

Pages: 437
31. Synapses between second and third-order neurons in the anterolateral pathways occur in the:
    A) dorsal root ganglia
    B) spinal cord
    C) medulla oblongata
    D) thalamus
    E) somatosensory cortex
    Answer: D

Pages: 440
32. Axons of upper motor neurons in the direct (pyramidal) pathways travel from the motor cortex to the midbrain via the:
    A) medial lemniscus
    B) corticobulbar tracts
    C) internal capsule
    D) cranial nerves III-VII
    E) hippocampus
    Answer: C

Pages: 440
33. Injury to upper motor neurons results in:
    A) kinesthesia
    B) spastic paralysis
    C) flaccid paralysis
    D) coma
    E) long-term potentiation
    Answer: B

Pages: 440
34. Injury to lower motor neurons results in:
    A) kinesthesia
    B) spastic paralysis
    C) flaccid paralysis
    D) coma
    E) long-term potentiation
    Answer: C

Pages: 441
35. **ALL** of the following are indirect (extrapyramidal) motor tracts **EXCEPT** the:
    A) rubrospinal tract
    B) tectospinal tract
    C) vestibulospinal tract
    D) corticobulbar tract
    E) lateroreticulospinal tract
    Answer: D

Pages: 442
36. A hereditary disease whose symptoms result from loss of acetylcholine-releasing neurons and GABA-releasing neurons in the basal ganglia is:
A) Parkinson's disease
B) Huntington's chorea
C) cerebral palsy
D) tabes dorsalis
E) narcolepsy
Answer: B

Pages: 447
37. Long-term potentiation is a phenomenon that occurs in:
A) formation of short-term memories
B) transition from NREM to REM sleep
C) conversion of short-term to long-term memory
D) destruction of neurons by the rubella virus
E) recovery of reflex activity following spinal shock
Answer: C

Pages: 430
38. Crude localization and identification of the type of sensory input entering the brain is made by the:
A) thalamus
B) hypothalamus
C) cerebellum
D) medulla oblongata
E) pons
Answer: A

Pages: 430
39. The distinct quality that makes one sensation different from others is called its:
A) perception
B) receptive field
C) receptor potential
D) generator potential
E) modality
Answer: E

Pages: 430
40. Conversion of a stimulus into a generator potential is called:
A) stimulation
B) perception
C) conduction
D) transduction
E) translation
Answer: D

41. First-order sensory neurons conduct impulses from:
    A) the spinal cord to the brain stem
    B) the thalamus to the somatosensory cortex
    C) the CNS to an effector
    D) a receptor to the CNS
    E) one part of the spinal cord to another
    Answer: D

42. The special senses of hearing, vision, smell, and taste are sensed by structures classified by location as:
    A) exteroreceptors (only)
    B) interoreceptors (only)
    C) proprioceptors (only)
    D) both exteroreceptors and interoreceptors
    E) both interoreceptors and proprioceptors
    Answer: A

43. **ALL** of the following are **TRUE** for receptor potentials **EXCEPT**:
    A) they are produced in receptors serving the special senses of vision and hearing
    B) they trigger action potentials in the receptors in which they occur
    C) they may be depolarizations
    D) they may be hyperpolarizations
    E) they regulate release of neurotransmitter from the receptors in which they occur
    Answer: B

44. **ALL** of the following are **TRUE** for generator potentials **EXCEPT**:
    A) they are produced in receptors for tactile sensations and pain
    B) they may trigger action potentials in the receptors in which they occur
    C) they always cause depolarization of the receptors in which they occur
    D) they sometimes cause hyperpolarization in the receptors in which they occur
    E) they occur in first-order sensory neurons
    Answer: D

45. Which of the following would have the densest concentration of cutaneous receptors?
    A) back of hand
    B) umbilical region of abdomen
    C) tip of tongue
    D) buttocks
    E) tips of toes
    Answer: C

Pages: 434
46. The reason visceral pain is referred to surface structures is that:
    A) there are no nociceptors in internal organs, so visceral sensory neurons must stimulate somatic sensory neurons to get a response
    B) Substance P is produced only in second-order sensory neurons that are used for both sensory and somatic sensations
    C) the affected organ has been removed, so the brain localizes the pain in the nearest remaining structure
    D) neurotransmitters released at the site of injury diffuse through the interstitial fluid to surface structures, triggering impulses there
    E) sensory neurons for both visceral pain and surface structures enter the same segment of the spinal cord, and surface sensations are better localized by the brain
    Answer: E

Pages: 434
47. A patient complains of pain in the right shoulder blade. This pain is most likely indicating problems with the:
    A) heart
    B) urinary bladder
    C) gallbladder
    D) kidneys
    E) brain
    Answer: C

Pages: 435
48. Local anesthetics, such as Novocaine, work by:
    A) blocking release of chemicals that stimulate nociceptors
    B) blocking conduction of impulses along first-order sensory neurons
    C) altering cerebral perception of pain
    D) breaking down Substance P in synaptic clefts
    E) blocking Substance P receptors on second-order sensory neurons
    Answer: B

Pages: 435
49. The receptors for changes in length of skeletal muscle are anchored to:
    A) actin myofilaments
    B) Z discs
    C) tendons
    D) articular capsules
    E) endomysium and perimysium
    Answer: E

Pages: 435
50. The type la fibers of a muscle spindle are stimulated when:
    A) intrafusal muscle fibers stretch
    B) it is necessary to stimulate extrafusal muscle fibers
    C) they are stimulated by gamma motor neurons
    D) they are stimulated by alpha motor neurons
    E) extrafusal muscle fibers release the appropriate neurotransmitter
    Answer: A

Pages: 436
51. The tendon organs of proprioception consist of:
    A) intrafusal muscle wrapped by a type Ia sensory fiber
    B) type Ib sensory neurons penetrating a capsule of connective tissue enclosing a few collagen fibers
    C) free nerve endings
    D) layers of connective tissue enclosing a dendrite
    E) bundles of myelinated axons
    Answer: B

Pages: 435
52. Muscle spindles are located:
    A) interspersed within skeletal muscle fibers
    B) at the junction of tendon and muscle
    C) at the junction of tendon and bone
    D) within articular capsules
    E) in the cerebellum
    Answer: A

Pages: 436
53. Tendon organs are located:
    A) interspersed within skeletal muscle fibers
    B) at the junction of tendon and muscle
    C) at the junction of tendon and bone
    D) within articular capsules
    E) in the cerebellum
    Answer: B

Pages: 436
54. Joint kinesthetic receptors are located:
    A) interspersed within skeletal muscle fibers
    B) at the junction of tendon and muscle
    C) at the junction of tendon and bone
    D) within articular capsules
    E) in the cerebellum
    Answer: D

Pages: 439
55. Tabes dorsalis is, in part, a loss of sensory function associated with:
    A) German measles (rubella)
    B) Alzheimer's disease
    C) Parkinson's disease
    D) cerebral palsy
    E) syphilis
    Answer: E

Pages: 440
56. Cell bodies for upper motor neurons are located in the:
    A) cerebral cortex
    B) cerebellum
    C) brain stem
    D) anterior gray horns of the spinal cord
    E) connective tissues surrounding skeletal muscles
    Answer: A

Pages: 440
57. Cell bodies of some lower motor neurons are located in the:
    A) cerebral cortex
    B) cerebellum
    C) basal ganglia
    D) anterior gray horns of the spinal cord
    E) connective tissues surrounding skeletal muscles
    Answer: D

Pages: 440
58. Lower motor neurons whose cell bodies are in nuclei in the brain stem stimulate:
    A) upper motor neurons
    B) association neurons
    C) movements of the face and head
    D) movements of limbs
    E) parts of the cerebellum
    Answer: C

Pages: 440
59. Axons of lower motor neurons whose cell bodies are in nuclei in the brain stem leave the brain via the:
    A) cranial nerves
    B) internal capsule
    C) medial lemniscus
    D) anterior spinothalamic tracts
    E) anterior spinocerebellar tracts
    Answer: A

Pages: 441
60. The route of conduction of motor impulses controlling the vocal cords is:
    A) cerebellum to internal capsule, to anterior corticospinal tract, to anterior root of cervical spinal nerve
    B) postcentral gyrus to medial lemniscus, to anterolateral tract, to dorsal root of cervical spinal nerve
    C) motor area of cortex to medial lemniscus, to anterolateral tract, to dorsal root of cervical spinal nerve
    D) motor area of cortex to internal capsule, to corticobulbar tract, to cranial nerve
    E) motor area of cortex to internal capsule, to lateral corticospinal tract, to anterior root of cervical spinal nerve
    Answer: D

Pages: 440
61. Decussation of axons of upper motor neurons occurs in the:
    A) lateral corticospinal tracts
    B) internal capsule
    C) medial lemniscus
    D) basal ganglia
    E) precentral gyrus
    Answer: A

Pages: 441
62. Upper motor neurons of the anterior corticospinal tracts synapse with association neurons or lower motor neurons in the:
    A) postcentral gyrus
    B) internal capsule
    C) cerebellar peduncles
    D) anterior gray horn of the spinal cord
    E) anterior white columns of the spinal cord
    Answer: D

Pages: 440
63. A small cerebral hemorrhage in the internal capsule will most likely result in:
    A) loss of kinesthesia
    B) anesthesia in skin of affected dermatomes
    C) paralysis on the opposite side of the body
    D) paralysis on the same side of the body
    E) phantom pain
    Answer: C

Pages: 449
64. The stage of deep sleep dominated by delta waves on an EEG, and during which events such as bed-wetting and sleepwalking occur is:
    A) REM sleep
    B) Stage 1 NREM sleep
    C) Stage 2 NREM sleep
    D) Stage 3 NREM sleep
    E) Stage 4 NREM sleep
    Answer: E

Pages: 449
65. In an adult's normal sleep period, REM sleep totals about:
    A) 5-10 minutes
    B) one hour
    C) 90-120 minutes
    D) 50% of the total sleep period
    E) close to 100% of the total sleep period
    Answer: C

True-False

Write T if the statement is true and F if the statement is false.

Pages: 432
1. Information concerning body position and equilibrium is transmitted by interoreceptors.
    Answer: False

Pages: 432
2. Some mechanoreceptors respond to changes in blood pressure.
    Answer: True

Pages: 432
3. Pain receptors are called nociceptors.
    Answer: True

Pages: 432
4. Tonic receptors are rapidly adapting receptors.
    Answer: False

Pages: 433
5. Tactile sensations are all detected by mechanoreceptors.
    Answer: True

Pages: 433
6. Nociceptors are rapidly adapting receptors.
    Answer: False

Pages: 436
7. Joint kinesthetic receptors are located at the junction of a tendon with a muscle.
    Answer: False

Pages: 437
8. Third-order sensory neurons conduct nerve impulses from receptors into the spinal cord or brain stem.
    Answer: False

Pages: 437
9. The medial lemniscus is a projection tract that extends from the medulla to the thalamus.
    Answer: True

Pages: 439
10. The spinocerebellar tracts are the major routes for subconscious proprioception input into the cerebellum.
    Answer: True

Pages: 442
11. Destruction of certain basal ganglia connections results in abnormal muscle rigidity.
    Answer: True

12. In Alzheimer's Disease, memory loss seems to be related to a decrease in the amount of dopamine produced in the brain.
Answer: False

13. As a person ages, the average amount of time spent sleeping increases.
Answer: False

14. Impulses for chronic pain are usually conducted along myelinated Type A fibers.
Answer: False

15. Cutaneous receptors consist of free nerve endings or dendrites with an epithelial or connective tissue capsule.
Answer: True

Short Answer

Write the word or phrase that best completes each statement or answers the question.

1. A sense organ transduces a stimulus into a _____.
Answer: generator potential

2. Nerve impulses are conducted from a receptor to the central nervous system by _____ neurons.
Answer: (first-order) sensory/afferent

3. The medical term for pain relief is _____.
Answer: analgesia

4. Parkinson's Disease is associated with deterioration of neural connections between the _____ of the midbrain and the _____ of the cerebrum.
Answer: substantia nigra; basal ganglia

5. Long-term potentiation is a phenomenon believed to occur in a region of the cerebrum known as the _____.
Answer: hippocampus

6. The daily sleep-wake cycle is known as _____ rhythm.
Answer: circadian

Pages: 437
7. The ability to recognize by "feel" the size, shape, and texture of an object is called
   _____.
   Answer: stereognosis

Pages: 430
8. _____ is the conscious or unconscious awareness of external or internal stimuli.
   Answer: Sensation

Pages: 430
9. Each specific type of sensation is called a sensory _____.
   Answer: modality

Pages: 430
10. The region of a sensory neuron that is able to respond to an appropriate stimulus is called the
    _____.
    Answer: receptive field

Pages: 432
11. Receptors that provide information concerning body position and movement are classified as
    _____.
    Answer: proprioceptors

Pages: 432
12. Receptors located in the blood vessel walls and visceral organs are classified as
    visceroreceptors or _____.
    Answer: interoreceptors

Pages: 432
13. If a stimulus directly increases or decreases the exocytosis of neurotransmitter from synaptic
    vesicles, it is said to elicit a _____.
    Answer: receptor potential

Pages: 432
14. A decrease in sensitivity to a long-term stimulus is called _____.
    Answer: adaptation

Pages: 432
15. Tactile sensations, thermal sensations, and pain sensations are classified as _____,
    because the receptors are located in the skin or underlying connective tissue.
    Answer: cutaneous

Pages: 433
16. All tactile sensations are detected by receptors classified as _____.
    Answer: mechanoreceptors

Pages: 434
17. Visceral pain that is perceived as localized in the skin served by the same segment of the
    spinal cord is called _____.
    Answer: referred pain

18. Pain that is perceived as occurring in an amputated limb is called _____.
Answer: phantom pain

19. Surgical severing of spinal posterior nerve roots to relieve chronic pain is called _____.
Answer: rhizotomy

20. Intrafusal muscle fibers and the connective tissue enclosing them make up the proprioceptors known as _____.
Answer: muscle spindles

21. Proprioceptors that monitor the force of muscle contraction are the _____.
Answer: tendon organs

22. Receptors that detect changes in temperature are called _____.
Answer: thermoreceptors

23. Most dreaming occurs during the _____ stage of sleep.
Answer: REM

24. Awakening from sleep involves increased activity in fibers known as the _____ that project from the brain stem through the thalamus to the cerebral cortex.
Answer: reticular activating system

25. The _____ of the cerebral cortex is the major control region for initiation of voluntary movement.
Answer: primary motor area (precentral gyrus)

26. For a sensation to arise, four events typically occur: stimulation, transduction, conduction, and _____.
Answer: translation

27. An imbalance of neurotransmitter activity - too little dopamine and too much acetylcholine - is thought to bring about the symptoms of _____.
Answer: Parkinson's disease

28. Paralysis of all four limbs is referred to as _____.
Answer: quadriplegia

Pages: 448
29. The final stage of brain failure that is characterized by total unresponsiveness to all external stimuli is called _____.
Answer: coma

Pages: 446
30. The medical term for jerky, uncoordinated movements is _____.
Answer: ataxia

Matching

Choose the item from Column 2 that best matches each item in Column 1.

1.	Type I cutaneous mechanoreceptors (Merkel discs)	discriminative touch
2.	Type II cutaneous mechanoreceptors (end organs of Ruffini)	heavy, continuous touch
3.	lamellated (Pacinian) corpuscles	pressure & high-frequency vibration
4.	thermoreceptors	warm & cold stimuli
5.	nociceptors	pain
6.	muscle spindles	rate & degree of change in muscle length
7.	tendon organs	force of muscle contraction
8.	joint kinesthetic receptors	acceleration & deceleration of joint movement
9.	photoreceptors	light
10.	chemoreceptors	smell & taste stimuli

Essay

Write your answer in the space provided or on a separate sheet of paper.

Pages: 432

1. List the functional classes of receptors and the types of stimuli received by each.

   Answer: Mechanoreceptors detect mechanical pressure or stretching (including touch, pressure, vibration, proprioception, hearing, equilibrium, and blood pressure); thermoreceptors detect temperature changes; nociceptors detect pain; photoreceptors detect light; chemoreceptors detect chemicals in mouth, nose, and body fluids.

Pages: 431

2. List and briefly describe the events that occur in order to perceive a sensation.

   Answer: 1) Stimulation - sensory neuron responds to appropriate modality in its receptive field
   2) Transduction - conversion of stimulus into generator potential by receptor
   3) Conduction - transmission of nerve impulse to CNS via sensory neuron(s)
   4) Translation/Integration - interpretation of sensation by CNS

Pages: 432

3. Describe the differences between generator potentials and receptor potentials.

   Answer: 1) generator potential at threshold triggers nerve impulse in receptor in which it occurs; receptor potential triggers release of neurotransmitter from receptor in which it occurs
   2) generator potentials are always depolarizations; receptor potentials may be either depolarizations or hyperpolarizations

Pages: 440-442

4. Compare the direct and indirect pathways for control and coordination of movement.

   Answer: Direct pathways - for precise, movements; upper motor neurons in motor cortex through internal capsule to medulla; most axons cross over and terminate in nuclei of cranial nerves or anterior gray horn of spinal cord; lower motor neurons extend to effectors; may be association neurons between upper and lower neurons
   Indirect pathways - involve motor cortex, basal ganglia, limbic system, thalamus, cerebellum, reticular formation, nuclei in brain stem; provide excitatory & inhibitory input to lower motor neurons

Pages: 446-447

5. Compare short-term memory vs. long-term memory with regard to specific changes that are though to occur in the brain.

   Answer: Short-term memory may depend on forming new synapses and reverberating circuits. Long-term memory is thought to involve high-frequency stimulation within the hippocampus at glutamate synapses. Nitric oxide and ACh may be involved. Neurons develop new presynaptic terminals, larger synaptic end bulbs, and more dendritic branches. Enhanced facilitation occurs. Possibly DNA and RNA are involved.

Pages: 448-449
6. Describe the events in the stages of sleep.
   Answer: 1) Stage 1/NREM - 1-7 min. at start of cycle; alpha & theta waves on EEG; general relaxation
   2) Stage 2/NREM - light sleep; dream fragments; slow eye rolling; sleep spindles on EEG
   3) Stage 3/NREM - moderately deep sleep; very relaxed; decreased body temp. and blood pressure; sleep spindles & delta waves on EEG
   4) Stage 4/NREM - very deep sleep; very relaxed; hard to wake; delta waves on EEG
   5) REM - dreaming sleep; rapid eye movement; cycles every 90 min. with NREM; penile erections in males

Pages: 445
7. Briefly describe the four aspects of cerebellar function.
   Answer: 1) Monitoring intentions - what movements are planned? input from motor cortex & basal ganglia
   2) Monitoring actual movement - input from proprioceptors via spinal tracts & vestibular apparatus
   3) Comparison of actual movement vs. intentions
   4) Providing corrective feedback - modifies brain output for successful completion of desired movement

Pages: 433
8. Two children were born in England who lacked nociceptors. What effects would this have on the lives of these children? Explain your answer.
   Answer: Lack of pain sensations would require careful monitoring by parents and the children themselves for tissue damage (or the potential thereof...). The question might be expanded to include speculation of the social effects of this condition.

Pages: 447
9. A well-known musician suffered nearly complete destruction of the hippocampus. Predict the effects of this damage. Explain your answer.
   Answer: Students should describe role of hippocampus on conversion of short-term to long-term memory. Patient would be unable to form new memories. Question could be expanded to include potential social effects of this problem.

Pages: 442
10. A patient is exhibiting tremors of the hands while sitting still. Propose where the damage might be. Explain your answer.
    Answer: The most likely answer would probably be damage to basal ganglia, particular those portions involved in inhibition of unnecessary movements. Other answers may be acceptable.

Chapter 16   The Special Senses

Multiple-Choice

Choose the one alternative that best completes the statement or answers the question.

Pages: 466
1. In the process of forming an image on the retina, convergence occurs to allow:
   A) a change in shape of the lens
   B) three dimensional image formation
   C) refraction of light rays
   D) focusing of light through the center of the lens
   E) production of new photoreceptors
   Answer: B

Pages: 459
2. The conjunctiva covers the:
   A) lens
   B) sclera
   C) cornea
   D) retina
   E) optic nerve
   Answer: B

Pages: 461
3. The retina is held in place by the:
   A) optic disc
   B) vitreous body
   C) ciliary muscle
   D) bipolar neurons
   E) iris
   Answer: B

Pages: 461
4. Photoreceptors are located in the:
   A) choroid
   B) cornea
   C) iris
   D) sclera
   E) retina
   Answer: E

Pages: 461
5. The function of the ciliary processes of the eye is to:
   A) produce aqueous humor
   B) produce the vitreous body
   C) produce tears
   D) change the shape of the lens
   E) respond to red and green light
   Answer: A

Pages: 459
6. The mucous membrane that lines the eyelids is the:
   A) cornea
   B) conjunctiva
   C) choroid
   D) sclera
   E) iris
   Answer: B

Pages: 472
7. Ceruminous glands are located in the:
   A) spiral organ
   B) ciliary body
   C) external auditory meatus
   D) middle ear
   E) auditory tube
   Answer: C

Pages: 473
8. The auditory (Eustachian) tube connects the:
   A) middle ear and inner ear
   B) external ear and middle ear
   C) middle ear and nasopharynx
   D) cochlea and vestibule
   E) inner ear and primary auditory area of the brain
   Answer: C

Pages: 473
9. The ossicles are the major structures of the:
   A) external ear
   B) middle ear
   C) vestibule
   D) cochlea
   E) auditory regions of the cerebrum
   Answer: B

Pages: 480
10. The primary function of the utricle and saccule is to:
    A) transduce sound waves into generator potentials
    B) monitor static equilibrium
    C) monitor dynamic equilibrium
    D) cause movements in the ossicles
    E) produce endolymph in the cochlea
    Answer: B

11. The receptors in the utricle and saccule are stimulated when:
    A) endolymph flows over the spiral organ
    B) the otoliths bend stereocilia in response to gravity
    C) the tympanic membrane vibrates
    D) perilymph bends hair cells
    E) the stapes pushes into the oval window
    Answer: B

12. The scala tympani and scala vestibuli are part of the:
    A) middle ear
    B) vestibule
    C) cochlea
    D) posterior chamber of the eye
    E) anterior chamber of the eye
    Answer: C

13. The first synapse in the olfactory pathway occurs in the:
    A) olfactory epithelium on the nasal conchae
    B) olfactory bulbs
    C) olfactory tract
    D) frontal lobe of the cerebral cortex
    E) epithelium of the nasopharynx
    Answer: B

14. Gustatory hairs are the receptors for:
    A) smell
    B) taste
    C) color vision
    D) night vision
    E) dynamic equilibrium
    Answer: B

15. Circumvallate and fungiform papillae contain receptors for:
    A) smell
    B) taste
    C) color vision
    D) night vision
    E) dynamic equilibrium
    Answer: B

Pages: 454
16. Generator potentials are produced in:
    A) olfactory hairs
    B) gustatory hairs
    C) rods
    D) hair cells of the spiral organ
    E) all types of receptors
    Answer: A

Pages: 457
17. Which of the following best describes the sensitivity of the tongue to the basic tastes?
    A) tip most sensitive to sour, back to salty, sides to bitter and sweet
    B) tip most sensitive to bitter, back to sweet and sour, sides to salty
    C) tip most sensitive to salty, back to sweet and sour, sides to bitter
    D) tip most sensitive to bitter, back to salty, sides to sweet and sour
    E) tip most sensitive to sweet and salty, back to bitter, sides to sour
    Answer: E

Pages: 457
18. The threshold is lowest for which of the primary tastes?
    A) sweet
    B) sour
    C) salty
    D) bitter
    E) all taste receptors have equal thresholds
    Answer: D

Pages: 454
19. Olfactory sensations reach the brain via:
    A) cranial nerve I
    B) cranial nerve II
    C) cranial nerve VII
    D) cranial nerve VIII
    E) spinal nerves C1
    Answer: A

Pages: 457
20. Most impulses related to gustatory sensations arising on the tongue are conveyed to the brain via:
    A) cranial nerve I
    B) cranial nerve II
    C) cranial nerve VII
    D) cranial nerve VIII
    E) spinal nerves C1
    Answer: C

Pages: 457
21. The primary gustatory area is located in the:
    A) frontal lobe of the cerebral cortex
    B) occipital lobe of the cerebral cortex
    C) limbic system
    D) parietal lobe of the cerebral cortex
    E) olfactory bulbs
    Answer: D

Pages: 459
22. "Bloodshot eyes" are the result of dilation of blood vessels in the:
    A) lens
    B) cornea
    C) sclera
    D) conjunctiva
    E) iris
    Answer: D

Pages: 459
23. The curved, avascular, transparent, fibrous coat covering the iris is the:
    A) cornea
    B) sclera
    C) lens
    D) uvea
    E) pupil
    Answer: A

Pages: 459
24. The dense connective tissue forming the "white" of the eye is the:
    A) cornea
    B) sclera
    C) conjunctiva
    D) uvea
    E) palpebrae
    Answer: B

Pages: 461
25. The iris is part of the:
    A) lens
    B) fibrous tunic
    C) uvea
    D) ganglion cell layer
    E) photoreceptor layer
    Answer: C

Pages: 461
26. The function of the ciliary muscle is to:
    A) secrete aqueous humor
    B) produce the vitreous body
    C) transduce certain light rays into receptor potentials
    D) control the diameter of the pupil
    E) alter the shape of the lens for near or far vision
    Answer: E

Pages: 461
27. The ciliary body extends from the:
    A) optic disc to the ora serrata
    B) anterior margin of the retina to the sclerocorneal junction
    C) pupil to the outer edge of the iris
    D) point at which the optic nerve exits, across the outer surface to the edge of the cornea
    E) medial surface of the orbit to the nasal cavity
    Answer: B

Pages: 461
28. The pupil is:
    A) the colored part of the eye you can see when you look at someone
    B) the white of the eye
    C) a hole in the center of the iris
    D) the site of clearest color vision
    E) the site of greatest density of rods
    Answer: C

Pages: 468
29. Which of the following occurs when bright light stimulates the eye?
    A) parasympathetic fibers stimulate contraction of the circular muscles of the iris to decrease pupil diameter
    B) parasympathetic fibers stimulate contraction of the circular muscles of the iris to increase pupil diameter
    C) sympathetic fibers stimulate contraction of the radial muscles of the iris to decrease pupil diameter
    D) sympathetic fibers stimulate contraction of the radial muscles of the iris to increase pupil diameter
    E) sympathetic fibers stimulate contraction of the circular muscles of the iris to increase pupil diameter
    Answer: A

Pages: 468
30. Which of the following occurs when dim light stimulates the eye?
    A) parasympathetic fibers stimulate contraction of the circular muscles of the iris to decrease pupil diameter
    B) parasympathetic fibers stimulate contraction of the circular muscles of the iris to increase pupil diameter
    C) sympathetic fibers stimulate contraction of the circular muscles of the iris to decrease pupil diameter
    D) sympathetic fibers stimulate contraction of the radial muscles of the iris to decrease pupil diameter
    E) sympathetic fibers stimulate contraction of the radial muscles of the iris to increase pupil diameter
    Answer: E

Pages: 461
31. The central retinal artery and vein exit the eyeball at the:
    A) ora serrata
    B) canal of Schlemm
    C) pupil
    D) optic disk
    E) central fovea
    Answer: D

Pages: 461
32. The function of melanin in the choroid and pigment epithelium of the retina is to:
    A) transduce blue, green, and yellow/orange light into generator potentials
    B) transduce dim blue light into receptor potentials
    C) prevent reflection and scattering of light within the eye
    D) reflect color onto the iris
    E) act as a neurotransmitter between photoreceptor neurons and bipolar neurons
    Answer: C

Pages: 467
33. **ALL** of the following are **TRUE** regarding the photoreceptors of the eye **EXCEPT**:
    A) there are more cones than rods in the retina
    B) rods contain the photopigment rhodopsin
    C) cones respond best to bright light
    D) rods respond best to dim light
    E) there are no photoreceptors in the optic disk
    Answer: A

Pages: 462
34. The area of highest visual acuity (resolution) on the retina is the:
    A) optic disk
    B) uvea
    C) ora serrata
    D) lacrimal puncta
    E) central fovea
    Answer: E

Pages: 462
35. The optic nerve is made up of the:
    A) axons of the rods and cones
    B) axons of the bipolar cells
    C) axons of the ganglion cells
    D) dendrites of the bipolar cells
    E) dendrites of the ganglion cells
    Answer: C

Pages: 463
36. The function of the suspensory ligaments of the eye is to:
    A) move the eyeball from side to side
    B) rotate the eyeball
    C) hold the retina in place
    D) hold the lens in place
    E) change the diameter of the pupil
    Answer: D

Pages: 463
37. A cataract is:
    A) a lesion on the cornea
    B) detachment of the retina from the back of the eye
    C) loss of transparency of the lens
    D) loss of photopigments from the photoreceptors
    E) damage to photoreceptors due to high intraocular pressure
    Answer: C

Pages: 461
38. Nutrients are provided to the posterior surface of the retina by blood vessels in the darkly
    pigmented portion of the vascular tunic known as the:
    A) choroid
    B) ciliary body
    C) iris
    D) sclera
    E) macula lutea
    Answer: A

Pages: 463
39. Glaucoma is the result of:
    A) loss of the vitreous body
    B) loss of transparency of the lens
    C) high intraocular pressure
    D) vitamin A deficiency
    E) loss of elasticity of the lens
    Answer: C

40. The spaces in the eye that are anterior to the lens are normally filled with:
    A) vitreous body
    B) aqueous humor
    C) air
    D) blood
    E) photopigments
    Answer: B

41. The space in the eye that is posterior to the lens is normally filled with:
    A) vitreous body
    B) aqueous humor
    C) air
    D) blood
    E) photopigments
    Answer: A

42. **ALL** of the following are **TRUE** regarding aqueous humor **EXCEPT**:
    A) it creates intraocular pressure
    B) it helps nourish the lens and cornea
    C) it drains into the scleral venous sinus (canal of Schlemm)
    D) it is produced by the ciliary processes during embryonic life, and is never replaced
    E) it fills the anterior cavity
    Answer: D

43. How does the cornea obtain nutrients?
    A) via blood vessels within the cornea
    B) via aqueous humor
    C) via the vitreous body
    D) via blood vessels in the sclera
    E) The cornea does not require nutrients because it is not a living tissue.
    Answer: B

44. A scratch on the cornea most directly interferes with:
    A) accommodation
    B) refraction
    C) constriction of the pupil
    D) convergence
    E) transduction
    Answer: B

45. For sharpest visual acuity, light rays must be refracted so that they:
    A) change wavelength to fall within the visible range
    B) stimulate constriction of the pupil
    C) turn photopigments in the lens into their colorless form
    D) hit the melanin in the choroid
    E) fall directly on the central fovea
    Answer: E

46. Relaxation of the ciliary muscle results in:
    A) flattening of the lens for distance vision
    B) flattening of the lens for near vision
    C) bulging of the lens for distance vision
    D) bulging of the lens for near vision
    E) increased diameter of the pupil
    Answer: A

47. Which of the following occurs when trying to focus on a close object?
    A) relaxation of the ciliary muscle to flatten the lens
    B) relaxation of the ciliary muscle to make the lens more convex
    C) contraction of the ciliary muscle to flatten the lens
    D) contraction of the ciliary muscle to make the lens more convex
    E) contraction of the ciliary muscle to increase the diameter of the pupil
    Answer: D

48. The first step in visual transduction is:
    A) activation of retinal isomerase
    B) bleaching of the photopigment
    C) isomerization of the photopigment
    D) initiation of a receptor potential
    E) initiation of a generator potential
    Answer: C

49. When your mother told you to eat your carrots, she was giving you good advice because carrots contain carotenoids necessary for the formation of:
    A) the opsin that responds to yellow-orange light
    B) the retinal of photopigments
    C) melanin
    D) aqueous humor
    E) cyclic GMP
    Answer: B

50. Light excites the bipolar cells that synapse with rods by:
    A) increasing the amount of cyclic GMP
    B) opening light-gated sodium ion channels in the bipolar neurons
    C) depolarizing rods, thus increasing release of photopigments from synaptic vesicles
    D) hyperpolarizing rods, thus decreasing release of the inhibitory neurotransmitter glycine
    E) stimulating isomerization of cis-retinal which opens chemically-gated sodium ion channels
    Answer: D

51. Activation of the enzyme **transducin** results in hyperpolarization of rods because it:
    A) activates phosphodiesterase, which breaks down cyclic GMP, thus closing sodium ion channels
    B) stimulates regeneration of rhodopsin
    C) stimulates production of cyclic GMP, which increases production of the inhibitory neurotransmitter glycine
    D) opens chemically-gated chloride ion channels
    E) alters the wavelength of light entering the rods
    Answer: A

52. The function of horizontal cells in the retina is to:
    A) spread the wave of depolarization to all cells in the retina
    B) inhibit bipolar cells in areas lateral to excited rods and cones to help make the visual scene clearer
    C) attach the retina to the back of the eye
    D) produce melanin
    E) provide feedback to the rods and cones to adjust their sensitivity to particular wavelengths of light
    Answer: B

53. Which of the following lists the route of impulse transmission in the correct order?
    A) photoreceptors, ganglion cells, bipolar cells, optic nerve, optic chiasm, optic tract
    B) bipolar cells, ganglion cells, photoreceptors, optic nerve, optic tract, optic chiasm
    C) photoreceptors, bipolar cells, ganglion cells, optic nerve, optic tract, optic chiasm
    D) photoreceptors, bipolar cells, ganglion cells, optic nerve, optic chiasm, optic tract
    E) bipolar cells, ganglion cells, photoreceptors, optic nerve, optic chiasm, optic tract
    Answer: D

54. The middle ear is normally filled with:
    A) air
    B) blood
    C) endolymph
    D) perilymph
    E) cerumen
    Answer: A

55. The malleus articulates with the:
    A) incus and stapes
    B) tympanic membrane and stapes
    C) stapes and oval window
    D) tympanic membrane and incus
    E) incus and oval window
    Answer: D

56. Endolymph fills the:
    A) external auditory meatus
    B) auditory canal
    C) middle ear
    D) bony labyrinth
    E) membranous labyrinth
    Answer: E

57. The cochlear duct is separated from the scala tympani by the:
    A) vestibular membrane
    B) basilar membrane
    C) tectorial membrane
    D) tympanic membrane
    E) oval window
    Answer: B

58. A delicate, flexible gelatinous membrane projecting over and in contact with the hair cells of the spiral organ is the:
    A) vestibular membrane
    B) basilar membrane
    C) tectorial membrane
    D) tympanic membrane
    E) oval window
    Answer: C

59. When the stapes moves back and forth, it **directly** causes:
    A) vibration of the tympanic membrane
    B) movement of the malleus
    C) formation of waves in the perilymph as the oval window moves in and out
    D) movement of hair cells against the tectorial membrane
    E) increased production of endolymph
    Answer: C

Pages: 478
60. Which of the following events of hearing occurs first?
    A) The endolymph moves the dendrites in the spiral organ.
    B) The malleus moves the incus.
    C) The tympanic membrane vibrates.
    D) The stapes pushes into the oval window.
    E) The incus moves the stapes.
    Answer: C

Pages: 475
61. The spiral organ is located in the:
    A) cochlea
    B) semicircular canals
    C) utricle
    D) saccule
    E) middle ear
    Answer: A

Pages: 478
62. Mechanically-gated channels in the hair cells of the spiral organ open when:
    A) endolymph flows between the bony and membranous labyrinths
    B) cyclic GMP activity decreases, decreasing the activity of glycine
    C) microvilli on the cell membranes bend in response to vibration of the basilar membrane
    D) otoliths roll out of the way as endolymph flows over them
    E) the round window bulges in and out, causes bending of the hair cells around it
    Answer: C

Pages: 482
63. The sense of continued movement after you have stopped spinning around is due to:
    A) slow conduction of visual impulses to the brain
    B) continued flow of endolymph over the hair cells of the spiral organ
    C) continued flow of endolymph over the hair cells of the semicircular ducts
    D) prolonged vibration of the tympanic membrane
    E) buildup of air bubbles in the endolymph of the vestibular apparatus
    Answer: C

Pages: 475
64. High-frequency sounds induce:
    A) maximal vibrations in the part of the basilar membrane at the apex of the cochlea
    B) maximal vibrations in the part of the basilar membrane at the base of the cochlea
    C) larger waves in the endolymph than low-frequency sounds
    D) smaller waves in the endolymph than low-frequency sounds
    E) generator potentials in the cochlear branch of cranial nerve VIII
    Answer: B

65. When mechanically-gated ion channels open in the hair cells of the spiral organ, the hair cells:
   A) depolarize because sodium ions enter the cell
   B) hyperpolarize because potassium and calcium ions leave the cell
   C) depolarize because potassium and calcium ions enter the cell
   D) hyperpolarize because sodium ions leave the cell
   E) hyperpolarize because chloride ions enter the cell
   Answer: C

True-False

Write T if the statement is true and F if the statement is false.

Pages: 454
1. Olfactory receptors are replaced approximately every thirty days.
   Answer: True

Pages: 456
2. A chemical must be dissolved in saliva to be tasted.
   Answer: True

Pages: 457
3. The tip of the tongue is the region most sensitive to bitter substances.
   Answer: False

Pages: 459
4. The conjunctiva covers the cornea.
   Answer: False

Pages: 459
5. The choroid of the eyeball is highly vascular.
   Answer: True

Pages: 461
6. Contraction of circular muscles of the iris results in dilation of the pupil.
   Answer: False

Pages: 462
7. The optic disk is the area of the retina with the greatest visual acuity.
   Answer: False

Pages: 461
8. Light must pass through the ganglion and bipolar cell layers of the retina before reaching the photoreceptor layer.
   Answer: True

Pages: 463
9. The anterior cavity is filled with aqueous humor.
   Answer: True

Pages: 463
10. The vitreous body is being continually produced and reabsorbed.
    Answer: False

Pages: 465
11. Images focused on the retina are inverted.
    Answer: True

Pages: 474
12. The stapes fits into the round window.
    Answer: False

Pages: 478
13. Hair cells in the spiral organ are mechanoreceptors that are stimulated by bending.
    Answer: True

Pages: 478
14. Bending of hair cells in the spiral organ results in a generator potential.
    Answer: False

Pages: 480
15. The utricle and saccule contain the receptors for static equilibrium.
    Answer: True

Short Answer

Write the word or phrase that best completes each statement or answers the question.

Pages: 481
1. The hair cells of the utricle and saccule are covered in part by a layer of calcium carbonate crystals called _____.
    Answer: otoliths (otoconia)

Pages: 481
2. The _____ have the primary role in maintenance of dynamic equilibrium.
    Answer: semicircular ducts

Pages: 463
3. _____ is an abnormally high intraocular pressure due to buildup of aqueous humor.
    Answer: Glaucoma

Pages: 455
4. The receptors for gustatory sensations are located in the _____.
    Answer: taste buds

Pages: 459
5. A bactericidal enzyme present in lacrimal fluid is _____.
    Answer: lysozyme

6. The fibrous tunic of the eyeball consists of the _____ and the _____.
   Answer: cornea; sclera

7. The shape of the lens is altered for near or far vision by the _____.
   Answer: ciliary muscle

8. The hole in the center of the iris is the _____.
   Answer: pupil

9. Photoreceptors called _____ are most important for seeing shades of gray in dim light, while _____ provide color vision in bright light.
   Answer: rods; cones

10. The area in the exact center of the posterior portion of the retina whose name translates literally as "yellow spot" is the _____.
    Answer: macula lutea

11. The leading cause of blindness is a loss of transparency of the lens known as a _____.
    Answer: cataract

12. The anterior and posterior chambers of the anterior cavity of the eye are separated by the

    _____.
    Answer: iris

13. The anterior and posterior cavities of the eye are separated by the _____.
    Answer: lens

14. Intraocular pressure is produced mainly by _____.
    Answer: aqueous humor

15. The jellylike substance filling the posterior cavity of the eye is the _____.
    Answer: vitreous body

16. Bending of light rays as they pass through different media is called _____.
    Answer: refraction

17. The increase in curvature of the lens for near vision is called _____.
    Answer: accommodation

Pages: 465
18. With aging, the lens loses its elasticity, and therefore, its ability to change shape. This condition is known as _____.
Answer: presbyopia

Pages: 466
19. The medial movement of the eyeballs so that both are directed toward the object being viewed is called _____.
Answer: convergence

Pages: 467
20. The photopigment in rods is _____.
Answer: rhodopsin

Pages: 467
21. The light-absorbing portion of all visual photopigments is _____, which is derived from vitamin _____.
Answer: retinal; A

Pages: 473
22. The external auditory canal and the middle ear are separated by the _____.
Answer: tympanic membrane

Pages: 473
23. The auditory ossicles are the _____, the _____, and the _____.
Answer: malleus; incus; stapes

Pages: 473
24. The middle ear and the nasopharynx are connected by the _____.
Answer: auditory (Eustachian) tube

Pages: 474
25. The three areas of the bony labyrinth are the _____, the _____, and the _____.
Answer: cochlea; vestibule; semicircular canals

Pages: 474
26. The membranous labyrinth of the vestibule consists of two sacs called the _____ and the _____.
Answer: utricle; saccule

Pages: 475
27. The three channels of the cochlea are the _____, the _____, and the _____.
Answer: scala tympani; scala vestibuli; cochlear duct

Pages: 475
28. Sound intensity is measured in units called _____.
Answer: decibels

Pages: 481
29. Movements of the _____ of hair cells in the utricle and saccule initiate depolarizing receptor potentials.
Answer: sterocilia

Pages: 474
30. The bony labyrinth is filled with a fluid called _____.
Answer: perilymph

Matching

Choose the item from Column 2 that best matches each item in Column 1.

1. avascular, transparent part   cornea   conjunctiva
   of fibrous tunic

2. the "white" of the eye   sclera

3. transparent, avascular   lens
   protein structure that
   fine-tunes focusing of
   light rays

4. circular and radial muscles   iris
   that help regulate the
   amount of light entering
   the posterior cavity

5. produces aqueous humor and   ciliary body
   alters shape of lens

6. jellylike substance in   vitreous body   aqueous humor
   posterior cavity that helps
   hold retina flush against
   eyeball

7. highly vascular, darkly   choroid
   pigmented portion of uvea

8. photoreceptors for vision   rods
   in dim light

9. photoreceptors for color   cones
   vision in bright light

10. the hole through which   pupil   optic disc
    light passes to reach the
    lens, posterior cavity, and
    retina

Essay

Write your answer in the space provided or on a separate sheet of paper.

Pages: 459-463
1. Outline the structure of the eyeball.
   Answer: 1) Fibrous tunic - avascular; white sclera of dense CT; transparent cornea
   2) Vascular tunic - vascular, pigmented choroid lining sclera; ciliary body from ora serrata to outer edge of cornea; iris and pupil
   3) Retina - photoreceptors here, plus bipolar and ganglion cells; macula lutea at center w/ central fovea (no rods); optic disk where optic nerve leaves
   4) Aqueous humor in anterior cavity in front of lens; vitreous body in posterior cavity behind lens
   5) Lens suspended behind pupil

Pages: 464-466
2. Describe the process of image formation on the retina.
   Answer: 1) Refraction - bending of light as medium changes to focus light on central fovea
   2) Accommodation of lens for near/distance vision - shape changes to make light focus on retina
   3) Constriction of pupil - ANS reflex to prevent scattering of light through edges of lens
   4) Convergence of eyes - to focus both eyes on same object and provide binocular vision Images focused on retina upside-down and mirror image; brain translates

Pages: 468
3. Describe the mechanism by which photoreceptors work.
   Answer: 1) Isomerization - cis to trans conversion of retinal as light hits; sodium channels close; cell hyperpolarizes; less glutamate released
   2) Bleaching - trans-retinal separates from opsin
   3) Conversion of trans back to cis via retinal isomerase
   4) Regeneration - as cis-retinal binds to opsin

Pages: 472-475
4. Outline the structure of the ear.
   Answer: 1) External ear - auricle (flap of elastic cartilage covered by skin); external auditory canal = 1" tube in temporal bone; tympanic membrane = eardrum
   2) Middle ear - air-filled via auditory tube; contains ossicles - malleus attached to eardrum and incus, stapes attached to incus and fits into oval window
   3) Inner ear - structures of cochlea, vestibule, and semicircular canals should be described (see pp. 472-475)

Pages: 478
5. Describe the events of hearing.
   Answer: This long answer is outlined on page 478.

Pages: 479-482

6. Differentiate between static and dynamic equilibrium. Describe the structures and physiological mechanisms involved in receiving and transducing vestibular sensations.

    Answer: · Static equilibrium - maintenance of body position relative to gravity; hair cells in maculae of utricle & saccule bend as otolithic membrane slides forward due to gravity; receptor potential transmitted to cranial nerve VIII to pons

    · Dynamic equilibrium - maintenance of body position in response to movement; endolymph flowing over hair cells in cristae of semicircular ducts cause bending; receptor potentials passed to cranial nerve VIII to pons

Pages: 464-470

7. Blindness can occur for many reasons. Using your knowledge of the structure of the eye and the processing of light stimuli, predict where some of the potential trouble spots might be. Explain your answers.

    Answer: Examples could include: loss of transparency of cornea or lens (light can't pass); detachment of retina (loss of photoreceptive and conductive function); no photoreceptors or pigments (no transduction); damage to optic nerve (no conduction); damage to visual pathway or visual cortex (no conduction or translation)

Pages: 478-479

8. Deafness can occur for many reasons. Using your knowledge of the structure of the ear and the processing of sound stimuli, predict where some of the potential trouble spots might be. Explain your answers.

    Answer: Examples could include: blockage of external auditory meatus (sound doesn't reach eardrum); hole in eardrum (incomplete vibrations); fusion of ossicles or paralysis of associated muscles (no transmission to inner ear); damage to hair cells of spiral organ (no receptors); damage to cranial nerve VIII (no conduction); brain damage in auditory areas (no translation)

Pages: 466

9. Discuss the problems caused by abnormal refraction in the eyeball and the usual correction for each.

    Answer: 1) Myopia - image focused in front of retina due to elongated eyeball or thickened lens; correct with concave lens to diverge light rays

    2) Hypermetropia - image focused behind retina due to shortened eyeball or thin lens; corrected with convex lens to converge light rays

    3) Astigmatism - due to irregularities in cornea or lens prevent accurate focusing of light on retina; correction depends on specific problem

Pages: 456-457

10. Describe the physiological mechanisms involved in perceiving the taste of a particular food.

    Answer: Food dissolved in saliva; chemicals stimulate gustatory hairs of one or more of the primary tastes; resulting receptor potential causes release of neurotransmitter; info travels over cranial nerves VII, IX, and X to medulla, then the limbic system, hypothalamus, thalamus, and parietal lobe of cortex; for complete "taste" olfactory hairs must be stimulated also

Multiple-Choice

Choose the one alternative that best completes the statement or answers the question.

Pages: 494
1. In the autonomic nervous system, all preganglionic fibers release the neurotransmitter:
   A) acetylcholine
   B) norepinephrine
   C) serotonin
   D) dopamine
   E) epinephrine
   Answer: A

Pages: 494
2. In the autonomic nervous system, most sympathetic postganglionic fibers release the neurotransmitter:
   A) acetylcholine
   B) norepinephrine
   C) serotonin
   D) dopamine
   E) GABA
   Answer: B

Pages: 492
3. Terminal ganglia are where:
   A) the cell bodies for sympathetic preganglionic fibers are located
   B) parasympathetic preganglionic fibers synapse with parasympathetic postganglionic fibers
   C) the cell bodies of sensory neurons are located
   D) sympathetic preganglionic fibers synapse with sympathetic postganglionic fibers
   E) sympathetic preganglionic fibers synapse with parasympathetic postganglionic fibers
   Answer: B

Pages: 493
4. The adrenal medulla receives stimulation from:
   A) sympathetic preganglionic fibers
   B) sympathetic postganglionic fibers
   C) parasympathetic preganglionic fibers
   D) parasympathetic postganglionic fibers
   E) all cranial nerves
   Answer: A

Pages: 489
5. Some sympathetic preganglionic neurons have their cell bodies in the:
   A) brain stem
   B) sacral region of the spinal cord
   C) lumbar region of the spinal cord
   D) adrenal medulla
   E) sympathetic effectors
   Answer: C

Pages: 494
6. An adrenergic neuron produces the neurotransmitter:
   A) serotonin
   B) GABA
   C) norepinephrine
   D) acetylcholine
   E) glycine
   Answer: C

Pages: 488
7. Which of the following is an example of an effector for a visceral efferent neuron?
   A) quadriceps femoris
   B) diaphragm
   C) extrinsic eye muscles
   D) smooth muscle in the wall of the small intestine
   E) All of the above are correct except the quadriceps femoris.
   Answer: D

Pages: 489
8. **ALL** of the following are **TRUE** for postganglionic neurons **EXCEPT**:
   A) They have nicotinic receptors on their membranes.
   B) Their cell bodies are in the gray horns of the spinal cord.
   C) Their axons are unmyelinated.
   D) They are stimulated by preganglionic fibers.
   E) They stimulate visceral effectors.
   Answer: B

Pages: 488
9. Which of the following is a somatic efferent response?
   A) changes in the size of the pupil
   B) dilation of blood vessels
   C) adjustment of the rate and force of heartbeat
   D) contractions within the wall of the gastrointestinal tract
   E) contraction of the rectus abdominis muscle
   Answer: E

Pages: 488
10. Which of the following statements is **TRUE** regarding the autonomic nervous system?
    A) Its sensory input includes impulses from interoreceptors.
    B) Its preganglionic fibers release acetylcholine or norepinephrine.
    C) Its postganglionic fibers release only acetylcholine.
    D) Its effectors include some skeletal muscles.
    E) The sympathetic division is the "energy conservation-restoration" system.
    Answer: A

Pages: 488
11. The efferent portion of the autonomic nervous system has two principal divisions, the:
    A) sensory and motor
    B) preganglionic and postganglionic
    C) CNS and PNS
    D) sympathetic and parasympathetic
    E) spinal nerves and cranial nerves
    Answer: D

Pages: 488
12. The cell bodies of preganglionic neurons in the ANS are located in the:
    A) dorsal root ganglia
    B) prevertebral ganglia
    C) brain stem or spinal cord
    D) visceral effectors
    E) major spinal nerve plexuses
    Answer: C

Pages: 489
13. Cell bodies of preganglionic neurons of the parasympathetic division are located in the nuclei of **ALL** of the following cranial nerves **EXCEPT**:
    A) III
    B) V
    C) VII
    D) IX
    E) X
    Answer: B

Pages: 491
14. The cell bodies of parasympathetic preganglionic neurons are located in the lateral gray horns of spinal segments:
    A) T1-T12
    B) T2-T10
    C) L1-L5
    D) T1-L5
    E) S2-S4
    Answer: E

Pages: 491-492
15. **ALL** of the following are acceptable names for groups of autonomic ganglia **EXCEPT** the:
    A) collateral ganglia
    B) terminal ganglia
    C) prevertebral ganglia
    D) dorsal root ganglia
    E) intramural ganglia
    Answer: D

Pages: 492
16. Postganglionic sympathetic fibers from the superior cervical ganglia are distributed to **ALL** of the following **EXCEPT** the:
    A) blood vessels of the face
    B) smooth muscle of the eye
    C) cartilage of the nose
    D) sweat glands of the head
    E) major salivary glands
    Answer: C

Pages: 493
17. The exception to the usual pattern of two efferent neurons in an autonomic motor pathway is the pathway to the:
    A) adrenal medulla
    B) brain
    C) colon
    D) reproductive organs
    E) rectum
    Answer: A

Pages: 493
18. The smooth muscle of the eye is innervated by parasympathetic postganglionic fibers from the:
    A) ciliary ganglia
    B) lingual ganglia
    C) otic ganglia
    D) pterygopalatine ganglia
    E) submandibular ganglia
    Answer: A

Pages: 497
19. Which of the following receive only parasympathetic stimulation?
    A) brain
    B) heart
    C) lacrimal glands
    D) small intestine
    E) stomach
    Answer: C

Pages: 496
20. Which of the following responses is initiated by the sympathetic nervous system?
    A) decreased heart rate
    B) constriction of pupils
    C) splitting of glycogen to glucose by the liver
    D) constriction of bronchioles
    E) decreased blood pressure
    Answer: C

Pages: 491-492
21. Which of the following pairs of ganglia receive sympathetic preganglionic fibers?
    A) vertebral chain and terminal ganglia
    B) collateral and intramural ganglia
    C) terminal and dorsal root ganglia
    D) paravertebral and prevertebral ganglia
    E) ciliary and otic ganglia
    Answer: D

Pages: 491
22. The sympathetic ganglia that lie in a vertical row on either side of the vertebral column are called the:
    A) prevertebral ganglia
    B) paravertebral ganglia
    C) terminal ganglia
    D) collateral ganglia
    E) dorsal root ganglia
    Answer: B

Pages: 491
23. The sympathetic ganglia that lie close to large abdominal arteries are the:
    A) prevertebral ganglia
    B) paravertebral ganglia
    C) terminal ganglia
    D) sympathetic trunk ganglia
    E) dorsal root ganglia
    Answer: A

Pages: 493
24. Sympathetic preganglionic fibers that pass through sympathetic trunk ganglia to synapse in prevertebral ganglia are called:
    A) white rami communicantes
    B) gray rami communicantes
    C) splanchnic nerves
    D) sympathetic chains
    E) anterior roots
    Answer: C

Pages: 492
25. Sympathetic preganglionic fibers extend from the CNS to autonomic ganglia via the:
A) dorsal roots and gray rami communicantes
B) dorsal roots and white rami communicantes
C) dorsal roots and sympathetic chains
D) anterior roots and gray rami communicantes
E) anterior roots and white rami communicantes
Answer: E

Pages: 492
26. Which of the following is a sympathetic ganglion?
A) middle cervical ganglion
B) ciliary ganglion
C) otic ganglion
D) submandibular ganglion
E) pterygopalatine ganglion
Answer: A

Pages: 493
27. Which of the following is a parasympathetic ganglion?
A) hypogastric ganglion
B) middle cervical ganglion
C) celiac ganglion
D) inferior mesenteric ganglion
E) submandibular ganglion
Answer: E

Pages: 492
28. What is the usual distribution of sympathetic trunk ganglia?
A) 7 cervical, 12 thoracic, 5 lumbar, 5 sacral
B) 3 cervical, 11 thoracic, 4 lumbar, 4 sacral
C) 7 cervical, 12 thoracic, 4 lumbar, 4 sacral
D) 4 cervical, 8 thoracic, 4 lumbar, 4 sacral
E) 5 cervical, 10 thoracic, 5 lumbar, 5 sacral
Answer: B

Pages: 492
29. The sympathetic trunk ganglion located at the level of vertebra C6 and the cricoid cartilage of the larynx is the:
A) superior cervical ganglion
B) middle cervical ganglion
C) inferior cervical ganglion
D) ciliary ganglion
E) pterygopalatine ganglion
Answer: B

Pages: 492
30. The sympathetic trunk ganglion located at the level of vertebra C7 and the first rib is the:
    A) superior cervical ganglion
    B) middle cervical ganglion
    C) inferior cervical ganglion
    D) ciliary ganglion
    E) pterygopalatine ganglion
    Answer: C

Pages: 492
31. The sympathetic trunk ganglion located at the level of vertebra C2 and an internal carotid artery is the:
    A) superior cervical ganglion
    B) middle cervical ganglion
    C) inferior cervical ganglion
    D) ciliary ganglion
    E) pterygopalatine ganglion
    Answer: A

Pages: 492
32. Sympathetic stimulation to the heart is provided by fibers from the:
    A) hypogastric ganglion
    B) celiac ganglion
    C) inferior cervical ganglion
    D) pterygopalatine ganglion
    E) submandibular ganglion
    Answer: C

Pages: 492
33. Sympathetic stimulation to the salivary glands is provided by fibers from the:
    A) superior cervical ganglion
    B) middle cervical ganglion
    C) celiac ganglion
    D) pterygopalatine ganglion
    E) submandibular ganglion
    Answer: A

Pages: 492
34. The structure containing sympathetic postganglionic fibers that connect sympathetic trunk ganglia to the spinal nerves is the:
    A) sympathetic chain
    B) anterior root
    C) posterior root
    D) white ramus communicans
    E) gray ramus communicans
    Answer: E

Pages: 493
35. Axon collaterals of sympathetic preganglionic fibers that extend between sympathetic trunk ganglia form the:
   A) anterior roots
   B) posterior roots
   C) sympathetic chains
   D) white rami communicantes
   E) gray rami communicantes
   Answer: C

Pages: 493
36. Splanchnic nerves from the thoracic area terminate in the:
   A) inferior cervical ganglion
   B) celiac ganglion
   C) superior mesenteric plexus
   D) terminal ganglia near the heart and lungs
   E) dorsal root ganglia associated with thoracic spinal nerves
   Answer: B

Pages: 493
37. The solar plexus is another name for the:
   A) superior cervical ganglion
   B) junction of sympathetic preganglionic fibers and the adrenal medulla
   C) celiac ganglion
   D) ciliary ganglion
   E) junction of the vagus nerve with thoracic visceral effectors
   Answer: C

Pages: 493
38. Sympathetic postganglionic fibers innervating the stomach, spleen, and liver arise from the:
   A) ciliary plexus
   B) celiac plexus
   C) superior mesenteric plexus
   D) inferior mesenteric plexus
   E) hypogastric plexus
   Answer: B

Pages: 493
39. The greater splanchnic nerve terminates in the:
   A) celiac ganglion
   B) superior mesenteric ganglion
   C) inferior mesenteric ganglion
   D) adrenal medulla
   E) lumbar plexus
   Answer: A

Pages: 493
40. The lesser splanchnic nerve terminates in the:
    A) celiac ganglion
    B) superior mesenteric ganglion
    C) inferior mesenteric ganglion
    D) adrenal medulla
    E) lumbar plexus
    Answer: B

Pages: 493
41. The lumbar splanchnic nerve terminates in the:
    A) celiac ganglion
    B) superior mesenteric ganglion
    C) inferior mesenteric ganglion
    D) adrenal medulla
    E) lumbar plexus
    Answer: C

Pages: 493
42. Parasympathetic stimulation to the smooth muscle cells in the eyeball is provided by fibers arising from the:
    A) superior cervical ganglion
    B) middle cervical ganglion
    C) otic ganglion
    D) ciliary ganglion
    E) celiac ganglion
    Answer: D

Pages: 493
43. Parasympathetic stimulation to the nasal mucosa, pharynx, and lacrimal glands is provided by fibers arising from the:
    A) superior cervical ganglion
    B) middle cervical ganglion
    C) inferior cervical ganglion
    D) pterygopalatine ganglion
    E) submandibular ganglion
    Answer: D

Pages: 493
44. Parasympathetic stimulation to the parotid salivary glands is provided by fibers arising from the:
    A) middle cervical ganglion
    B) inferior cervical ganglion
    C) submandibular ganglion
    D) pterygopalatine ganglion
    E) otic ganglion
    Answer: E

Pages: 493
45. Parasympathetic preganglionic fibers traveling with cranial nerve III terminate in the:
A) superior cervical ganglion
B) inferior cervical ganglion
C) ciliary ganglia
D) otic ganglia
E) submandibular ganglia
Answer: C

Pages: 493
46. Some parasympathetic preganglionic fibers traveling with cranial nerve VII terminate in the:
A) inferior cervical ganglion
B) ciliary ganglia
C) otic ganglia
D) celiac ganglion
E) pterygopalatine ganglia
Answer: E

Pages: 493
47. Parasympathetic preganglionic fibers traveling with cranial nerve IX terminate in the:
A) middle cervical ganglion
B) inferior cervical ganglion
C) ciliary ganglia
D) otic ganglia
E) celiac ganglion
Answer: D

Pages: 493
48. Parasympathetic stimulation to the liver, stomach, and gallbladder is provided by fibers traveling with the:
A) vagus nerve
B) greater splanchnic nerve
C) lesser splanchnic nerve
D) white rami communicantes
E) gray rami communicantes
Answer: A

Pages: 493
49. The ureters and urinary bladder receive parasympathetic stimulation from postganglionic fibers that have been stimulated by the:
A) greater splanchnic nerve
B) lesser splanchnic nerve
C) vagus nerve
D) pelvic splanchnic nerves
E) adrenal medulla
Answer: D

Pages: 497
50. A major organ that receives sympathetic stimulation, but not parasympathetic stimulation, is the:
A) heart
B) liver
C) stomach
D) lung
E) kidney
Answer: E

Pages: 494
51. Acetylcholine is released into the synaptic cleft by the process of:
A) primary active transport
B) simple diffusion
C) exocytosis
D) filtration
E) receptor-mediated endocytosis
Answer: C

Pages: 494
52. Acetylcholine exerts its effects on postsynaptic cells when it:
A) is broken down by enzymes in the synaptic cleft, and the end-products diffuse into the postsynaptic cell
B) binds to specific receptors on the postsynaptic cell membrane and changes the permeability of the membrane to particular ions
C) diffuses into the postsynaptic cell and changes the pH of the intracellular fluid
D) actively transports ions from the synaptic cleft into the postsynpatic cell
E) blocks specific receptors to which other neurotransmitters could attach
Answer: B

Pages: 494
53. Blood-borne norepinephrine and epinephrine are broken down by enzymes in the:
A) synpatic cleft
B) postsynaptic cells they stimulate
C) liver
D) brain
E) kidneys
Answer: C

Pages: 494
54. Norepinephrine and epinephrine enter the bloodstream when sympathetic stimulation is provided to the:
A) adrenal medulla
B) liver
C) brain
D) heart
E) kidneys
Answer: A

Page 339 - The Autonomic Nervous System

55. Because the principal active ingredient in tobacco is nicotine, you might expect smoking to enhance the effects of:
   A) acetylcholine on parasympathetic visceral effectors
   B) acetylcholine on any postganglionic neurons
   C) norepinephrine on the heart and blood vessels
   D) norepinephrine on the limbic system
   E) norepinephrine on most sympathetic visceral effectors
   Answer: B

56. Cholinergic sympathetic postganglionic neurons stimulate the:
   A) adrenal medulla
   B) basal ganglia
   C) sweat glands
   D) heart
   E) walls of the gastrointestinal tract
   Answer: C

57. Norepinephrine principally stimulates:
   A) alpha receptors only
   B) beta receptors only
   C) muscarinic receptors
   D) nicotinic receptors
   E) alpha and beta receptors equally
   Answer: A

58. Loss of control over bladder and bowel functions in situations involving so-called paradoxical fear is due to:
   A) the fight-or-flight response
   B) failure of the sympathetic nervous system to respond
   C) failure of the parasympathetic nervous system to respond
   D) inability to produce adequate amounts of acetylcholine to maintain muscle tone
   E) massive activation of the parasympathetic nervous system
   Answer: E

59. You have just discovered that your pants were unzipped the entire time you gave a speech to your class. Which of the following responses would you most likely experience?
   A) increased parasympathetic stimulation to the iris
   B) increased parasympatheitc stimulation to the stomach
   C) increased sympathetic stimulation to the heart
   D) decreased sympathetic stimulation to the bronchioles
   E) decreased sympathetic stimulation to the small intestine
   Answer: C

Pages: 496
60. During a fight-or-flight response, you would expect dilation of blood vessels in **ALL** of the following areas **EXCEPT** the:
A) skin
B) skeletal muscles
C) cardiac muscle
D) adipose tissue
E) liver
Answer: A

Pages: 497
61. You are just about to perform a clinical procedure for the first time, and your palms begin to sweat. This is due to:
A) increased sympathetic stimulation of sweat glands possessing alpha receptors
B) increased sympathetic stimulation of sweat glands possessing beta receptors
C) increased parasympathetic stimulation of sweat glands possessing nicotinic receptors
D) increased parasympathetic stimulation of sweat glands possessing muscarinic receptors
E) increased sympathetic stimulation of sweat glands possessing muscarinic receptors
Answer: A

Pages: 497
62. Increased sympathetic stimulation increases the secretion of **ALL** of the following **EXCEPT**:
A) glucagon
B) renin
C) epinephrine
D) insulin
E) norepinephrine
Answer: D

Pages: 498
63. Fibers of the ANS traveling with cranial nerve X will stimulate which of the following reponses?
A) increased secretion from salivary glands
B) increased heart rate
C) increased activity of sweat glands on the palms and soles
D) decreased force of contraction of cardiac muscle
E) dilation of blood vessels in skeletal muscles of the torso
Answer: D

Pages: 498
64. Which of the following is stimulated by the parasympathetic nervous system?
A) dilation of the bronchioles
B) erection of the penis
C) increased gluconeogenesis
D) dilation of the pupil
E) increased secretion of sweat glands in the palms and soles
Answer: B

Pages: 497
65. Which of the following is stimulated by the sympathetic nervous system?
    A) increased glycogenolysis in the liver
    B) increased secretion of insulin
    C) movement of fatty acids into adipose tissue for storage
    D) increased bile secretion
    E) increased secretion of digestive enzymes from the pancreas
    Answer: A

True-False

Write T if the statement is true and F if the statement is false.

Pages: 498
1. There are two sets of efferent neurons in an autonomic motor pathway.
   Answer: True

Pages: 489
2. The axons of postganglionic autonomic fibers are myelinated.
   Answer: False

Pages: 489
3. The cell bodies of some sympathetic preganglionic neurons are located in the brain stem.
   Answer: False

Pages: 489
4. The cell bodies of some parasympathetic preganglionic neurons are located in the brain stem.
   Answer: True

Pages: 491
5. Prevertebral (collateral) ganglia receive parasympathetic preganglionic fibers.
   Answer: False

Pages: 492
6. Terminal (intramural) ganglia receive parasympathetic preganglionic fibers.
   Answer: True

Pages: 492
7. A single sympathetic preganglionic fiber may synapse with twenty or more postganglionic fibers.
   Answer: True

Pages: 493
8. Most of the total craniosacral outflow is carried by cranial nerve VII.
   Answer: False

Pages: 494
9. Cholinergic neurons release norepinephrine or epinephrine.
   Answer: False

10. All preganglionic neurons release the neurotransmitter acetylcholine.
Answer: True

11. COMT and MAO are neurotransmitters released by sympathetic postganglionic fibers.
Answer: False

12. All postganglionic neurons have nicotinic receptors.
Answer: True

13. Activation of nicotinic receptors always leads to depolarization, while activation of muscarinic receptors always leads to hyperpolarization.
Answer: False

14. In a fight-or-flight response, blood sugar (glucose) increases.
Answer: True

15. Increased sympathetic stimulation increases the rate at which food is moved through the gastrointestinal tract.
Answer: False

Short Answer

Write the word or phrase that best completes each statement or answers the question.

1. Effector tissues for autonomic motor neurons are _____, _____, and _____.
Answer: cardiac muscle; smooth muscle; glands

2. An autonomic motor neuron that extends from the CNS to an autonomic ganglion is called a _____ neuron.
Answer: preganglionic

3. An autonomic motor neuron that extends from an autonomic ganglion to a visceral effector is called a _____ neuron.
Answer: postganglionic

4. Based on the locations of preganglionic cell bodies, the sympathetic division of the ANS is sometimes called the _____ division.
Answer: thoracolumbar

Pages: 491
5. Based on the locations of preganglionic cell bodies, the parasympathetic division of the ANS is sometimes called the _____ division.
Answer: craniosacral

Pages: 492
6. Preganglionic parasympathetic fibers synapse in _____ ganglia in or near visceral effectors.
Answer: terminal (intramural)

Pages: 491
7. Ganglia lying close to large abdominal arteries that receive sympathetic preganglionic fibers are the _____ ganglia.
Answer: prevertebral (collateral)

Pages: 491
8. _____ ganglia lie in a vertical row on either side of the vertebral column from the base of the skull to the coccyx.
Answer: Sympathetic trunk (paravertebral, vertebral chain)

Pages: 492
9. The celiac and superior mesenteric ganglia, which are named for the large abdominal arteries they are near, are examples of _____ ganglia.
Answer: prevertebral (collateral)

Pages: 492
10. Sympathetic preganglionic fibers that connect the anterior ramus of a spinal nerve with sympathetic trunk ganglia are collectively called the _____.
Answer: white rami communicantes

Pages: 492
11. The effector for sympathetic postganglionic fibers leaving the middle and inferior cervical ganglia is the _____.
Answer: heart

Pages: 492
12. Sympathetic postganglionic fibers that connect sympathetic trunk ganglia with spinal nerves are collectively called the _____.
Answer: gray rami communicantes

Pages: 493
13. Axon collaterals of the fibers of the white rami communicantes that extend through multiple sympathetic trunk ganglia are called _____.
Answer: sympathetic chains

Pages: 493
14. Sympathetic preganglionic fibers that pass through sympathetic trunk ganglia to terminate in prevertebral ganglia are called _____.
Answer: splanchnic nerves

15. Another name for the celiac ganglion is the _____.
Answer: solar plexus

16. Parasympathetic cranial outflow has five components: four pairs of ganglia and the plexuses associated with the _____ nerve.
Answer: vagus

17. The four pairs of ganglia that receive cranial parasympathetic outflow are the _____ ganglia, the pterygopalatine ganglia, the submandibular ganglia, and the _____ ganglia.
Answer: ciliary; otic

18. The sacral parasympathetic outflow consisting of preganglionic fibers from the anterior roots of spinal nerves S2-S4 collectively form the _____.
Answer: pelvic splanchnic nerves

19. Cholinergic neurons release the neurotransmitter _____.
Answer: acetylcholine

20. The enzyme that inactivates acetylcholine is _____.
Answer: acetylcholinesterase

21. Adrenergic neurons release the neurotransmitters _____ or _____.
Answer: norepinephrine; epinephrine

22. The enzymes that inactivate norepinephrine and epinephrine are _____ and _____.
Answer: COMT; MAO

23. _____ receptors for acetylcholine are present on both sympathetic and parasympathetic postganglionic neurons.
Answer: Nicotinic

24. _____ receptors for acetylcholine are present on all effectors innervated by parasympathetic postganglionic neurons.
Answer: Muscarinic

25. The two types of receptors for norepinephrine and epinephrine are _____ and _____ receptors.
Answer: alpha; beta

Pages: 496

26. Activation of the sympathetic division of the ANS also results in the release of hormones from the modified postganglionic cells of the _____.
Answer: adrenal medulla

Pages: 498

27. The major control and integration center of the ANS is the _____.
Answer: hypothalamus

Pages: 496

28. The phrase "energy conservation-restorative system" describes the activities of the _____ division of the ANS.
Answer: parasympathetic

Pages: 496

29. The "fight-or-flight response" is a term used to describe the physiological responses triggered by the _____ division of the ANS.
Answer: sympathetic

Pages: 496

30. A drug that acts as a "beta-blocker" is blocking the effects of the neurotransmitters _____ or _____.
Answer: norepinephrine; epinephrine

Matching

Choose the item from Column 2 that best matches each item in Column 1.

1. receive sympathetic preganglionic fibers; located along either side of vertebral column

   sympathetic trunk ganglia

2. receive sympathetic preganglionic fibers; located close to large abdominal arteries

   prevertebral ganglia

3. receive parasympathetic preganglionic fibers; located close to or within walls of visceral organs

   terminal ganglia

4. contain preganglionic fibers connecting anterior ramus of spinal nerve with sympathetic trunk ganglia

   white rami communicantes

5. contain postganglionic fibers connecting ganglion of sympathetic trunk to spinal nerve

gray rami communicantes

6. axon collaterals of preganglionic fibers that form fibers along which sympathetic trunk ganglia are strung

sympathetic chains

7. sympathetic preganglionic fibers that pass through sympathetic trunk ganglia and extend to prevertebral ganglia

splanchnic nerves

8. site of synapses between splanchnic nerves of the thoracic region and postganglionic cell bodies

celiac ganglion (solar plexus)

Essay

Write your answer in the space provided or on a separate sheet of paper.

Pages: 492-495

1. Explain why the sympathetic division of the ANS has more widespread and longer-lasting effects than the parasympathetic division.

   Answer: A single sympathetic preganglionic neuron synapses with 20 or more postganglionic neurons, vs. about five for parasympathetic. The sympathetic neurotransmitters are broken down more slowly than ACh, so postsynaptic cells are stimulated longer. The sympathetic division also stimulates release of catecholamines from the adrenal medulla, thus enhancing the sympathetic effects via the endocrine system. Many more visceral effectors have receptors for catecholamines than ACh.

Pages: 493

2. Describe the anatomical arrangement of the components of the parasympathetic division of the ANS.

   Answer: Cell bodies of preganglionic neurons are located in the brain stem and in the sacral region of the spinal cord. Preganglionic axons travel with cranial nerves III, VII, IX, and X, or with the anterior roots of sacral spinal nerves as pelvic splanchnic nerves. They synapse with postganglionic cell bodies in terminal ganglia in or near the walls of visceral effectors, thus postganglionic fibers are relatively short.

Pages: 492
3. Describe the anatomical arrangement of the components of the sympathetic division of the ANS.

Answer: Cell bodies of preganglionic neurons arise in either the thoracic or lumbar region of the spinal cord. Axons of these neurons extend (via the white rami communicantes) to sympathetic trunk or prevertebral ganglia, where they synapse with cell bodies of postganglionic neurons. Postganglionic axons extend to visceral effectors. Some preganglionic axons form splanchnic nerves as they pass through sympathetic trunk ganglia to prevertebral ganglia. Others send out collaterals that form sympathetic chains.

Pages: 494
4. An autonomic neuron releases the neurotransmitter acetylcholine. What can you tell about this neuron's role in the ANS? What possible characteristics can be determined about the postsynaptic cell? Explain your answers.

Answer: The neuron could be a preganglionic neuron in either division of the ANS, a parasympathetic postganglionic neuron, or one of a few types of sympathetic postganglionic neurons. The postsynaptic cell must possess either nicotinic or muscarinic receptors in order to respond to ACh. This effector cell could be either a postganglionic neuron of either division, a parasympathetic visceral effector, or one of a few sympathetic effectors.

Pages: 494-495
5. A postsynaptic cell has beta receptors on its membrane. Assuming that these beta receptors bind an ANS neurotransmitter, what are the possible general characteristics/activities of the postsynaptic cell? Explain your answer.

Answer: Possession of beta receptors indicates responsiveness to NE or epinephrine. The cell must be an effector in the sympathetic ANS, since these neurotransmitters are released only by sympathetic postganglionic neurons. The cell could be part of most any visceral effector, and whether the cell is excited or inhibited depends on the specific type of beta receptor and the cell that possesses it.

Pages: 494
6. Compare acetylcholine and norepinephrine with regard to their ANS sources and their effects on postsynaptic cells.

Answer: · Cholinergic neurons - all preganglionic neurons, all parasympathetic postganglionic neurons, some sympathetic postganglionic neurons - release ACh; postsynaptic cells with nicotinic receptors depolarized; those with muscarinic receptors are depolarized or hyperpolarized, depending on the cell.

· Adrenergic neurons - most sympathetic postganglionic neurons - release NE; NE affects primarily those postsynaptic cells with alpha receptors, which are generally excitatory, though cells with beta receptors may respond also. Excitation or inhibition may be the effect.

Pages: 498

7. List and briefly describe the functions of the components of an autonomic reflex arc. What are the general functions of such reflex arcs?

   Answer: 1) Receptor - senses changes in environment
   2) Sensory neuron - somatic or visceral; conducts impulse to CNS
   3) Association neuron - connects neurons within CNS
   4) Autonomic motor neurons - preganglionic conducts impulse from CNS to autonomic ganglion; postganglionic conducts impulse from ganglion to visceral effector
   5) Visceral effector - carries out response; cardiac or smooth muscle, or gland These reflexes help regulate homeostasis. [See p. 498 for examples.]

Pages: 496

8. Describe the physiological responses triggered by the sympathetic division of the ANS.

   Answer: Most sympathetic ("fight-or-flight") responses enhance changes seen in emergency and embarrassing situations and during exercise, while activities not necessary in such situations are inhibited. For specific responses, see p. 496.

Pages: 496

9. Describe the physiological responses triggered by the parasympathetic division of the ANS.

   Answer: "SLUD" responses include salivation, lacrimation, urination, and defecation. Parasympathetic responses enhance digestion and absorption of food. A decrease in heart rate is another important parasympathetic response.

Pages: 494

10. Predict the effects of a drug called "an MAO inhibitor." Explain your answer.

    Answer: An MAO inhibitor inhibits the enzyme monoamine oxidase, which breaks down epinephrine and norepinephrine. If the enzyme is inhibited, then these neurotransmitters accumulate, thus enhancing the effects of most of the sympathetic ANS.

Chapter 18   The Endocrine System

Multiple-Choice

Choose the one alternative that best completes the statement or answers the question.

Pages: 515
1. Hormones secreted from the neurohypophysis are synthesized by the:
   A) adenohypophysis
   B) thyroid gland
   C) neurohypophysis
   D) hypothalamus
   E) pineal gland
   Answer: D

Pages: 513
2. An increase in the secretion of GHRH results in:
   A) an increase in the secretion of human growth hormone
   B) a decrease in the secretion of human growth hormone
   C) an increase in the secretion of FSH and LH
   D) simultaneous stimulation of the sympathetic nervous system
   E) an increase in the secretion of GnRH
   Answer: A

Pages: 510
3. The primary target for prolactin is the:
   A) ovaries
   B) hypothalamus
   C) adrenal cortex
   D) uterus
   E) mammary glands
   Answer: E

Pages: 515
4. Which of the following hormones is **NOT** produced by the anterior pituitary gland?
   A) ACTH
   B) FSH
   C) CRH
   D) TSH
   E) hGH
   Answer: C

Pages: 514
5. The testes are a primary target for:
   A) FSH
   B) ACTH
   C) aldosterone
   D) corticosterone
   E) GnRH
   Answer: A

Pages: 514
6. Increasing the uptake of iodide by the thyroid gland and increasing the growth of the thyroid gland are two functions of:
    A) TRH
    B) TSH
    C) T_3
    D) thyroglobulin
    E) thyroid binding globulin
    Answer: B

Pages: 518
7. The main target for ADH is the:
    A) kidney
    B) hypothalamus
    C) uterus
    D) adrenal cortex
    E) neurohypophysis
    Answer: A

Pages: 515
8. Nerve impulses from the hypothalamus stimulate the release of hormones from the:
    A) anterior pituitary gland only
    B) posterior pituitary gland only
    C) thyroid gland
    D) adrenal cortex
    E) both the anterior and posterior pituitary gland
    Answer: B

Pages: 520
9. The thyroid gland is located:
    A) under the sternum
    B) behind and beneath the stomach
    C) in the sella turcica of the sphenoid bone
    D) in the neck, anterior to the trachea
    E) in the roof of the third ventricle of the brain
    Answer: D

Pages: 522
10. The primary effect of T_3 and T_4 is to:
    A) decrease blood glucose
    B) promote the release of calcitonin
    C) promote heat-generating reactions
    D) stimulate the uptake of iodide by the thyroid gland
    E) promote excretion of sodium ions in urine
    Answer: C

Pages: 525
11. The primary effect of calcitonin is to:
    A) increase blood glucose
    B) decrease blood glucose
    C) increase excretion of calcium ions in urine
    D) increase blood calcium
    E) decrease blood calcium
    Answer: E

Pages: 527
12. The stimulus for release of parathyroid hormone is:
    A) PRH
    B) low levels of calcium in the blood
    C) TSH
    D) nerve impulses from the hypothalamus
    E) calcitonin
    Answer: B

Pages: 533
13. Hormones that are "sympathomimetic" are produced by the:
    A) adrenal medulla only
    B) adrenal cortex only
    C) adenohypophysis
    D) neurohypophysis
    E) both the adrenal cortex and the adrenal medulla
    Answer: A

Pages: 533
14. Increased heart rate and force of contraction are effects of:
    A) ADH
    B) insulin
    C) cortisol
    D) epinephrine
    E) aldosterone
    Answer: D

Pages: 529
15. The outermost layer of the adrenal cortex is called the:
    A) zona reticularis
    B) zona glomerulosa
    C) zona fasciculata
    D) medulla
    E) follicular zone
    Answer: B

Pages: 531
16. An increase in blood glucose and an anti-inflammatory effect are important effects of:
    A) epinephrine
    B) glucagon
    C) corticosterone
    D) insulin
    E) ADH
    Answer: C

Pages: 529
17. A primary effect of mineralocorticoids is to promote:
    A) increased urine production
    B) excretion of potassium ions by the kidney
    C) excretion of sodium ions by the kidney
    D) decreased blood glucose
    E) increased secretion of ACTH
    Answer: B

Pages: 536
18. The primary target for glucagon is the:
    A) liver
    B) hypothalamus
    C) adrenal cortex
    D) pancreas
    E) kidney
    Answer: A

Pages: 533
19. The Islets of Langerhans are the endocrine portion of the:
    A) adrenal cortex
    B) adrenal medulla
    C) anterior pituitary gland
    D) posterior pituitary gland
    E) pancreas
    Answer: E

Pages: 539
20. An endocrine gland located in the roof of the third ventricle of the brain that may control the maturation and aging process is the:
    A) thymus
    B) adenohypophysis
    C) neurohypophysis
    D) pineal gland
    E) parathyroid gland
    Answer: D

Pages: 515
21. Antidiuretic hormone is secreted by the:
    A) kidney
    B) adrenal cortex
    C) adrenal medulla
    D) adenohypophysis
    E) neurohypophysis
    Answer: E

Pages: 512
22. An increase in the rate of protein synthesis, especially by the cells in the skeletal and muscular systems, is the main effect of:
    A) glucagon
    B) hGH
    C) aldosterone
    D) PTH
    E) GHRH
    Answer: B

Pages: 533
23. Alpha and/or beta receptors are required for a tissue to respond to:
    A) epinephrine
    B) corticosterone
    C) insulin
    D) aldosterone
    E) hGH
    Answer: A

Pages: 536
24. The only hormone that promotes anabolism of glycogen, fats, and proteins is:
    A) hGH
    B) insulin
    C) epinephrine
    D) aldsoterone
    E) corticosterone
    Answer: B

Pages: 507
25. When a hormone that uses a second messenger binds to a target cell, the next thing that happens is that:
    A) phosphodiesterase is activated
    B) a protein kinase is formed
    C) a gene is activated in the nucleus
    D) adenylate cyclase is activated by a G-protein
    E) voltage-regulated ion channels open in the cell membrane
    Answer: D

Pages: 507
26. Most hormones that use a second messenger are:
   A) proteins
   B) enzymes
   C) steroids
   D) nucleic acids
   E) biogenic amines
   Answer: A

Pages: 507
27. The compound that most often acts as a second messenger is:
   A) cholesterol
   B) phosphodiesterase
   C) cyclic AMP
   D) COMT
   E) CRH
   Answer: C

Pages: 506
28. When a steroid hormone binds to its target cell receptor, it:
   A) causes the formation of cyclic AMP
   B) is converted into cholesterol, which acts as a second messenger
   C) causes the formation of releasing hormones
   D) turns specific genes of the nuclear DNA on or off
   E) alters the membrane's permeability to G-proteins
   Answer: D

Pages: 510
29. Releasing and inhibiting hormones are produced by the:
   A) neurohypophysis to control the adenohypophysis
   B) adenohypophysis to control the neurohypophysis
   C) hypothalamus to control the adenohypophysis
   D) hypothalamus to control the neurohypophysis
   E) pineal gland to control the hypothalamus
   Answer: C

Pages: 536
30. Blood glucose is raised by **ALL** of the following **EXCEPT**:
   A) glucagon
   B) hGH
   C) epinephrine
   D) cortisol
   E) insulin
   Answer: E

Pages: 515
31. Endorphins are derived from the same parent compound as:
    A) ADH
    B) ACTH
    C) cortisol
    D) insulin
    E) hGH
    Answer: B

Pages: 514
32. GnRH is the primary stimulus for secretion of:
    A) FSH
    B) hGH
    C) TSH
    D) GHRH
    E) GHIH
    Answer: A

Pages: 517
33. The milk ejection reflex is an important effect of:
    A) prolactin
    B) estrogen
    C) progesterone
    D) oxytocin
    E) FSH
    Answer: D

Pages: 515
34. Vasopressin is another name for:
    A) antidiuretic hormone
    B) thyrotropin
    C) aldosterone
    D) growth hormone
    E) somatostatin
    Answer: A

Pages: 522
35. Thyroglobulin is:
    A) the major component of colloid inside follicles
    B) another name for thyroid hormone
    C) the major stimulus for release of thyroid hormones
    D) the protein that transports thyroxine in the blood
    E) the protein that transports TSH to the thyroid gland
    Answer: A

Pages: 522
36. During the production of T_3 and T_4, iodide is added to:
    A) cholesterol
    B) thyrotropin
    C) tyrosine
    D) TRH
    E) TBG
    Answer: C

Pages: 533
37. Chromaffin cells are:
    A) modified postganglionic sympathetic neurons
    B) producers of epinephrine
    C) located in the adrenal medulla
    D) cells that secrete more hormones in response to acetylcholine
    E) all of the above are correct
    Answer: E

Pages: 536
38. An increase in glycogenolysis by the liver is an important effect of:
    A) glucagon only
    B) insulin only
    C) PTH
    D) aldosterone
    E) both glucagon and insulin
    Answer: A

Pages: 531
39. Long-term therapy with steroid drugs, such as cortisone, can cause osteoporosis and muscle wasting because of:
    A) increased blood glucose
    B) increased protein catabolism
    C) bacterial breakdown of bone and muscle
    D) increased protein anabolism
    E) increased excretion of calcium ions in urine
    Answer: B

Pages: 532
40. Androgens are:
    A) female sex hormones
    B) releasing hormones that target the pituitary gland
    C) male sex hormones
    D) releasing hormones that target the gonads
    E) transport proteins that carry sex hormones in the bloodstream
    Answer: C

Pages: 533
41. Somatostatin is produced by the:
    A) anterior pituitary gland
    B) posterior pituitary gland
    C) alpha cells of the pancreas
    D) delta cells of the pancreas
    E) parafollicular cells of the thyroid gland
    Answer: D

Pages: 536
42. The primary stimulus for the release of insulin is:
    A) an elevated level of blood glucose
    B) a decreased level of blood glucose
    C) insulin releasing hormone
    D) insulin-like growth factors
    E) pancreatic stimulating hormone
    Answer: A

Pages: 540
43. Melatonin is a hormone produced in response to varying levels of daylight by the:
    A) neurohypophysis
    B) gonads
    C) pineal gland
    D) thymus
    E) skin
    Answer: C

Pages: 541
44. A hormone that counteracts the effects of both ADH and aldosterone is:
    A) thymosin
    B) atrial natriuretic peptide
    C) calcitonin
    D) somatostatin
    E) insulin
    Answer: B

Pages: 513
45. Insulin-like growth factors are necessary for the full effect of:
    A) insulin
    B) hGH
    C) somatostatin
    D) triiodothyronine
    E) glucagon
    Answer: B

46. Diabetes insipidus results from:
    A) hyposecretion of insulin
    B) hypersecretion of insulin
    C) hyposecretion of aldosterone
    D) hypersecretion of ADH
    E) hyposecretion of ADH
    Answer: E

47. Severe mental retardation, skeletal dwarfing, and low metabolic rate are signs of:
    A) hyposecretion of hGH (only) in childhood
    B) hyposecretion of T_3 and T_4 (only) in childhood
    C) hypersecretion of T_3 and T_4 in childhood
    D) hypersecretion of hGH in childhood
    E) hyposecretion of either hGH or T_3 and T_4 in childhood
    Answer: B

48. The primary stimulus for release of cortisol and corticosterone is:
    A) ACTH
    B) increased levels of blood glucose
    C) the renin-angiotensin pathway
    D) increased levels of sodium ions in the blood
    E) increased levels of calcium ions in the blood
    Answer: A

49. An autoimmune disease in which an antibody mimics the action of TSH is:
    A) myxedema
    B) cretinism
    C) acromegaly
    D) Cushing's disease
    E) Graves' disease
    Answer: E

50. A "moon face," heavy fat deposition on the torso, but spindly limbs, and a "buffalo hump" over the scapula suggests that a person has chronically high levels of:
    A) insulin
    B) hGH
    C) thyroid hormones
    D) cortisol
    E) estrogens
    Answer: D

51. Which of the following is **NOT** a steroid hormone?
    A) aldosterone
    B) ADH
    C) cortisol
    D) estrogen
    E) DHEA
    Answer: B

52. Suckling is an important stimulus for release of:
    A) oxytocin
    B) DHEA
    C) estrogen
    D) FSH
    E) LH
    Answer: A

53. Hypersecretion of hGH in adulthood results in:
    A) acromegaly
    B) giantism
    C) cretinism
    D) diabetes insipidus
    E) myxedema
    Answer: A

54. The adrenal glands are located:
    A) in the sella turcica of the sphenoid bone
    B) in the neck, behind the trachea
    C) on top of the kidneys
    D) just under the diaphragm
    E) in the roof of the third ventricle of the brain
    Answer: C

55. Which of the following would you expect to see in someone suffering from Graves' disease?
    A) high levels of TRH
    B) high levels of TSH
    C) high levels of thyroxine
    D) high levels of calcitonin
    E) all of the above except calcitonin
    Answer: C

56. A chemical grouping of hormones derived from arachidonic acid is the:
   A) eicosanoids
   B) biogenic amines
   C) proteins
   D) peptides
   E) steroids
   Answer: A

57. Which of the following hormones works by direct gene activation?
   A) ADH
   B) hGH
   C) insulin
   D) cortisol
   E) glucagon
   Answer: D

58. The function of adenylate cyclase is to:
   A) break down a protein hormone when it binds to its receptor
   B) turn on a G-protein
   C) catalyze the conversion of ATP to cAMP
   D) activate a protein kinase
   E) inactivate cAMP
   Answer: C

59. Somatotrophs of the adenohypophysis secrete:
   A) GHRH
   B) somatostatin
   C) ACTH
   D) GnRH
   E) hGH
   Answer: E

60. Dopamine acts as an inhibiting hormone for:
   A) somatotropin
   B) somatostatin
   C) TSH
   D) prolactin
   E) oxytocin
   Answer: D

61. Synthesis and secretion of insulin-like growth factors is promoted by:
    A) insulin only
    B) somatotropin
    C) thyroxine
    D) glucagon only
    E) both insulin and glucagon
    Answer: B

62. Protein anabolism is promoted by:
    A) insulin
    B) somatotropin
    C) triiodothyronine
    D) cortisol
    E) all of the above except cortisol
    Answer: E

63. Interleukin-I stimulates the secretion of:
    A) ACTH
    B) ADH
    C) somatostatin
    D) insulin
    E) thyroxine
    Answer: A

64. You have consumed a six-pack of beer in the course of an evening. The combination of fluid intake and alcohol act to inhibit the secretion of:
    A) ACTH
    B) vasopressin
    C) insulin
    D) TSH
    E) atrial natriuretic peptide
    Answer: B

65. If levels of PTH are high, one would expect to see:
    A) increased osteoblast activity
    B) increased excretion of calcium ions in urine
    C) increased excretion of phosphate ions in urine
    D) decreased absorption of vitamin D
    E) increased deposition of calcium ions in bone
    Answer: C

Pages: 519,530
66. Which of the following pairs of hormones are **NOT** antagonists?
    A) GHRH-somatostatin
    B) PTH-calcitonin
    C) insulin-glucagon
    D) aldosterone - atrial natriuretic peptide
    E) aldosterone - ADH
    Answer: E

Pages: 519
67. Stimulation of osmoreceptors results in secretion of:
    A) renin
    B) aldosterone
    C) ACTH
    D) ADH
    E) oxytocin
    Answer: D

Pages: 524
68. Exophthalmos is a characteristic sign of:
    A) myxedema
    B) cretinism
    C) Graves' disease
    D) acromegaly
    E) Cushing's disease
    Answer: C

Pages: 532
69. Levels of ACTH are high in Addison's disease because:
    A) levels of glucocorticoids are low
    B) levels of glucocorticoids are high
    C) excessive water is being lost with sodium ions
    D) there is a tumor of the corticotrophs in the adenohypophysis
    E) there is a hypothalamic tumor of cells that make CRH
    Answer: A

Pages: 534
70. F cells of the pancreas secrete pancreatic polypeptide, which regulates the release of:
    A) insulin
    B) glucagon
    C) somatostatin
    D) insulin-like growth factors
    E) pancreatic digestive enzymes
    Answer: E

Pages: 537
71. Destruction of the beta cells of the pancreas results in:
    A) Type I diabetes mellitus
    B) Type II diabetes mellitus
    C) diabetes insipidus
    D) hyperglycemia
    E) pheochromocytoma
    Answer: A

Pages: 539
72. Male secondary sex characteristics develop primarily in response to:
    A) estrogens
    B) progesterone
    C) relaxin
    D) LH
    E) testosterone
    Answer: E

Pages: 542
73. Tumor angiogenesis factors stimulate:
    A) differentiation of mature neurons
    B) growth of new blood vessels
    C) growth of connective tissue capsules around tumors
    D) differentiation of stem cells in bone marrow
    E) release of blood cell growth factors by the kidney
    Answer: B

Pages: 542
74. Epidermal growth factor is produced by:
    A) neurons
    B) neuroglia
    C) Schwann cells
    D) submaxillary salivary glands
    E) ovaries and testes
    Answer: D

Pages: 542
75. Proliferation of mesodermally derived cells, cell migration, and production of adhesion proteins are all stimulated by:
    A) platelet-derived growth factors
    B) tumor angiogenesis factors
    C) nerve growth factor
    D) transforming growth factor-beta
    E) fibroblast growth factor
    Answer: E

True-False

Write T if the statement is true and F if the statement is false.

Pages: 506
1. Steroid hormones exert their effects on target cells by triggering the formation of a second messenger.
   Answer: False

Pages: 507
2. Phosphodiesterase inactivates cyclic AMP.
   Answer: True

Pages: 508
3. When a hormone's ability to exert its effects on a target cell is dependent on previous or simultaneous exposure to another hormone, the interaction is called a synergistic effect.
   Answer: False

Pages: 510
4. Hypophysiotropic hormones control the secretion of hormones from the neurohypophysis.
   Answer: False

Pages: 511
5. GnRH controls the secretion of both FSH and LH.
   Answer: True

Pages: 515
6. The posterior pituitary gland is made up of axons whose cell bodies are in the hypothalamus.
   Answer: True

Pages: 519
7. More ADH is secreted in response to increased blood volume.
   Answer: False

Pages: 522
8. Iodide trapping by thyroid follicle cells is stimulated by TSH.
   Answer: True

Pages: 522
9. Thyroglobulin is a transport protein that carries thyroid hormones in the blood.
   Answer: False

Pages: 527
10. PTH increases the number and activity of osteoblasts.
    Answer: False

Pages: 530
11. The principal mechanism controlling secretion of aldosterone is via stimulation by ACTH.
    Answer: False

Pages: 531
12. Cortisol stimulates both protein catabolism and gluconeogenesis.
Answer: True

Pages: 536
13. The principal target for glucagon is the liver.
Answer: True

Pages: 537
14. Ketoacidosis may occur in Type I diabetics due to increased use of fatty acids to produce ATP.
Answer: True

Pages: 543
15. The reactions of the General Adaptation Syndrome collectively known as the alarm reaction are due to elevated levels of CRH, GHRH, and TRH.
Answer: False

Short Answer

Write the word or phrase that best completes each statement or answers the question.

Pages: 503
1. The specific cells affected by a particular hormone are called _____ cells.
Answer: target

Pages: 504
2. Local hormones that act on neighboring cells are called _____, while those that act on the same cell that secreted them are called _____.
Answer: paracrines; autocrines

Pages: 505
3. Prostaglandins and leukotrienes belong to the structural (chemical) class of hormones known as _____, which are derived from fatty acids.
Answer: eicosanoids

Pages: 506
4. Specific proteins that transport most steroid hormones in the blood are synthesized in the _____.
Answer: liver

Pages: 507
5. The role of cyclic AMP in the function of water-soluble hormones is to act as a(n) _____.
Answer: second messenger

Pages: 507
6. Cyclic AMP is inactivated by the enzyme _____.
Answer: phosphodiesterase

7. Receptors on the outer surface of target cell membranes are linked to adenylate cyclase molecules on the inner surface by molecules called _____.
   Answer: G-proteins

8. When two hormones complement each other's action and both are needed for full expression of the hormonal effects, the interaction is called a(n) _____ effect.
   Answer: synergistic

9. Release of hormones from the adenohypophysis is regulated by releasing and inhibiting hormones secreted by the _____.
   Answer: hypothalamus

10. Hormones that influence endocrine glands other than their source are called _____.
    Answer: tropins (tropic hormones)

11. GHIH, or _____, inhibits the release of hGH.
    Answer: somatostatin

12. The target organ for ACTH is the _____.
    Answer: adrenal cortex

13. The target organs for oxytocin are the _____ and the _____.
    Answer: uterus; mammary glands

14. Lower than normal water concentration in the blood is sensed by osmoreceptors in the _____, which activate the cells that synthesize and release the hormone _____.
    Answer: hypothalamus; ADH

15. Thyroid hormones (T_3 and T_4) are synthesized by attaching _____ atoms to the amino acid _____.
    Answer: iodine; tyrosine

16. The increase in body temperature resulting from increased metabolic rate is called the _____ effect of thyroid hormones.
    Answer: calorigenic

17. The parafollicular cells of the thyroid gland secrete _____.
    Answer: calcitonin

18. The principal target organ for aldosterone is the _____, where it stimulates the reabsorption of _____ ions and the excretion of _____ ions.
    Answer: kidney; sodium (chloride, bicarbonate); potassium (hydrogen)

Pages: 531
19. The glucocorticoid that is most abundant and that is responsible for most of glucocorticoid activity is _____.
    Answer: cortisol

Pages: 532
20. DHEA is the principal _____ secreted by the adrenal cortex.
    Answer: androgen

Pages: 533
21. Hormones that are sympathomimetic are secreted by the _____.
    Answer: adrenal medulla

Pages: 533
22. The alpha cells of the pancreas secrete _____, while the beta cells secrete _____, and the delta cells secrete _____.
    Answer: glucagon; insulin; somatostatin (GHIH)

Pages: 540
23. The hormone whose release from the pineal gland is governed by the daily dark-light cycle is _____.
    Answer: melatonin

Pages: 539
24. The ovaries produce the female sex hormones _____ and _____.
    Answer: estrogens; progesterone

Pages: 537
25. If a diabetic injects too much insulin, the principal symptom would be _____.
    Answer: hypoglycemia

Pages: 531
26. If levels of glucocorticoids are low, the regulatory negative feedback loop would dictate an increase in secretion of _____ and _____.
    Answer: CRH; ACTH

Pages: 530
27. The target organs for angiotensin II are the _____ and the _____.
    Answer: adrenal cortex; arterioles

Pages: 529
28. The adrenal cortex is subdivided into three zones that secrete different hormones - the zona glomerulosa, which secretes _____, the zona fasciculata, which secretes _____, and the zona reticularis, which secretes _____.
    Answer: mineralocorticoids; glucocorticoids; adrenocortical androgens

29. The effect of parathyroid hormone on the kidney is to increase the reabsorption of _____ and _____ ions and to promote excretion of _____ ions.
    Answer: calcium; magnesium; phosphate

30. An enlarged thyroid gland is called a _____.
    Answer: goiter

Matching

Choose the item from Column 2 that best matches each item in Column 1.

   Match the following:

   1. hGH   acromegaly

   2. thyroid hormones   myxedema

   3. PTH   tetany

   4. glucocorticoids   Cushing's disease

   5. catecholamines   pheochromocytoma

   6. insulin   diabetes mellitus

   7. ADH   diabetes insipidus

   8. melatonin   seasonal affective disorder

   Match the following...

   9. CRH   ACTH

   10. GnRH   FSH

   11. TRH   TSH

   12. high blood glucose   insulin

   13. TSH   triiodothyronine

   14. dehydration   ADH

   15. low blood calcium   PTH

   16. high blood calcium   calcitonin

17. angiotensin II   aldosterone

18. LH   testosterone

Essay

Write your answer in the space provided or on a separate sheet of paper.

Pages: 503
1. Compare the roles of the nervous and endocrine systems in controlling homeostasis.
   Answer: The nervous system controls homeostasis via release of neurotransmitters in response
            to nerve impulses, which affect muscle fibers, glands, & neurons. Time between
            stimulation and response (contraction or secretion) is usually brief, and the effects
            of short duration. The endocrine system controls homeostasis via hormones, which cause
            metabolic changes in target cells, which include virtually all body cells. Time
            between stimulus and response varies from seconds to days with long-lasting effects.

Pages: 502
2. List the seven broad areas into which the actions of hormones can be categorized.
   Answer: 1) regulate chemical composition and volume of extracellular fluid
            2) regulate metabolism and energy balance
            3) regulate contraction of smooth & cardiac muscle fibers and secretion from glands
            4) maintain homeostasis in spite of pathologies and trauma
            5) regulate certain immune functions
            6) play role in integration of growth & development
            7) contribute to basic reproductive processes

Pages: 504
3. Define up-regulation and down-regulation as they relate to hormone action. Under what
   circumstances would you expect to see these phenomena?
   Answer: These terms refer to changes in the number of receptors present on target cells
            depending on the concentration of a hormone. Down-regulation occurs when hormone
            levels are chronically high to decrease responsiveness, while up-regulation occurs in
            circumstances of hormone deficiency to increase sensitivity of target cells.

Pages: 506-507
4. Compare and contrast the mechanisms of action of lipid-soluble vs. water-soluble hormones.
   Answer: Upon reaching their targets, lipid-soluble hormones diffuse through the phospholipid
            bilayer of the target cell membrane and bind to receptors in the cytosol or nucleus.
            The activated receptor turns a gene on or off, thus regulating synthesis of a new
            protein. Water-soluble hormones bind to membrane receptors, activating a G-protein,
            which activates adenylate cyclase, which converts ATP to the 2nd messenger cAMP, which
            activates a protein kinase to regulate enzyme action.

Pages: 508
5. Compare and contrast permissive, synergistic, and antagonistic hormone interactions.
    Answer: · Permissive effect - effect of one hormone requires previous or simultaneous exposure
        to another hormone
        · Synergistic effect - two or more hormones complement each other's actions and both
        are needed for full expression of hormonal effects
        · Antagonistic effect - effect of one hormone opposed by another

Pages: 518
6. Describe the positive feedback loop that illustrates the role of oxytocin in the maintenance of
    labor. Why is this considered a positive feedback loop?
    Answer: Cervix stretched by baby's head; sensory impulses sent to hypothalamus, which sends
        impulses to the neurohypophysis to release OT; OT travels in blood to uterus &
        increases contraction of smooth muscle; baby pushed out; more cervical distension
        Loop is positive feedback because action of effectors enhances (rather than reverses)
        original stimulus.

Pages: 529
7. Explain why the hormones secreted by the adrenal cortex are considered to be essential for
    life.
    Answer: Mineralocorticoids have an important role in maintaining balance of sodium and
        potassium ions, which, in turn, affects balance of other electrolytes, pH, and water.
        Glucocorticoids have an important role in regulating glucose and protein metabolism
        and providing stress resistance.

Pages: 524
8. What is a goiter? Using the appropriate negative feedback loops in your answer, explain how
    goiters develop in both hyposecretion and hypersecretion disorders. In these hyposecretion and
    hypersecretion disorders, would you expect the levels of other hormones involved in the loops
    to be high or low? Why?
    Answer: A goiter is an enlarged thyroid gland. Hyposecretion goiters are usually due to
        insufficient iodide in the diet. Resulting low levels of thyroid hormones cause
        increased TRH and TSH until adequate thyroid activity is restored. Graves' disease
        causes hypersecretion of thyroid hormones by mimicking TSH. Thyroid enlarges, and
        production of thyroid hormones increases. TRH and natural TSH remain low due to
        negative feedback, but false TSH pushes thyroid activity.

Pages: 530
9. Describe in detail the negative feedback loop involving both the adrenal gland and the kidney
    in the control of blood pressure.
    Answer: Low blood pressure triggers release of renin from JG cells of kidney. In blood, renin
        converts angiotensinogen to angiotensin I, which is converted to angiotensin II by ACE
        in lung capillaries. Angiotensin II stimulates secretion of aldosterone by the adrenal
        cortex, which causes the kidneys to save sodium ions. Water is reabsorbed for osmotic
        balance, thus increasing blood volume and blood pressure.

Pages: 531
10. Describe/explain the anti-inflammatory effects of glucocorticoids.
  Answer: · decrease number of mast cells, reducing histamine release
      · stabilize lysosomal membranes to decrease release of destructive enzymes
      · decrease blood capillary permeability
      · decrease phagocytosis

Pages: 543
11. Describe the physiological responses that occur in the General Adaptation Syndrome.
  Answer: · Alarm reaction - sympathetic fight-or-flight response triggered by hypothalamus to increase circulation to essential organs, promote catabolism for ATP production, & decrease non-essential activities
      · Resistance reaction - stimulated by action of CRH, GHRH, and TRH to keep sufficient glucose available for ATP production, to maintain blood pressure and volume, and to maintain pH of extracellular fluid
      · Exhaustion - depletion of hormones kept high during resistance reaction resulting in failure to maintain homeostasis

Pages: 537
12. Compare the causes, effects, and treatments of type I (IDDM) vs. type II (NIDDM) diabetes mellitus.
  Answer: Type I most often develops before age 20, probably due to autoimmune destruction of beta cells of pancreas in genetically susceptible people; effects similar to starvation, since glucose can't enter cells; replace insulin.
      Type II - occurs most often in overweight people over 40; milder effects controllable via diet, exercise, weight loss; possibly due to decreased sensitivity of targets; boost beta cell production or give insulin.

Pages: 531
13. Many people are on long-term therapy with drugs in the glucocorticoid family. What would you expect the long-term effects of these drugs to be? Explain your answer.
  Answer: Chronically high levels of glucose due to increased gluconeogenesis; weakening of bones and muscles due to increased protein catabolism; redistribution of fat to face, back, abdomen; decreased resistance to stress and infection due to anti-inflammatory effects

Pages: 508
14. Explain the mechanism by which effects of low levels of protein hormones are amplified.
  Answer: One hormone molecule binds to membrane receptor, thus activating perhaps 100 G-proteins. Each G-protein activates one adenylate cyclase molecule, which results in production of perhaps thousands of cAMP molecules. Each cAMP activates a protein kinase, which acts on hundreds to thousands of substrate molecules.

Chapter 19   The Cardiovascular System: The Blood

Multiple-Choice

Choose the one alternative that best completes the statement or answers the question.

Pages: 554
1. Blood plasma is made up mostly of:
    A) formed elements
    B) water
    C) plasma proteins
    D) hemoglobin
    E) dissolved ions
    Answer: B

Pages: 564
2. The most abundant of the leukocytes is the:
    A) lymphocyte
    B) basophil
    C) monocyte
    D) neutrophil
    E) eosinophil
    Answer: D

Pages: 559
3. The function of hemoglobin is to:
    A) protect the DNA of erythrocytes
    B) produce red blood cells
    C) produce antibodies
    D) carry oxygen
    E) trigger the cascade of clotting reactions
    Answer: D

Pages: 564
4. The formed elements that are fragments of larger cells called metamegakaryocytes are:
    A) neutrophils
    B) lymphocytes
    C) erythrocytes
    D) thrombocytes
    E) plasma proteins
    Answer: D

Pages: 554
5. The most abundant of the plasma proteins are the:
    A) albumins
    B) hemoglobins
    C) gamma globulins
    D) clotting proteins
    E) alpha globulins
    Answer: A

Pages: 570
6. A person's ABO blood type is determined by antigens present on the:
   A) erythrocytes
   B) platelets
   C) leukocytes
   D) gamma globulins
   E) blood vessel walls
   Answer: A

Pages: 569
7. Fibrinolysis is:
   A) production of a blood clot
   B) dissolving of a blood clot
   C) prevention of blood loss
   D) vasoconstriction
   E) breakdown of hemoglobin
   Answer: B

Pages: 564
8. White blood cells that contain heparin and histamine are:
   A) neutrophils
   B) basophils
   C) eosinophils
   D) lymphocytes
   E) monocytes
   Answer: B

Pages: 562
9. Agranular leukocytes that are phagocytic are the:
   A) neutrophils
   B) monocytes
   C) lymphocytes
   D) eosinophils
   E) all of the above except eosinophils
   Answer: B

Pages: 558
10. "5 million per cubic millimeter" is a value falling within the normal adult range for the number of:
    A) platelets
    B) all leukocytes
    C) erythrocytes
    D) hemoglobin molecules
    E) neutrophils
    Answer: C

Pages: 556
11. In the adult, red bone marrow would normally be found in the:
    A) sternum
    B) diaphysis of the femur
    C) diaphysis of the humerus
    D) irregular bones of the face
    E) all of the above except the bones of the face
    Answer: A

Pages: 559
12. How long do erythrocytes normally stay in circulation?
    A) 10 days
    B) 120 days
    C) one year
    D) only a few hours
    E) one month
    Answer: B

Pages: 572
13. Which of the following would be **TRUE** for a person with Type B blood?
    A) He theoretically could donate to a Type O person.
    B) His own plasma contains anti-B antibodies.
    C) He must be Rh positive.
    D) He theoretically could donate blood to a Type AB person.
    E) He theoretically could receive blood from a Type AB person.
    Answer: D

Pages: 572
14. Which of the following would be **TRUE** for a normal person with anti-A antibodies circulating in his blood?
    A) He could be blood type A.
    B) He could be blood type B.
    C) He could be blood type AB.
    D) He could be either type B or type AB.
    E) It is impossible to know his possible blood types.
    Answer: B

Pages: 561
15. An increased number of reticulocytes in circulation indicates:
    A) an increased demand for erythrocytes
    B) an increased demand for neutrophils
    C) spontaneous clotting is occurring
    D) there must be a bone marrow problem
    E) a transfusion reaction has occurred
    Answer: A

Pages: 558
16. 15 g/100 ml would within the normal adult range for:
    A) amount of total plasma in whole blood
    B) total blood cell count
    C) hemoglobin levels
    D) glucose levels in plasma
    E) levels of clotting proteins in serum
    Answer: C

Pages: 570
17. Heparin works as an anticoagulant by:
    A) preventing the clumping of platelets
    B) acting as an antagonist to vitamin K
    C) working with AT-III to interfere with the action of thrombin
    D) binding calcium ions
    E) enhancing the production of tissue plasminogen activator
    Answer: C

Pages: 560
18. When red blood cells wear out, the iron is saved, and the remainder of the hemoglobin is:
    A) also saved
    B) excreted as bile pigments
    C) rearranged into gamma globulins
    D) broken down by plasmin
    E) used as an anticoagulant
    Answer: B

Pages: 568
19. The conversion of fibrinogen to fibrin is catalyzed by the enzyme:
    A) plasmin
    B) prothrombinase
    C) prothrombin
    D) thrombin
    E) t-PA
    Answer: D

Pages: 566
20. The initial stimulus for the vasoconstriction that occurs in hemostasis is:
    A) prothrombinase
    B) mechanical damage to the vessel
    C) thromboxane A2
    D) plasmin
    E) thrombin
    Answer: B

Pages: 570
21. The anticoagulant produced by mast cells is:
    A) CPD
    B) heparin
    C) histamine
    D) coumarin
    E) antithrombin III
    Answer: B

Pages: 566
22. Platelets initially stick to the wall of a damaged blood vessel because:
    A) exposed collagen fibers make a rough surface to which the platelets are attracted
    B) histamine causes vasoconstriction so that platelets can't fit through the opening
    C) fibrin threads act like glue to hold them there
    D) prothrombinase alters the electrical charge of the vessel wall
    E) the intracellular fluid released by damaged cells in the blood vessel wall has a higher viscosity than plasma
    Answer: A

Pages: 553
23. The total blood volume in an average adult is about:
    A) 8 liters
    B) one liter
    C) 3 liters
    D) 5 liters
    E) 10 liters
    Answer: D

Pages: 533
24. Which of the following values falls within the normal pH range for blood plasma?
    A) 7.00
    B) 7.25
    C) 7.30
    D) 7.42
    E) both 7.30 and 7.42 are within normal range
    Answer: D

Pages: 554
25. **ALL** of the following are important functions of plasma proteins **EXCEPT**:
    A) protection against bacteria and viruses
    B) maintenance of osmotic pressure
    C) protection against blood loss
    D) transportation of steroid hormones
    E) transportation of oxygen
    Answer: E

Pages: 557
26. Myeloblasts are the stem cells that give rise to:
    A) erythrocytes
    B) neutrophils
    C) monocytes
    D) lymphocytes
    E) platelets
    Answer: B

Pages: 563
27. Which of the following types of cells usually has the longest life-span?
    A) erythrocytes
    B) neutrophils
    C) monocytes
    D) lymphocytes
    E) platelets
    Answer: D

Pages: 561
28. Males tend to have a higher hematocrit than females because:
    A) their red blood cells are larger
    B) they have more solutes and proteins in their plasma
    C) testosterone stimulates synthesis of erythropoietin
    D) they usually have a higher dietary intake of factors needed for erythropoiesis
    E) they have red bone marrow in more locations than females
    Answer: C

Pages: 556
29. Erythropoietin is synthesized primarily by the:
    A) red bone marrow
    B) yellow bone marrow
    C) erythrocytes
    D) spleen
    E) kidneys
    Answer: E

Pages: 558
30. Oxygen is transported by red blood cells by binding to:
    A) specific receptors on the plasma membrane
    B) specific receptors within the nucleus of the red blood cell
    C) the beta polypeptide chain of the globin portion of hemoglobin
    D) the polypeptide chain of the heme portion of hemoglobin
    E) the iron ion in the heme portion of hemoglobin
    Answer: E

31. Carbaminohemoglobin forms when:
   A) a faulty gene causes substitution of amino acids in the globin portion of hemoglobin
   B) carbon dioxide replaces iron in the heme portion of hemoglobin
   C) carbon dioxide binds to the globin portion of hemoglobin for transport to the lungs
   D) hemoglobin is broken down for recycling and excretion
   E) red blood cells are in the intermediate erythroblast stage of development
   Answer: C

32. Sickle-cell anemia occurs when:
   A) a faulty gene results in the substitution of amino acids in the globin portion of hemoglobin
   B) intrinsic factor is lacking in gastric juice
   C) there is not enough iron in the diet
   D) industrial pollutants bind to hemoglobin, causing it to change conformation
   E) malaria parasites invade red blood cells
   Answer: A

33. The function of ferritin and hemosiderin is to:
   A) transport iron via facilitated diffusion from the small intestine into the bloodstream
   B) store iron within muscle fibers, liver, and spleen
   C) transport iron through the bloodstream
   D) catalyze the formation of bile pigments
   E) stimulate the formation of hemoglobin in developing erythrocytes
   Answer: B

34. Biliverdin and bilirubin are:
   A) pigments that form during the breakdown of the non-iron portion of heme
   B) compounds that transport iron in the bloodstream
   C) storage molecules for iron in the liver and spleen
   D) precursors to hemoglobin seen in immature red blood cells
   E) inactive forms of clotting factors secreted by platelets
   Answer: A

35. A normoblast is:
   A) any normal, but immature, formed element in circulation
   B) the stem cell from which all formed elements are derived
   C) a nucleated red blood cell found in red marrow, but rarely in blood
   D) the type of cell that produces plasma proteins
   E) an immature form of lymphocyte that produces gamma globulins
   Answer: C

Pages: 561
36. In the negative feedback loop that controls the rate of erythropoiesis, the stress is:
    A) a change in the normal percentages of each type of formed element
    B) increased levels of carbon dioxide in the red bone marrow
    C) increased levels of carbon dioxide in the cerebrospinal fluid
    D) decreased levels of oxygen in the kidney
    E) decreased solute concentration of blood circulating in the hypothalamus
    Answer: D

Pages: 561
37. In the negative feedback loop that controls the rate of erythropoiesis, the target for erythropoietin is the:
    A) kidney
    B) spleen
    C) proerythroblasts in red bone marrow
    D) erythrocytes in circulation
    E) liver
    Answer: C

Pages: 561
38. The function of intrinsic factor is to:
    A) stimulate production of prothrombinase
    B) stimulate differentiation of stem cells in red bone marrow
    C) activate erythropoietin in the bloodstream
    D) break down fibrin clots
    E) aid in absorption of vitamin B_{12} in the small intestine
    Answer: E

Pages: 564
39. Your lab slip reporting your blood work results states, "PMNs 72%." This indicates that:
    A) your hematocrit is very high
    B) your neutrophils are within normal range
    C) you don't have enough granular leukocytes
    D) you have too many immature red blood cells in circulation
    E) your platelets are making up an abnormally high percentage of formed elements
    Answer: B

Pages: 562
40. The names of the granular leukocytes are derived from:
    A) their hemoglobin content
    B) the different pigments present in each type of cell
    C) their functions
    D) their relative sizes
    E) their different responses to hematology stains
    Answer: E

Pages: 562
41. Leukocytes that stain reddish-orange with Wright's stain are:
    A) neutrophils
    B) basophils
    C) eosinophils
    D) lymphocytes
    E) monocytes
    Answer: C

Pages: 562
42. Leukocytes containing large granules that commonly obscure the nucleus and that stain blue-black when stained with Wright's stain are the:
    A) neutrophils
    B) basophils
    C) eosinophils
    D) lymphocytes
    E) monocytes
    Answer: B

Pages: 562
43. The non-granular leukocytes that move constantly from blood to fluids outside the blood vessels and back again are the:
    A) neutrophils
    B) eosinophils
    C) basophils
    D) lymphocytes
    E) all of the above
    Answer: D

Pages: 563
44. The term chemotaxis refers to:
    A) attraction of phagocytic cells to chemicals released by damaged tissues
    B) transport of iron and other chemicals in the blood by plasma proteins
    C) incorporation of iron atoms into developing hemoglobin molecules
    D) transport of gases by hemoglobin
    E) the mechanism by which hormones identify their target cells
    Answer: A

Pages: 563
45. Defensins are:
    A) oxidizing chemicals within neutrophils
    B) a form of gamma globulin
    C) chemicals released by damaged tissues to which phagocytes are attracted
    D) antimicrobial proteins within neutrophils
    E) chemicals contained in platelets that help form the platelet plug
    Answer: D

Pages: 564
46. Eosinophils are most commonly elevated during:
    A) prolonged periods of hypoxia
    B) infections with parasitic worms
    C) bacterial infections
    D) chronic internal bleeding
    E) multiple myeloma
    Answer: B

Pages: 564
47. The cells that produce antibodies are a form of:
    A) neutrophil
    B) monocyte
    C) T lymphocyte
    D) B lymphocyte
    E) erythrocyte
    Answer: D

Pages: 566
48. The role of ADP in hemostasis is to:
    A) immediately stimulate vascular spasm when smooth muscle in vessel walls is damaged
    B) stabilize fibrin clots
    C) make platelets sticky so that they will clump together
    D) initiate the clotting cascade
    E) activate prothrombinase
    Answer: C

Pages: 566
49. Thromboxane A2 is:
    A) an enzyme that stimulates the activity of fibroblasts to help repair damage to blood vessel walls
    B) a growth factor that stimulates the production of platelets
    C) a type of anticoagulant that blocks the action of antithrombin III
    D) an oxygen-storage molecule inside platelets and phagocytes
    E) a prostaglandin in platelets that causes vasoconstriction and platelet activation
    Answer: E

Pages: 568
50. Thromboplastin is:
    A) a protein released by damaged tissues in the extrinsic clotting pathway that initiates the formation of prothrombinase
    B) the inactive form of thrombin
    C) the inactive form of prothrombin
    D) an enzyme that breaks down fibrin
    E) the enzyme formed by both the intrinsic and extrinsic clotting pathways to activate thrombin
    Answer: A

51. The enzyme prothrombinase forms from the combination of:
    A) thromboplastin and calcium ions
    B) prothrombin and calcium ions
    C) clotting factors X and V in the presence of calcium ions
    D) clotting factors I-IV
    E) fibrin and t-PA
    Answer: C

52. Contact with collagen activates clotting factor:
    A) I
    B) II
    C) IV
    D) VIII
    E) XII
    Answer: E

53. The function of platelet-derived growth factor in the process of hemostasis is to stimulate:
    A) an increase in the rate of mitosis of platelets clumping at the site of injury
    B) the production of longer fibrin threads to make the clot more stable
    C) proliferation of vascular endothelial and smooth muscle cells
    D) increased production and release of prothrombin from the liver
    E) increased production and secretion of clotting factors by aggregated platelets
    Answer: C

54. People suffering from disorders that prevent absorption of fat from the intestine may suffer uncontrolled bleeding because:
    A) the fat-soluble vitamin K cannot be absorbed, so levels of prothrombin and other clotting factors drop
    B) fat droplets in the blood normally help platelets and other formed elements stick together during hemostasis
    C) vitamin B_{12} cannot be absorbed, so there are inadequate formed elements present to produce a clot
    D) fatty acids are necessary for the complete synthesis of fibrin
    E) fat prevents the complete inactivation of plasmin
    Answer: A

55. Syneresis is:
    A) the movement of leukocytes from the blood into interstitial fluid
    B) the differentiation of stem cells into more mature forms of cells
    C) the interruption in blood flow created by a moving blood clot that lodges in a small vessel
    D) release of clotting factors and vasoconstrictors from platelets
    E) the contraction of fibrin threads as platelets pull on them to pull the edges of a damaged vessel closer together
    Answer: E

Pages: 570
56. Low doses of aspirin are sometimes prescribed to reduce the risk of thrombus formation because aspirin prevents the formation of:
    A) thromboxane A2
    B) prostacyclin
    C) thromboplastin
    D) fibrinogen
    E) prothrombinase
    Answer: A

Pages: 571
57. A person with the genotype $I^A i$ has:
    A) A isoantigens (only) on red blood cells
    B) O isoantigens (only) on red blood cells
    C) anti-B isoantibodies in plasma (only)
    D) both A and O isoantigens on red blood cells
    E) both A isoantigens on RBCs and anti-B isoantibodies in plasma
    Answer: E

Pages: 572
58. Type O is considered the theoretical universal:
    A) recipient because there are no A or B isoantigens on RBCs
    B) donor because there are no A or B isoantigens on RBCs
    C) recipient because there are no anti-A or anti-B antibodies in plasma
    D) donor because there are no anti-A or anti-B antibodies in plasma
    E) donor because there are no A or B isoantigens on RBCs, nor are there anti-A or anti-B isoantibodies in plasma
    Answer: B

Pages: 573
59. The symptoms of hemolytic disease of the newborn occur because:
    A) the baby has a faulty gene that makes its hemoglobin unable to bind oxygen
    B) the baby begins making antibodies to its own A or B isoantigens
    C) anti-Rh antibodies produced by the mother pass the placenta into the bloodstream of the fetus
    D) the baby is premature and is unable to produce enough plasma proteins to keep osmotic pressure at the correct levels
    E) the mother took high doses of aspirin during the last trimester of development
    Answer: C

Pages: 573
60. The purpose for giving RhoGAM to women who have just delivered a child or who have had a miscarriage or abortion is to:
    A) stimulate contraction of uterine smooth muscle to expel all uterine contents
    B) trigger the clotting cascade to prevent excessive bleeding
    C) block the stimulation of pain receptors
    D) stimulate erythropoiesis to replace red blood cells lost during delivery
    E) block recognition of any fetal red blood cells by the mother's immune system
    Answer: E

61. The normal number of thrombocytes in circulation is:
    A) about 5 million per cubic millimeter
    B) about 7000 per cubic millimeter
    C) about 300,000 per cubic millmeter
    D) about 25% of all formed elements
    E) about 25,000 per cubic millimeter
    Answer: C

62. Pernicious anemia results from:
    A) destruction of bone marrow
    B) insufficient iron in the diet
    C) presence of antibodies that cause hemolysis
    D) abnormal hemoglobin structure
    E) lack of intrinsic factor in the stomach
    Answer: E

63. The antibiotic chloramphenicol may induce aplastic anemia, which means that:
    A) blood cells are abnormally shaped
    B) red bone marrow fails to produce new blood cells
    C) normal numbers of red blood cells are manufactured, but lack hemoglobin
    D) too many red blood cells are produced
    E) B lymphocytes are produced that make antibodies to a person's own red blood cells
    Answer: B

64. A disease in which mature leukocytes accumulate because they do not die at the end of their normal life spans is:
    A) acute leukemia
    B) chronic leukemia
    C) infectious mononucleosis
    D) thalassemia
    E) polycythemia
    Answer: B

65. MHC antigens are:
    A) determined by the combination of I-gene alleles
    B) viral antigens whose presence indicates infectious mononucleosis
    C) proteins encoded by genes present in nucleated cells that must be matched for successful tissue transplantation
    D) the antigens that are attacked in hemolytic disease of the newborn
    E) the molecules to which emigrating leukocytes attach before leaving the bloodstream
    Answer: C

Pages: 564
66. Which of the following indicates a normal differential count in a healthy adult?
    A) 50% neutrophils, 30% lymphocytes, 15% monocytes, 4% eosinophils, 1% basophils
    B) 65% lymphocytes, 20% neutrophils, 10% monocytes, 4% eosinophils, 1% basophils
    C) 65% neutrophils, 25% lymphocytes, 6% eosinophils, 2% monocytes, 2% basophils
    D) 65% neutrophils, 25% lymphocytes, 6% monocytes, 3% eosinophils, 1% basophils
    E) 50% lymphocytes, 30% neutrophils, 10% eosinophils, 8% monocytes, 2% basophils
    Answer: D

Pages: 569
67. The most common type of hemophilia is a hereditary deficiency of:
    A) clotting factor VIII
    B) heparin
    C) intrinsic factor
    D) thromboplastin
    E) normal hemoglobin
    Answer: A

Pages: 564
68. If 100 leukocytes are counted in a normal adult blood sample, how many monocytes would you expect to see?
    A) 60-70
    B) 20-25
    C) 3-8
    D) 2-4
    E) one
    Answer: C

Pages: 568
69. Which of the following events occurs first in the intrinsic pathway of the blood clotting cascade?
    A) activation of factor IX
    B) formation of thrombin
    C) formation of prothrombinase
    D) activation of factor XII
    E) activation of factor XIII
    Answer: D

Pages: 568
70. Which of the following events occurs first in the extrinsic and common pathways of the blood clotting cascade?
    A) combination of factors IV, V and X
    B) activation of factor I
    C) activation of factor X
    D) activation of factor XII
    E) activation of factor VII
    Answer: E

True-False

Write T if the statement is true and F if the statement is false.

Pages: 553
1. The normal temperature of blood is 37° C (98.6° F).
   Answer: False

Pages: 554
2. Normally, blood is about 45% formed elements and 55% plasma.
   Answer: True

Pages: 554
3. Most plasma proteins are synthesized by the bone marrow.
   Answer: False

Pages: 554
4. Gamma globulins are the most abundant type of plasma protein.
   Answer: False

Pages: 556
5. In children, erythropoiesis normally occurs in the red bone marrow, the liver, and the spleen.
   Answer: False

Pages: 558
6. Mature erythrocytes cannot reproduce.
   Answer: True

Pages: 559
7. Mature erythrocytes remain in circulation for about 120 days.
   Answer: True

Pages: 561
8. The usual stimulus for release of erythropoietin is nerve impulses from the sympathetic nervous system.
   Answer: False

Pages: 561
9. A value of 45% would be a normal adult hematocrit.
   Answer: True

Pages: 561
10. Red blood cells enter the bloodstream in the form of reticulocytes.
    Answer: True

Pages: 562
11. Lymphocytes are phagocytic cells.
    Answer: False

12. The average adult white blood cell count is about 5,000,000 per cubic millimeter.
Answer: False

13. Basophils develop into mast cells when they leave the bloodstream.
Answer: True

14. T cells, B cells, and natural killer cells are the major types of lymphocytes.
Answer: True

15. Lymphocytes are normally the most abundant type of leukocyte.
Answer: False

Short Answer

Write the word or phrase that best completes each statement or answers the question.

1. The buffy coat of centrifuged whole blood is formed by the _____.
Answer: leukocytes

2. Plasma minus its clotting proteins is called _____.
Answer: serum

3. Gamma globulins are also called _____.
Answer: antibodies (immunoglobulins)

4. The most common of the plasma proteins are the _____.
Answer: albumins

5. Gamma globulins are produced by _____.
Answer: plasma cells

6. The process by which formed elements of the blood are produced is called _____.
Answer: hem(at)opoiesis

7. Erythropoietin is a hormone produced mainly by the _____.
Answer: kidneys

Pages: 559
8. Carbaminohemoglobin forms when hemoglobin combines with _____.
    Answer: carbon dioxide

Pages: 558
9. Red blood cells are highly specialized to transport _____.
    Answer: oxygen

Pages: 560
10. During the breakdown of red blood cells, iron is removed from hemoglobin and transported in the bloodstream by a plasma protein called _____.
    Answer: transferrin

Pages: 560
11. During the breakdown of red blood cells, the non-iron portion of heme is converted into a green pigment called _____, then into an orange pigment called _____.
    Answer: biliverdin; bilirubin

Pages: 561
12. The percentage of red blood cells in whole blood is called the _____.
    Answer: hematocrit

Pages: 561
13. Intrinsic factor aids in the absorption of _____ by the small intestine.
    Answer: vitamin B_{12}

Pages: 562
14. The three types of granular leukocytes are the _____, the _____, and the _____.
    Answer: neutrophils; basophils; eosinophils

Pages: 563
15. The process by which phagocytes are attracted to chemicals released by microbes or inflamed tissue is called _____.
    Answer: chemotaxis

Pages: 564
16. Granular leukocytes that are thought to produce enzymes that are antagonistic to the mediators of allergic reactions are the _____.
    Answer: eosinophils

Pages: 565
17. The stoppage of bleeding is called _____.
    Answer: hemostasis

Pages: 569
18. The enzyme _____ digests the fibrin threads of a blood clot.
    Answer: plasmin

Pages: 570
19. A common anticoagulant that acts as an antagonist to vitamin K is _____.
    Answer: warfarin (Coumarin)

20. Clotting in an unbroken blood vessel is called _____.
    Answer: thrombosis

21. The rupture of red blood cells is called _____.
    Answer: hemolysis

22. Selectin and integrins are examples of _____ molecules that assist in the _____ of white blood cells from the bloodstream into the interstitial fluid.
    Answer: adhesion; emigration

23. The target cells for erythropoietin are _____, which are located in the _____.
    Answer: proerythroblasts; red bone marrow

24. The stimulus for release of erythropoietin is _____ in the kidney.
    Answer: hypoxia

25. White blood cells that produce histamine and heparin are the _____.
    Answer: basophils

26. Agranular leukocytes that develop into macrophages outside the bloodstream are the _____.
    Answer: monocytes

27. Among the various ethnic groups in the United States, the two most common blood types are type _____ and type _____.
    Answer: O; A

28. People with type A blood have isoantigens called _____ on their red blood cells and isoantibodies called _____ in their plasma.
    Answer: A antigens; anti-B antibodies

29. During the platelet release reaction, the prostaglandin _____ activates platelets and acts as a vasoconstrictor.
    Answer: thromboxane A2

30. Fibrinogen is converted into fibrin by the enzyme _____.
    Answer: thrombin

Matching

Choose the item from Column 2 that best matches each item in Column 1.

1. transports most oxygen and some carbon dioxide in blood

erythrocytes

2. fragments of cells that form a plug and release vasoconstrictors during hemostasis

thrombocytes

3. phagocytic granular leukocytes

neutrophils

4. intensify inflammatory response via release of histamine and serotonin

basophils

5. combat effects of histamine in allergic reactions

eosinophils

6. develop into plasma cells that secrete antibodies

B lymphocytes

7. attack viruses, cancer cells, and transplanted cells

T lymphocytes

8. develop into macrophages outside of circulation

monocytes

9. form of red blood cell that enter circulation from the bone marrow; serve as indicators of rate of erythropoiesis

reticulocytes

10. young neutrophils with rod-shaped nuclei

band cells

Essay

Write your answer in the space provided or on a separate sheet of paper.

Pages: 553
1. List and briefly describe the functions of blood.
   Answer: · Transportation - carries oxygen and carbon dioxide between cells and lungs; also transports nutrients, wastes, hormones, and heat
   · Regulation - helps maintain pH via buffers, body temperature via properties of water in plasma, and water balance via plasma proteins
   · Protection - via clotting, phagocytes, and complement

Pages: 558
2. Describe the structure of a typical mature erythrocyte, and relate that structure to the cell's function.
   Answer: · Cells 7-8 μ m in diameter; biconcave disks; no nucleus or mitochondria; 280 million molecules of hemoglobin
   · lack of nucleus gives flexibility for flow through small vessels; biconcave shape increases surface area; no mitochondria means no aerobic respiration, so no oxygen used; much hemoglobin allows much oxygen to be carried

Pages: 564
3. What is a differential white blood cell count, and why would it be performed? What are the normal values for a differential count? What is the usual reason for a high neutrophil count?
   Answer: A differential count counts WBCs by type, and calculates the percentage of the total WBC count to determine problems indicated by high or low levels of each type based on cell function. Neutrophils (60-70%, high in bacterial infection); lymphocytes (20-25%); monocytes (3-8%); eosinophils (2-4%); basophils (0.5-1%)

Pages: 574
4. Define anemia, and describe the possible mechanisms by which anemia develops.
   Answer: Anemia is a decreased oxygen carrying capacity of blood. This may be caused by too few RBCs (loss due to poisons, parasites, or immune destruction or decreased production) or by too little normal hemoglobin due to nutritional deficiencies (Fe, amino acids, various vitamins) or genetic problems (e.g., sickle cell anemia).

Pages: 567
5. Why does damaged endothelium present an increased risk of blood clotting?
   Answer: Blood may come in contact with collagen in the surrounding basal lamina, which activates clotting factor XII, which ultimately leads to the formation of fibrin clots. Platelets are also damaged by contact with damaged endothelium.

Pages: 566
6. Describe the activities of platelets during hemostasis.
   Answer: Platelets stick to damaged blood vessel walls (adhesion) and become activated. Contents of platelet granules (thromboxane A2, serotonin, ADP, etc.) are released, which activate other platelets and enhance vasoconstriction and platelet aggregation. Large numbers of platelets form a plug at the site of damage.

Pages: 567
7. Describe the process of coagulation of blood.
    Answer: 1) Formation of prothrombinase via extrinsic pathway in which thromboplastin from
    damaged cells outside vessels triggers the cascade, or via the intrinsic pathway in
    which damaged endothelial cells in vessel walls trigger cascade.
    2) Prothrombinase and calcium ions catalyze conversion of inactive prothrombin
    (produced by liver, present in blood) to active thrombin.
    3) Thrombin, in positive feedback, increases prothrombinase formation and activation
    of platelets, and converts soluble fibrinogen into insoluble fibrin threads around
    which formed elements are trapped.

Pages: 561
8. Describe the negative feedback loop that controls the rate of erythropoiesis. Under what
    circumstances would you expect the rate of erythropoiesis to be increased? How would it be
    possible to tell if the rate of erythropoiesis is elevated?
    Answer: Hypoxia in kidney leads to secretion of erythropoietin, which targets proerythroblasts
    in red marrow to mature into reticulocytes, which enter circulation to increase oxygen
    carrying capacity of blood. The rate should be increased in any form of anemia
    (reduced oxygen carrying capacity of blood) or when oxygen levels in external
    environment are low (e.g., high altitudes). High levels of reticulocytes in
    circulation indicate an increase in erythropoiesis.

Pages: 571
9. A woman who is blood type A has a baby whose blood type is O. Her husband insists that the
    child cannot be his because he is blood type B. Is his claim justified or not? What are the
    possible blood types of the baby's maternal grandparents? Explain your answer.
    Answer: The baby could be the husband's because the woman could be genotype $I^A i$ and the
    husband could be genotype $I^B i$, and each could have contributed the "i" allele,
    resulting in the child's O blood type. In order for the woman to be Type A with the
    heterozygous genotype, her parents could have both had the same genotype as the woman,
    or one could have been $I^A I^A$ and the other $I^A i$, or one could have been $I^A I^A$
    and the other ii.

Pages: 571
10. A patient who is blood type B is inadvertently transfused with type AB blood. Explain the
    specific interactions, if any, that will occur between the donor's and recipient's blood.
    Answer: The anti-A isoantibodies in the recipient's blood will bind to the A antigens on the
    donated red blood cells, causing them to undergo hemolysis. The donated blood contains
    neither anti-A nor anti-B antibodies, so the recipient's cells are not affected by the
    immune interaction. Ruptured RBCs release their hemoglobin, which may damage the
    kidneys.

Pages: 561-567
11. When malaria parasites invade red blood cells, they cause the RBCs to become sticky, and
    ultimately cause them to rupture. Based on this information, what would you expect some of the
    symptoms of malaria to be? Explain your answers.
    Answer: It is not necessary for answers to be absolutely correct, only to show appropriate
    thought processes. Students may suggest 1) intravascular clotting leading to ischemia,
    2) hemolytic anemia leading to increased erythropoiesis and high reticulocyte count
    and possible kidney failure, 3) increased activity of phagocytes in liver and spleen
    to handle ruptured RBCs. Other answers may be acceptable based on student reasoning.

Multiple-Choice

Choose the one alternative that best completes the statement or answers the question.

Pages: 588
1. Blood flows from the pulmonary veins into the:
    A) pulmonary arteries
    B) right atrium
    C) lungs
    D) left atrium
    E) left ventricle
    Answer: D

Pages: 587
2. Blood flows from the right atrium into the:
    A) right ventricle
    B) superior vena cava
    C) left atrium
    D) pulmonary trunk
    E) pulmonary veins
    Answer: A

Pages: 583
3. The bicuspid valve is located between the:
    A) right ventricle and the aorta
    B) right ventricle and the pulmonary trunk
    C) left atrium and the left ventricle
    D) right and left atria
    E) right and left ventricles
    Answer: C

Pages: 585
4. There is a semilunar valve between the:
    A) right ventricle and the aorta
    B) right ventricle and the pulmonary trunk
    C) left atrium and left ventricle
    D) right atrium and right ventricle
    E) left ventricle and the pulmonary trunk
    Answer: B

Pages: 587
5. Blood flows from the superior vena cava into the:
    A) right atrium
    B) inferior vena cava
    C) left atrium
    D) aorta
    E) pulmonary trunk
    Answer: A

6. The circumflex artery and the anterior interventricular artery are branches of the:
A) left coronary artery
B) right coronary artery
C) great cardiac vein
D) coronary sinus
E) pulmonary trunk
Answer: A

7. Blood flows from the great cardiac vein into the:
A) left coronary artery
B) aorta
C) superior vena cava
D) pulmonary trunk
E) coronary sinus
Answer: E

8. Which of the following lists the elements of the heart's conduction system in the correct order?
A) SA node, AV bundle, bundle branches, AV node, conduction myofibers
B) AV node, SA node, AV bundle, bundle branches, conduction myofibers
C) SA node, AV node, AV bundle, bundle branches, conduction myofibers
D) conduction myofibers, AV bundle, bundle branches, AV node, SA node
E) SA node, AV bundle, AV node, bundle branches, conduction myofibers
Answer: C

9. During the normal cardiac cycle, the ventricles contract when they are directly stimulated by the:
A) SA node
B) AV node
C) conduction myofibers
D) baroreceptors
E) vagus nerve
Answer: C

10. During the normal cardiac cycle, the atria contract when they are directly stimulated by the:
A) SA node
B) AV node
C) conduction myofibers
D) baroreceptors
E) vagus nerve
Answer: A

Pages: 594
11. The atrioventricular valves close when the:
    A) SA node fires
    B) atria contract
    C) vagus nerve stimulates them
    D) ventricles relax
    E) ventricles contract
    Answer: E

Pages: 581
12. The myocardium is made of:
    A) smooth muscle
    B) cardiac muscle
    C) skeletal muscle
    D) endothelium
    E) dense connective tissue
    Answer: B

Pages: 581
13. The serous membrane covering the heart is called the:
    A) mediastinum
    B) ligamentum arteriosum
    C) endocardium
    D) myocardium
    E) pericardium
    Answer: E

Pages: 583
14. The function of the chordae tendineae is to:
    A) pull the walls of the ventricles inward during contraction
    B) open the semilunar valves
    C) open the AV valves
    D) prevent inversion of the AV valves during ventricular systole
    E) hold the heart in place within the mediastinum
    Answer: D

Pages: 590
15. The initiation of the heart beat is the responsibility of the:
    A) cardiovascular center
    B) baroreceptors
    C) vagus nerve
    D) SA node
    E) fossa ovalis
    Answer: D

Pages: 591
16. The decrease in speed of conduction from the AV node through the AV bundle results in:
   A) failure of the ventricles to contract
   B) adequate time for the ventricles to fill
   C) delayed opening of the atrioventricular valves
   D) the sensation of a skipped beat
   E) a decrease in the rate of blood flow from the atria to the ventricles
   Answer: B

Pages: 597
17. If acetylcholine is applied to the heart, but cardiac output is to remain constant, which of
   the following would have to happen?
   A) stroke volume must increase
   B) venous return must decrease
   C) force of contraction must decrease
   D) rate of conduction of impulses through the AV bundle must increase
   E) the oxygen content of blood in the coronary circulation must increase
   Answer: A

Pages: 597
18. The Frank-Starling Law of the Heart states that:
   A) the heart is dependent upon the autonomic nervous system for a stimulus to contract
   B) the heart contracts to the fullest extent possible for the conditions, or not at all
   C) cardiac output equals heart rate times stroke volume
   D) the absolute refractory period for the heart must be longer than the duration of contraction
      for efficient heart functioning
   E) a greater force of contraction can occur if the heart muscle is stretched first
   Answer: E

Pages: 597
19. If the cardiostimulatory center is stimulated and the cardioinhibitory center is inhibited,
   assuming all other conditions remain constant, then:
   A) heart rate decreases
   B) cardiac output decreases
   C) heart rate increases
   D) stroke volume increases
   E) both heart rate and cardiac output increase
   Answer: E

Pages: 608
20. A strong stimulus applied to the heart during the relative refractory period results in:
   A) immediate death
   B) spontaneous opening of the heart valves
   C) the sensation of a skipped beat
   D) no apparent effect
   E) complete emptying of the ventricles (ESV = 0)
   Answer: C

Pages: 589

21. Afferent nerve fibers associated with the heart transmit information about:
    A) pain
    B) blood pressure
    C) heart rate
    D) blood viscosity
    E) force of contraction
    Answer: A

Pages: 587

22. From the coronary sinus, blood normally flows next into the:
    A) aorta
    B) pulmonary trunk
    C) right atrium
    D) great cardiac vein
    E) left atrium
    Answer: C

Pages: 591

23. The action potential normally travels most slowly from the:
    A) SA node to the AV node
    B) AV node through the AV bundle
    C) AV node to the SA node
    D) conduction myofibers to the cardiac muscle cells
    E) the conduction rate is constant throughout the cardiac conduction system
    Answer: B

Pages: 591

24. A heartbeat is normally initiated when:
    A) a nerve impulse arrives from the vasomotor center in the brain
    B) a critical volume of blood fills the ventricles
    C) enough sodium and calcium ions leak into the cells of the SA node to reverse their resting potentials
    D) enough potassium ions leak out of the cells of the SA node to reverse their resting potentials
    E) the chordae tendineae recoil after being stretched
    Answer: C

Pages: 596

25. The typical heart sounds are made by the:
    A) vibration of the chordae tendineae
    B) flow of blood into the coronary arteries
    C) opening of the valves
    D) closing of the valves
    E) recoil of the aorta and pulmonary trunk
    Answer: D

Pages: 587
26. Blood in pulmonary capillaries that has just picked up oxygen will go next into:
    A) pulmonary arteries to the left atrium
    B) pulmonary arteries to the right atrium
    C) pulmonary veins to the left atrium
    D) pulmonary veins to the right atrium
    E) pulmonary veins to the superior vena cava
    Answer: C

Pages: 593
27. On an ECG, depolarization of the ventricles is represented by the:
    A) P wave
    B) T wave
    C) QRS complex
    D) P-Q interval
    E) S-T segment
    Answer: C

Pages: 593
28. On an ECG, depolarization of the atria is represented by the:
    A) P wave
    B) T wave
    C) QRS complex
    D) P-Q interval
    E) S-T segment
    Answer: A

Pages: 593
29. On an ECG, repolarization of the ventricles is represented by the:
    A) P wave
    B) T wave
    C) QRS complex
    D) P-Q interval
    E) S-T segment
    Answer: B

Pages: 580
30. The apex of the heart is formed by the:
    A) arch of the aorta
    B) upper border of the right atrium
    C) upper border of the left atrium
    D) tip of the right ventricle
    E) tip of the left ventricle
    Answer: E

31. Increased firing of impulses by the sympathetic nervous system would cause:
    A) an increase in force of contraction of the heart
    B) a shorter absolute refractory period in the SA node
    C) vasoconstriction in coronary circulation
    D) an increase in cardiac output
    E) all of the above except vasoconstriction in coronary circulation
    Answer: E

32. The pericardial cavity normally contains:
    A) air
    B) blood
    C) lymph
    D) serous fluid
    E) areolar connective tissue
    Answer: D

33. The function of intercalated discs is to:
    A) initiate the heart beat
    B) anchor the heart in place within the mediastinum
    C) prevent the inversion of valves
    D) provide a mechanism for rapid conduction of action potentials among myofibers
    E) provide an anchoring point for chordae tendineae
    Answer: D

34. Cardiac muscle fibers remain depolarized longer than skeletal muscle fibers because:
    A) voltage-gated sodium ion channels close more quickly to trap sodium ions inside longer
    B) calcium ions enter the cytosol from the extracellular fluid to contribute more positive charges slightly after the sodium ions have entered
    C) voltage-gated potassium ion channels open at the same time as sodium ion channels, allowing more positively charged potassium ions to enter
    D) it takes longer to reach threshold, and the duration of depolarization is directly proportional to the time it takes to reach threshold
    E) because the intercalated discs are very thick relative to the rest of the sarcolemma, it takes longer for potassium ions to exit the cell to cause repolarization
    Answer: B

35. Repolarization of cardiac myofibers occurs due to:
    A) opening of voltage-gated fast sodium ion channels
    B) opening of voltage-gated slow calcium ion channels
    C) opening of voltage-gated potassium ion channels
    D) release of calcium ions from the sarcoplasmic reticulum
    E) opening of chemically-gated potassium ion channels responsive to acetylcholine
    Answer: C

Pages: 598
36. The force of cardiac muscle contraction is influenced primarily by the:
    A) number of calcium ions entering the cells through slow channels
    B) rate at which sodium ions diffuse into the cells
    C) number of calcium ions that can be stored in the sarcoplasmic reticulum
    D) duration of the absolute refractory period
    E) up- and down-regulation of beta adrenergic receptors on the cells
    Answer: A

Pages: 594
37. The dicrotic wave on the aortic pressure curve is due to:
    A) the force of the chordae tendineae pulling on the cusps of the aortic semilunar valve
    B) the closing of the AV valves during ventricular systole
    C) rebound of blood off the closed cusps of the aortic semilunar valves
    D) stretching of the elastic tissue in the wall of the aorta
    E) diversion of blood into the coronary arteries
    Answer: C

Pages: 594
38. Atrioventricular valves open when:
    A) the chordae tendineae contract
    B) they are stimulated by the AV node
    C) ventricular pressure falls below atrial pressure
    D) atrial pressure falls below ventricular pressure
    E) the papillary muscles contract
    Answer: C

Pages: 594
39. Diastasis occurs when:
    A) all valves are closed
    B) the AV valves first open
    C) the semilunar valves first open
    D) the AV valves have opened, but atrial systole has not yet occurred
    E) ventricular pressure rises above atrial pressure
    Answer: D

Pages: 596
40. The second heart sound (dupp) is created by the:
    A) closing of the atrioventricular valves
    B) opening of the atrioventricular valves
    C) closing of the semilunar valves
    D) opening of the semilunar valves
    E) vibration of the chordae tendineae during ventricular systole
    Answer: C

41. The two ventricles develop different pressures because:
    A) they hold different volumes of blood
    B) the diameters of the vessels they are pumping into are different
    C) one ventricle wall is made up of more cardiac muscle than the other
    D) the valves on one side of the heart don't close as securely as those on the other side
    E) the rate of ventricular filling is different
    Answer: C

42. All the valves are closed during:
    A) atrial systole
    B) ventricular ejection
    C) rapid ventricular filling
    D) diastasis
    E) isovolumetric ventricular contraction
    Answer: E

43. Narrowing of a heart valve is called:
    A) stenosis
    B) insufficiency
    C) heart block
    D) cardiac tamponade
    E) infarction
    Answer: A

44. If heart rate increases to very high levels, then:
    A) the autonomic nervous system will release more epinephrine to the SA node to stabilize the heart rate
    B) stroke volume increases to keep cardiac output constant
    C) the oxygen content of blood falls to levels insufficient to maintain cardiac activity
    D) end-diastolic volume drops because ventricular filling time is so short
    E) end-systolic volume increases because the valves are open for only a short time
    Answer: D

45. A substance that acts as a negative inotropic agent will:
    A) increase stroke volume by promoting inflow of calcium ions during contraction of cardiac muscle fibers
    B) decrease stroke volume by preventing inflow of calcium ions during contraction of cardiac muscle fibers
    C) decrease stroke volume by promoting inflow of calcium ions during contraction of cardiac muscle fibers
    D) increase stroke volume by preventing inflow of calcium ions during contraction of cardiac muscle fibers
    E) decrease stroke volume by preventing outflow of calcium ions during contraction of cardiac muscle fibers
    Answer: B

Pages: 598
46. A substance that acts as a positive inotropic agent is:
    A) acetylcholine
    B) epinephrine
    C) any calcium channel blocker
    D) any beta blocker
    E) carbon dioxide
    Answer: B

Pages: 598
47. The term **afterload** refers to:
    A) end-diastolic volume
    B) end-systolic volume
    C) the pressure that must be overcome before semilunar valves can open
    D) the pressure in blood vessels necessary to cause the semilunar valves to close
    E) the maximum possible cardiac output above resting cardiac output
    Answer: C

Pages: 598
48. Pulmonary edema results from:
    A) failure of the left ventricle
    B) failure of the right ventricle
    C) narrowing of the pulmonary trunk
    D) development of an ectopic pacemaker
    E) low blood pressure in the pulmonary arteries
    Answer: A

Pages: 599
49. The cardiac accelerator nerves extend from the:
    A) medulla oblongata to the thoracic region of spinal cord
    B) thoracic region of the spinal cord to the SA node and other areas of the heart
    C) medulla oblongata to the heart
    D) baroreceptors to the cardiovascular center in the medulla oblongata
    E) origin of the vagus nerve to the heart
    Answer: B

Pages: 599
50. Increased stimulation of the heart by the cardiac accelerator nerves causes:
    A) stimulation by acetylcholine of muscarinic receptors on the SA node and cardiac muscle fibers of the ventricles
    B) stimulation by norepinephrine of the SA node and of the beta receptors on the cardiac muscle fibers of the ventricles
    C) stimulation by norepinephrine of the SA node, but no effect on ventricular myofibers
    D) stimulation by acetylcholine of nicotinic receptors on the SA node and cardiac muscle fibers of the ventricles
    E) stimulation by norepinephrine of the SA node and of alpha receptors on the cardiac muscle fibers of the ventricles
    Answer: B

Pages: 599
51. Stimulation of the heart by autonomic nerve fibers traveling with the vagus nerve causes:
    A) increased heart rate and increased ventricular contractility
    B) decreased heart rate and decreased ventricular contractility
    C) increased heart rate and no change in ventricular contractility
    D) decreased heart rate and no change in ventricular contractility
    E) decreased heart rate and increased ventricular contractility
    Answer: D

Pages: 604
52. In the fetal heart, the foramen ovale is located:
    A) in the wall of the aorta
    B) in the interventricular septum
    C) in the interatrial septum
    D) between the openings of the superior and inferior venae cavae
    E) between the pulmonary trunk and the aorta
    Answer: C

Pages: 588
53. Shortly after birth, the ductus arteriosus closes leaving the remnant known as the:
    A) ligamentum arteriosum
    B) ductus venosus
    C) fossa ovalis
    D) truncus arteriosus
    E) bulbus cordis
    Answer: A

Pages: 592
54. Which of the following lists the electrical activity of the heart in the correct sequence?
    A) plateau, rapid depolarization, repolarization, refractory period
    B) refractory period, repolarization, rapid depolarization, plateau
    C) repolarization, rapid depolarization, refractory period, plateau
    D) rapid depolarization, plateau, repolarization, refractory period
    E) rapid depolarization, repolarization, plateau, refractory period
    Answer: D

Pages: 594
55. If the heart rate is about 75 beats per minute, each cardiac cycle takes about:
    A) 10 seconds
    B) 0.1 seconds
    C) 7.5 milliseconds
    D) 5 milliseconds
    E) 0.8 seconds
    Answer: E

Pages: 601

56. Lack of regular exercise increases the risk of heart disease because:
   A) the valves lose tone, and don't close securely
   B) the threshold of the SA node is increased, so the duration of the cardiac cycle is lengthened
   C) pumping efficiency decreases because as stroke volume decreases, heart rate must increase to maintain cardiac output
   D) blood flows too slowly over time to provide adequate oxygen to the myocardium
   E) respiratory rate is too low to take in sufficient oxygen for optimal oxygenation of the blood
   Answer: C

Pages: 602

57. A decreased risk of heart disease is associated most directly with:
   A) increased levels of low-density lipoproteins in plasma
   B) increased levels of high-density lipoproteins in plasma
   C) increased levels of very low-density lipoproteins in plasma
   D) decreased levels of triglycerides in plasma
   E) decreased levels of high-density lipoproteins in plasma
   Answer: B

Pages: 602

58. The function of high-density lipoproteins is to:
   A) deliver cholesterol to cells that need it
   B) transport triglycerides from liver cells to adipose cells
   C) remove excess cholesterol from body cells and transport it to the liver
   D) convert LDLs to VLDLs
   E) transport cholesterol from the gastrointestinal tract into the blood
   Answer: C

Pages: 602

59. The function of low-density lipoproteins is to:
   A) deliver cholesterol to cells that need it
   B) transport triglycerides from liver cells to adipose cells
   C) remove excess cholesterol from body cells and transport it to the liver
   D) convert HDLs to VLDLs
   E) transport cholesterol from the gastrointestinal tract into the blood
   Answer: A

Pages: 604

60. Contractions of the primitive heart normally begin:
   A) as soon as the egg is fertilized
   B) when labor contractions begin
   C) at the end of the first trimester
   D) at the end of the second trimester
   E) by day 22 of gestation
   Answer: E

61. Cellular injury caused by formation of oxygen free radicals occurs in:
    A) reperfusion damage
    B) rheumatic heart disease
    C) valvular stenosis
    D) first degree AV block
    E) coarctation of the aorta
    Answer: A

62. Normal resting cardiac output for an average adult is approximately:
    A) 70 ml/min
    B) one liter/min
    C) 2 liters/min
    D) 5 liters/min
    E) 10 liters/min
    Answer: D

63. The pain in the chest that usually accompanies myocardial ischemia is called:
    A) cor pulmonale
    B) cardiac tamponade
    C) angina pectoris
    D) atherosclerosis
    E) rheumatic heart disease
    Answer: C

64. What is happening during the plateau phase of cardiac depolarization?
    A) Voltage-gated potassium ion channels open to allow potassium ions to diffuse out.
    B) Sodium ions block voltage-gated calcium ion channels.
    C) Acetylcholine is hyperpolarizing the SA node.
    D) Nothing - all ion gates and valves are closed.
    E) Calcium ions enter the cytosol of cardiac myofibers from the sarcoplasmic reticulum and
       through voltage-gated slow channels.
    Answer: E

65. The longest phase of the cardiac cycle is:
    A) diastasis
    B) ventricular systole
    C) the relaxation period
    D) rapid ventricular filling
    E) atrial systole
    Answer: C

True-False

Write T if the statement is true and F if the statement is false.

Pages: 581
1. The pericardial cavity normally contains blood.
   Answer: False

Pages: 581
2. The myocardium is responsible for the pumping action of the heart.
   Answer: True

Pages: 583
3. The tricupsid valve is located between the left ventricle and the aorta.
   Answer: False

Pages: 585
4. The pulmonary semilunar valve is connected to papillary muscles by chordae tendineae.
   Answer: False

Pages: 585
5. Atrioventricular valves open when ventricular pressure is higher than atrial pressure.
   Answer: False

Pages: 587
6. The pulmonary veins return oxygenated blood to the left atrium.
   Answer: True

Pages: 588
7. The circumflex artery is a branch of the right coronary artery.
   Answer: False

Pages: 589
8. The coronary sinus normally contains air.
   Answer: False

Pages: 590
9. The AV bundle transmits impulses from the SA node to the AV node.
   Answer: False

Pages: 593
10. Cardiac muscle fiber membranes stay depolarized longer than those of skeletal muscle due to influx of calcium ions through voltage-gated channels.
    Answer: True

Pages: 592
11. Cardiac muscle fibers contract more strongly if inflow of calcium ions through voltage-gated slow calcium ion channels is enhanced.
    Answer: True

Pages: 593
12. The P wave and T wave of an ECG represent the opening of the AV and semilunar valves, respectively.
Answer: False

Pages: 594
13. Isovolumetric contraction and isovolumetric relaxation of the ventricles occur when both AV valves and both semilunar valves are closed.
Answer: True

Pages: 596
14. There should be no blood remaining in the ventricles following the conclusion of ventricular systole.
Answer: False

Pages: 596
15. The first heart sound is created by blood turbulence associated with closure of the AV valves as ventricular systole begins.
Answer: True

Short Answer

Write the word or phrase that best completes each statement or answers the question.

Pages: 579
1. The organs and tissues included in the cardiovascular system are the _____, _____, and _____.
Answer: heart; blood vessels; blood

Pages: 580
2. The _____ side of the heart is the pump for pulmonary circulation.
Answer: right

Pages: 580
3. The area of the thoracic cavity in which the heart is located is called the _____.
Answer: mediastinum

Pages: 581
4. The visceral layer of the serous pericardium is also called the _____.
Answer: epicardium

Pages: 581
5. Transverse thickenings of the sarcolemma of adjacent cardiac muscle fibers that contain gap junctions and desmosomes are called _____.
Answer: intercalated discs

Pages: 585
6. There are semilunar valves between the right ventricle and the _____ and between the left ventricle and the _____.
Answer: pulmonary trunk; aorta

Pages: 587
7. The left atrium receives blood from the _____.
   Answer: pulmonary veins

Pages: 589
8. The medical term for reduction in blood flow is _____.
   Answer: ischemia

Pages: 590
9. Cardiac excitation normally begins in the _____.
   Answer: sinoatrial (SA) node

Pages: 591
10. When a site other than the normal cardiac pacemaker takes over the job of initiation of the heart beat, that site is referred to as a(n) _____ focus.
    Answer: ectopic

Pages: 593
11. The QRS complex of an ECG represents _____.
    Answer: ventricular depolarization

Pages: 594
12. The term _____ refers to the contraction of the heart; the term _____ refers to the relaxation of the heart.
    Answer: systole; diastole

Pages: 596
13. The act of listening to the heart sounds with a stethoscope is called _____.
    Answer: auscultation

Pages: 597
14. The amount of blood ejected per beat from a ventricle is called the _____.
    Answer: stroke volume

Pages: 597
15. Cardiac output equals _____ times _____.
    Answer: stroke volume; heart rate

Pages: 597
16. The ratio between maximum achievable cardiac output and resting cardiac output is the _____.
    Answer: cardiac reserve

Pages: 599
17. Blood pressure is monitored by _____ in the walls of the aortic arch and the carotid arteries.
    Answer: baroreceptors

Pages: 604
18. In the fetus, blood flows directly from the right atrium to the left atrium through the

    _____.
    Answer: foramen ovale

Pages: 605
19. Proliferation of smooth muscle cells and accumulation of fatty substances in the walls of
    arteries characterize the condition known as _____.
    Answer: atherosclerosis

Pages: 581
20. Buildup of fluid or blood in the pericardial cavity leads to compression of the heart known as

    _____.
    Answer: cardiac tamponade

Pages: 582
21. The superior chambers of the heart are the _____, and the inferior chambers of the heart
    are the _____.
    Answer: atria; ventricles

Pages: 582
22. The external groove that separates the upper and lower chambers of the heart is the _____.
    Answer: coronary sulcus

Pages: 583
23. Irregular ridges and folds in the internal walls of the ventricles are known as the _____.
    Answer: trabeculae carnae

Pages: 583
24. The right atrioventricular valve is also known as the _____ valve, and the left
    atrioventricular valve is also known as the _____ valve.
    Answer: tricuspid; bicuspid (mitral)

Pages: 587
25. Deoxygenated blood is returned into the heart via the _____, the _____, and the

    _____.
    Answer: superior vena cava; inferior vena cava; coronary sinus

Pages: 588
26. The left coronary artery divides into the _____ branch and the _____ branch.
    Answer: anterior interventricular; circumflex

Pages: 588
27. The right coronary artery divides into the _____ branch and the _____ branch.
    Answer: posterior interventricular; marginal

Pages: 589
28. The principal tributaries carrying blood into the coronary sinus are the _____ and the

    _____.
    Answer: great cardiac vein; middle cardiac vein

Pages: 589

29. The medical term for a heart attack is a(n) _____.
    Answer: myocardial infarction

Pages: 590

30. Action potentials are transmitted from the bundle branches to myocardial cells of the ventricles by cells called _____.
    Answer: conduction myofibers (Purkinje fibers)

Matching

Choose the item from Column 2 that best matches each item in Column 1.

1. receives blood from the superior vena cava, inferior vena cava, and coronary sinus	right atrium
2. receives blood from pulmonary veins	left atrium
3. pumps blood into the pulmonary trunk	right ventricle
4. pumps blood into the aorta	left ventricle
5. located between right atrium and right ventricle	tricuspid valve
6. located between the left atrium and the left ventricle	bicuspid valve
7. opens when left ventricular pressure rises above 80 mm Hg	aortic semilunar valve
8. opens when right ventricular pressure rises above 15 mm Hg	pulmonary semilunar valve
9. located in atrial wall near opening of superior vena cava	SA node
10. located in interatrial septum	AV node

Essay

Write your answer in the space provided or on a separate sheet of paper.

Pages: 593

1. Name the waves and intervals present on an electrocardiogram, and state what electrical activity is represented by each. What general cardiac problems can be identified from an ECG?

    Answer: P wave represents atrial depolarization; QRS complex represents onset of ventricular depolarization; T wave represents ventricular repolarization; P-Q interval represents conduction time from SA node through all elements of the conduction system; S-T segment represents plateau phase of ventricular depolarization; Q-T interval represents time from start of ventricular depolarization to end of ventricular repolarization

    ECG can indicate abnormalities in conduction system, heart enlargement, or cardiac damage.

Pages: 594

2. Describe the sequence of pressure changes and the effects of those pressure changes that occur during a normal cardiac cycle.

    Answer: Decreasing ventricular pressure below pressure in aorta and pulmonary trunk causes semilunar valves to close while AV valves still closed; decreasing ventricular pressure below atrial pressure causes AV valves to open and blood to flow into ventricles; atrial systole increases pressure to push final 25% of blood into ventricles; ventricular systole increases ventricular pressure to close AV valves while semilunar valves still closed; further increase in ventricular pressure opens semilunar valves; blood flows out until relaxation begins.

Pages: 597

3. State the Frank-Starling law of the heart, and explain its relationship to cardiac output using the appropriate formula for determining cardiac output.

    Answer: A greater preload (EDV, stretch) on cardiac muscle fibers just before they contract increases the force of contraction. Because preload directly affects stroke volume, as EDV increases, so does stroke volume; thus force of contraction remains appropriate for volume in ventricles. Because CO = SV X HR, as stroke volume increases, so does cardiac output (assuming heart rate stays constant).

Pages: 598

4. Describe the role of the autonomic nervous system in regulating heart rate, and explain the effect on cardiac output using the appropriate formula for determining cardiac output.

    Answer: The cardiovascular center in the medulla oblongata initiates increases or decreases in heart rate by altering the rate of depolarization of the SA node. Sympathetic stimulation increases heart rate via norepinephrine and epinephrine. Parasympathetic stimulation decreases heart rate via acetylcholine. Because CO = SV X HR, as heart rate increases, so does cardiac output (to a point), and as heart rate decreases, so does cardiac output (assuming stroke volume stays constant).

Pages: 597
5. Define **stroke volume** and discuss how stroke volume can be altered. Explain the effect of changes in stroke volume on cardiac output using the appropriate formula for determining cardiac output.

   Answer: Stroke volume is the volume of blood ejected by the ventricle per contraction. SV is affected by preload (EDV) (stretches cardiac muscle), myocardial contractility (force of contraction at a given preload), and afterload (pressure needed to open the semilunar valves). Force of contraction increases if EDV increases (Frank-Starling), or if inflow of calcium ions into cardiac muscle increases. Increased afterload decreases stroke volume. Because CO = SV X HR, changes in SV have a direct effect on CO.

Pages: 600
6. Identify the five risk factors for development of heart disease that can be modified, and explain how each increases the risk.

   Answer: 1) high blood cholesterol-narrows vessels and increases risk of clots, which decreases blood flow

   2) high blood pressure-increased afterload increases end-systolic volume, leading to weakening of heart muscle associated with congestive heart failure

   3) cigarette smoking-nicotine vasoconstricts and increases secretion of catecholamines from adrenal medulla

   4) obesity-more blood vessels to new tissue increases blood pressure due to increased total circuit length

   5) lack of exercise-decreased force of contraction leads to decreased stroke vol.; heart rate must increase to maintain output

Pages: 592
7. Explain the importance of calcium ions to cardiac physiology.

   Answer: Calcium ions are responsible for the plateau phase of depolarization. As calcium ions enter cytosol from ECF through voltage-gated slow channels and from SR, depolarization is maintained for 250 msec, thus allowing ventricular filling. Calcium also binds to troponin to allow sliding of actin and myosin filaments during contraction, and changes in calcium ion levels alter force of contraction, and thus stroke volume.

Pages: 602
8. Explain the relationship between plasma lipids and heart disease.

   Answer: High blood cholesterol promotes formation of fatty plaques in artery walls causing narrowing of lumen and increasing risk of clotting. Triglycerides, cholesterol, and lipoproteins (HDLs, LDLs, VLDLs) must be in appropriate balance to maintain cardiovascular health. High HDL levels decrease risk of heart disease by removing cholesterol for elimination. LDLs may deposit cholesterol in and around smooth muscle fibers in artery walls. VLDLs may be converted to LDLs and are elevated in high fat diets.

Pages: 606
9. Coarctation of the aorta is a congenital defect in which a segment of the aorta is too narrow. Predict what the heart would need to do to compensate for this problem. Explain your answer.

   Answer: Answers need not be absolutely correct - just thoughtful. Students should suggest that blocked flow leads to increased pressure; heart must increase force of contraction and stroke volume to push blood through semilunar valve and into aorta (increased afterload); chronically high blood pressure may lead to congestive heart failure. Other answers may be acceptable depending on rationale given.

Pages: 603

10. Explain how regular exercise reduces the risk of heart disease.

   Answer: Maximum cardiac output and maximum oxygen delivery are increased by increasing stroke volume, while decreasing heart rate, thus increasing pumping efficiency. Accompanying weight loss reduces blood pressure. Also, HDLs are increased and triglycerides decreased, both helping to decrease risk of plaque buildup. Increased fibrinolytic activity reduces risk of intravascular clotting.

Chapter 21   The Cardiovascular System: Blood Vessels
and Hemodynamics

Multiple-Choice

Choose the one alternative that best completes the statement or answers the question.

Pages: 650
1. Blood flows from the sigmoid sinuses into the:
    A) Circle of Willis
    B) internal jugular veins
    C) superior sagittal sinus
    D) internal carotid arteries
    E) inferior sagittal sinus
    Answer: B

Pages: 639
2. The basilar artery is formed by the union of the:
    A) internal jugular veins
    B) vertebral arteries
    C) internal carotid arteries
    D) posterior cerebral arteries
    E) middle cerebral arteries
    Answer: B

Pages: 658
3. The hepatic portal vein is formed by the union of the superior mesenteric vein and the:
    A) hepatic artery
    B) inferior mesenteric vein
    C) splenic vein
    D) pancreatic vein
    E) hepatic vein
    Answer: C

Pages: 658
4. Before the hepatic portal vein enters the liver, it receives blood from the:
    A) hepatic artery
    B) cystic vein
    C) hepatic veins
    D) inferior vena cava
    E) superior mesenteric artery
    Answer: B

Pages: 626

5. The function of baroreceptors is to monitor changes in:
   A) heart rate
   B) stroke volume
   C) peripheral resistance
   D) blood pressure
   E) blood viscosity
   Answer: D

Pages: 626

6. Baroreceptors are located in the:
   A) wall of the right ventricle
   B) medulla oblongata
   C) SA node
   D) walls of the aorta and carotid arteries
   E) walls of the capillaries
   Answer: D

Pages: 626

7. Baroreceptors send nerve impulses to the:
   A) medulla oblongata only
   B) chemoreceptors only
   C) SA node only
   D) walls of muscular arteries only
   E) both the medulla oblongata and the SA node
   Answer: A

Pages: 624

8. The vasomotor center directly controls:
   A) heart rate by stimulating the SA node
   B) stroke volume by regulating total blood volume
   C) peripheral resistance by changing the diameter of blood vessels
   D) peripheral resistance by altering blood viscosity
   E) total blood volume by regulating release of ADH from the posterior pituitary gland
   Answer: C

Pages: 612

9. The tunica interna of a blood vessel is made of:
   A) smooth muscle
   B) cardiac muscle
   C) skeletal muscle
   D) endothelium
   E) dense connective tissue
   Answer: D

Pages: 612
10. Which of the following is considered an elastic artery?
    A) aorta
    B) brachiocephalic
    C) common carotid
    D) superior mesenteric
    E) all of the above except the superior mesenteric
    Answer: E

Pages: 613
11. Valves are present in:
    A) arteries
    B) arterioles
    C) veins
    D) capillaries
    E) all of the above except capillaries
    Answer: C

Pages: 612
12. The layer of a blood vessel wall that determines the diameter of the lumen is the:
    A) adventitia
    B) tunica externa
    C) tunica interna
    D) tunica media
    E) vasa vasorum
    Answer: D

Pages: 627
13. A deficiency of ADH would result in:
    A) reduced venous return
    B) a drop in systemic blood pressure
    C) reduced stroke volume
    D) decreased cardiac output
    E) all of the above
    Answer: E

Pages: 615
14. Thoroughfare channels are:
    A) wider and more permeable capillaries found in secreting organs
    B) low resistance pathways at the distal ends of metarterioles that provide bypass circulation around capillary beds
    C) all of the capillaries making up a capillary bed in a particular organ
    D) those vessels that have the fastest rate of blood flow
    E) special capillaries whose walls are impermeable to all solutes and colloids, such that no fluid or solutes are lost from blood flowing through them
    Answer: B

Pages: 627
15. Increased levels of epinephrine cause:
    A) a decrease in systemic blood pressure due to net vasodilation
    B) an increase in systemic blood pressure due to net vasoconstriction
    C) a decrease in systemic blood pressure due to a decrease in force of contraction in cardiac
       muscle
    D) an increase in systemic vascular resistance due to an increase in the rate of erythropoiesis
    E) a decrease in blood pressure due to increased movement of fluid from plasma to the
       interstitial fluid
    Answer: B

Pages: 622
16. An increase in venous return most directly affects:
    A) blood pressure
    B) systemic vascular resistance
    C) stroke volume
    D) blood viscosity
    E) heart rate
    Answer: C

Pages: 622
17. The viscosity of blood most directly affects:
    A) venous return
    B) stroke volume
    C) systemic vascular resistance
    D) heart rate
    E) net filtration pressure
    Answer: C

Pages: 627
18. Increased levels of aldosterone cause:
    A) an increase in venous return because more water is reabsorbed into the blood from the kidneys
       to increase total blood volume
    B) increased vasodilation because of direct hormone action on the tunica media of arterioles
    C) decreased heart rate because of the decrease in the sodium ion gradient in the SA node
    D) a decrease in blood pressure because of increased fluid loss from the kidneys
    E) an increase in blood viscosity because of the addition of water and sodium ions to the plasma
    Answer: A

Pages: 626
19. Decreased levels of oxygen in the blood in the carotid sinus would cause:
    A) stimulation of baroreceptors
    B) stimulation of the cardioinhibitory center
    C) stimulation of the vasomotor center by peripheral chemoreceptors
    D) reduced rate of conduction from the SA node to the AV node
    E) a decrease in venous return due to vasoconstriction of the superior vena cava
    Answer: C

20. Decreased firing of action potentials by baroreceptors results in:
    A) stimulation of the cardioinhibitory center
    B) inhibition of the cardiostimulatory center
    C) greater sympathetic stimulation of blood vessels by the vasomotor center
    D) a decrease in vasomotor tone
    E) a decrease in the rate of erythropoiesis
    Answer: C

21. If hemorrhage occurs, then:
    A) venous return increases
    B) release of aldosterone is inhibited
    C) ADH stimulates vasoconstriction
    D) epinephrine stimulates overall vasodilation
    E) ANP stimulates excretion of sodium ions
    Answer: C

22. Blood flow increases if:
    A) vasodilation increases
    B) sympathetic stimulation to vessels with alpha adrenergic receptors increases
    C) blood viscosity increases
    D) net filtration pressure increases
    E) parasympathetic stimulation to the heart increases
    Answer: A

23. At rest, the largest portion of total blood volume is in the:
    A) elastic arteries
    B) muscular arteries
    C) systemic veins and venules
    D) capillaries
    E) chambers of the heart
    Answer: C

24. Starling's law of the capillaries states that:
    A) blood flows more slowly through capillaries than arteries or veins because of their smaller diameter
    B) the volume of fluid reabsorbed at the venous end of a capillary is nearly equal to the volume of fluid filtered out at the arterial end
    C) if blood pressure is low, blood is diverted around large capillary beds
    D) if oxygen levels are low, blood is diverted into large capillary beds
    E) blood pressure in capillaries equals cardiac output divided by resistance
    Answer: B

25. Net filtration pressure equals:
    A) (blood hydrostatic pressure + blood colloid osmotic pressure) - (interstitial fluid hydrostatic pressure + interstitial fluid osmotic pressure)
    B) (blood hydrostatic pressure + interstitial fluid osmotic pressure) - (blood colloid osmotic pressure + interstitial fluid hydrostatic pressure)
    C) (blood hydrostatic pressure - blood colloid osmotic pressure) + (interstitial fluid hydrostatic pressure - interstitial fluid osmotic pressure)
    D) (blood hydrostatic pressure + interstitial fluid hydrostatic pressure) - (blood colloid osmotic pressure + interstitial fluid osmotic pressure)
    E) (blood hydrostatic pressure - interstitial fluid hydrostatic pressure) + (blood colloid osmotic pressure - interstitial fluid osmotic pressure)
    Answer: B

26. Of the pressures involved in determining net filtration pressure, the highest pressure at the arterial end of a capillary is usually:
    A) interstitial fluid hydrostatic pressure
    B) interstitial fluid osmotic pressure
    C) blood colloid osmotic pressure
    D) blood hydrostatic pressure
    E) blood hydrostatic pressure and blood colloid osmotic pressure are always equally high
    Answer: D

27. Of the pressures involved in determining net filtration pressure, the highest pressure at the venous end of a capillary is usually:
    A) interstitial fluid hydrostatic pressure
    B) interstitial fluid osmotic pressure
    C) blood hydrostatic pressure
    D) blood colloid osmotic pressure
    E) blood hydrostatic pressure and blood colloid osmotic pressure are always equally high
    Answer: D

28. Most fluid and proteins that escape from blood vessels to the interstitial fluid are normally:
    A) excreted via the urinary system
    B) reabsorbed via the urinary system
    C) returned to the blood via the hepatic portal system
    D) returned to the blood via the lymphatic system
    E) absorbed into tissue cells
    Answer: D

29. If the volume of interstitial fluid increases, then:
    A) interstitial fluid hydrostatic pressure increases
    B) interstitial fluid hydrostatic pressure decreases
    C) net filtration pressure increases
    D) blood is diverted into thoroughfare channels
    E) both net filtration pressure and interstitial fluid hydrostatic pressure increase
    Answer: A

Pages: 618
30. If plasma proteins are lost due to kidney disease, then which of the following pressure changes occur as a direct result?
    A) blood hydrostatic pressure increases
    B) blood colloid osomotic pressure increases
    C) interstitial fluid hydrostatic pressure decreases
    D) blood colloid osmotic pressure decreases
    E) interstitial fluid osmotic pressure decreases
    Answer: D

Pages: 618
31. If lymph channels are blocked, then in areas drained by the blocked vessels:
    A) interstitial fluid hydrostatic pressure increases
    B) blood hydrostatic pressure increases
    C) blood colloid osmotic pressure increases
    D) interstitial fluid hydrostatic pressure decreases
    E) interstitial fluid osmotic pressure decreases
    Answer: A

Pages: 618
32. Blood flow is defined as the:
    A) velocity at which blood moves through a vessel
    B) total blood volume in blood vessels at a particular point in time
    C) volume of blood flowing through a tissue in a given time period
    D) amount of blood ejected from the ventricles per heart beat
    E) degree of friction exerted by blood on the walls of vessels
    Answer: C

Pages: 620
33. In an adult at rest, normal circulation time is about:
    A) 30 seconds
    B) one minute
    C) two minutes
    D) five minutes
    E) ten minutes
    Answer: B

Pages: 620
34. Blood flows most slowly through:
    A) elastic arteries because of the elastic tissue in the walls
    B) superior and inferior venae cavae because of their large diameters and thin walls
    C) superior and inferior venae cavae because of the low pressure
    D) capillaries because of their small diameters
    E) capillaries because their total cross-sectional area is largest
    Answer: E

Pages: 621
35. Blood pressure would be highest in which of the following vessels?
    A) superior sagittal sinus
    B) brachiocephalic trunk
    C) inferior vena cava
    D) right subclavian vein
    E) blood pressure is the same in all vessels
    Answer: B

Pages: 620
36. Cardiac output equals:
    A) stroke volume X resistance
    B) mean arterial blood pressure X resistance
    C) stroke volume/resistance
    D) mean arterial blood pressure/resistance
    E) mean arterial blood pressure X heart rate
    Answer: D

Pages: 622
37. The diameter of blood vessels most directly affects:
    A) venous return
    B) blood viscosity
    C) resistance
    D) heart rate
    E) stroke volume
    Answer: C

Pages: 622
38. The resistance in large arteries is:
    A) high because of the large volume of blood they contain
    B) high because of the rapid velocity of flow
    C) high because of the elastic recoil of the tunica media
    D) low because of the elastic recoil of the tunica media
    E) low because most of the blood volume in the vessel does not come in contact with the vessel
       wall
    Answer: E

Pages: 622
39. Venous return decreases if:
    A) levels of aldosterone increase
    B) respiratory depth increases
    C) systemic blood pressure increases
    D) right atrial pressure increases
    E) levels of epinephrine increase
    Answer: D

Pages: 626
40. Marey's law of the heart describes the relationship between:
    A) heart rate and blood pressure
    B) heart rate and stroke volume
    C) cardiac output and blood pressure
    D) blood pressure and resistance
    E) cardiac output and resistance
    Answer: A

Pages: 626
41. The Bainbridge reflex increases heart rate and force of contraction in response to:
    A) increased net filtration pressure
    B) increased blood viscosity
    C) hypoxia in the carotid sinus
    D) hypercapnia in the aorta
    E) increased venous pressure
    Answer: E

Pages: 626
42. The carotid sinus reflex is most involved in controlling:
    A) systemic blood pressure
    B) normal blood pressure in the brain
    C) venous blood pressure
    D) heart rate
    E) systemic vascular resistance
    Answer: B

Pages: 626
43. The aortic reflex is involved primarily in controlling:
    A) systemic blood pressure
    B) normal blood pressure in the brain
    C) venous blood pressure
    D) heart rate
    E) systemic vascular resistance
    Answer: A

Pages: 626
44. Baroreceptors for the carotid sinus reflex send afferent impulses to the brain via the:
    A) vagus nerve
    B) cardiac accelerator nerves
    C) glossopharyngeal nerve
    D) phrenic nerve
    E) vasomotor nerves
    Answer: C

Pages: 627

45. Atrial natriuretic peptide helps regulate blood pressure by:
   A) increasing blood volume, thus raising blood pressure
   B) increasing blood viscosity, thus increasing resistance
   C) reducing blood volume, thus lowering blood pressure
   D) reducing blood viscosity, thus lowering resistance
   E) causing direct vasoconstriction of arterioles, thus increasing resistance
   Answer: C

Pages: 628

46. To say that smooth muscle in arteriole walls exhibits a **myogenic response** means that:
   A) more smooth muscle cells are added as blood pressure remains chronically high
   B) the cells possess beta adrenergic receptors for response to epinephrine
   C) the cells respond to sympathetic cholinergic stimulation
   D) the cells contract only in response to stimulation by vasomotor nerves
   E) the cells contract more forcefully when stretched by increased blood flow
   Answer: E

Pages: 627

47. Blood volume is regulated primarily by the actions of the hormones:
   A) aldosterone and ADH
   B) aldosterone and epinephrine
   C) ADH and epinephrine
   D) epinephrine and norepinephrine
   E) ADH and PTH
   Answer: A

Pages: 628

48. One way in which pulmonary and systemic circulations differ is that:
   A) blood hydrostatic pressure is much higher in pulmonary circulation
   B) blood colloid osmotic pressure is much lower in pulmonary circulation
   C) net filtration pressure is higher in pulmonary circulation
   D) pulmonary vessels constrict in response to low oxygen levels, while systemic vessels dilate
   E) pulmonary capillaries are much more permeable than systemic capillaries
   Answer: D

Pages: 628

49. The term **syncope** refers to:
   A) an irregular heart beat that reduces cardiac output
   B) a sudden, temporary loss of consciousness followed by spontaneous recovery
   C) the increase in heart rate in response to a drop in blood pressure
   D) the sounds heard when measuring blood pressure
   E) spontaneous contraction of smooth muscle in the tunica media in response to stretching
   Answer: B

50. The term **shock** used with regard to the cardiovascular system refers to:
    A) the rupturing of an aneurysm
    B) spontaneous establishment of pacemaker activity in the AV node following failure of the SA node
    C) a sudden, temporary loss of consciousness followed by spontaneous recovery
    D) failure of the cardiovascular system to deliver sufficient oxygen and nutrients to meet metabolic needs
    E) those effects of the fight-or-flight response that relate to the cardiovascular system
    Answer: D

51. Coolness of skin occurring as a symptom of circulatory shock results from:
    A) reduced blood flow to thermoregulatory centers in the hypothalamus
    B) slower blood flow through vessels in the skin, thus reducing friction
    C) sympathetic stimulation causing vasoconstriction in the skin
    D) failure of oxygen-requiring metabolic reactions that generate heat
    E) heat loss through skin pores that widen when blood flow is reduced
    Answer: C

52. The hypoxia that develops during circulatory shock stimulates:
    A) release of renin from the kidney
    B) release of ADH from the posterior pituitary
    C) increased firing of action potentials by aortic baroreceptors
    D) a myogenic response in the tunica media
    E) opening of precapillary sphincters
    Answer: E

53. During Stage I (compensated) shock, production of **ALL** of the following hormones would be increased **EXCEPT**:
    A) atrial natriuretic peptide
    B) angiotensin II
    C) epinephrine
    D) aldosterone
    E) ADH
    Answer: A

54. During Stage II (decompensated) shock, intravascular clotting results from:
    A) increased vasodilation allowing more blood into contact with damaged vessel walls
    B) decreased rate of blood flow allowing platelet aggregation
    C) spontaneous conversion of fibrinogen to fibrin due to oxygen deficiency in plasma
    D) release of clotting factors from platelets in response to decreased plasma pH
    E) buildup of ADP that occurs when there is insufficient oxygen available to convert ADP to ATP
    Answer: B

Pages: 612
55. The function of the vasa vasorum is to:
    A) provide collateral circulation
    B) connect fetal and maternal circulation
    C) regulate the opening and closing of precapillary sphincters
    D) provide blood flow to the cells in the walls of large vessels
    E) prevent backflow of blood in low-pressure veins
    Answer: D

Pages: 631
56. Acidosis develops during Stage II (decompensated) shock because:
    A) oxygen is an important regulator of pH, and oxygen levels are reduced
    B) as more aldosterone is released, the sodium ions that are reabsorbed make the plasma more acidic
    C) as aldosterone and ADH cause an increase in fluid retention, the change in concentration of solutes alters plasma pH
    D) most of the hormones released during shock are acidic compounds
    E) without oxygen, excessive lactic acid is produced from metabolic reactions
    Answer: E

Pages: 631
57. Resting pulse rate in an average adult is:
    A) 40-50 beats per minute
    B) 50-60 beats per minute
    C) 70-80 beats per minute
    D) 90-100 beats per minute
    E) at least 120 beats per minute
    Answer: C

Pages: 632
58. The first sound heard when measuring blood pressure corresponds to:
    A) blood pressure at ventricular contraction
    B) blood pressure at ventricular relaxation
    C) the closing of the semilunar valves
    D) the depolarization of the SA node
    E) the opening of the AV valves
    Answer: A

Pages: 612
59. An anastomosis is:
    A) the union of two or more vessels supplying the same body region
    B) a weak place in a blood vessel wall
    C) fatty plaque that builds up on vessels walls
    D) a blood vessel that provides oxygen and nutrients to cells in a vessel wall
    E) a special capillary that is very permeable
    Answer: A

Pages: 656
60. A superficial vein in the leg that is frequently used for prolonged administration of IV fluids is the:
A) femoral
B) popliteal
C) great saphenous
D) common iliac
E) peroneal
Answer: C

Pages: 646
61. Blood flows into the common iliac arteries from the:
A) external iliac arteries
B) femoral arteries
C) celiac artery
D) abdominal aorta
E) superior mesenteric artery
Answer: D

Pages: 637
62. Which of the following vessels branches off the ascending aorta?
A) celiac artery
B) right coronary artery
C) right subclavian artery
D) right common carotid artery
E) all of the above except the celiac artery
Answer: B

Pages: 642
63. The superior mesenteric artery supplies blood to the:
A) liver
B) stomach
C) diaphragm
D) kidneys
E) small intestine
Answer: E

Pages: 639
64. Blood flows into the left common carotid artery from the:
A) arch of the aorta
B) brachiocephalic trunk
C) right common carotid artery
D) left subclavian artery
E) left internal carotid artery
Answer: A

Pages: 642
65. The thoracic aorta becomes the abdominal aorta at the level of the:
    A) diaphragm
    B) intervertebral disc between T4 and T5
    C) seventh cervical vertebra
    D) celiac artery
    E) kidneys
    Answer: A

Pages: 638
66. Vessels that are part of the cerebral arterial circle include the:
    A) superior and inferior sagittal sinuses
    B) external carotid arteries
    C) internal carotid arteries
    D) vertebral arteries
    E) all of the above except the sagittal sinuses
    Answer: C

Pages: 646
67. The anterior and posterior tibial arteries are branches of the:
    A) femoral artery
    B) peroneal artery
    C) internal iliac artery
    D) external iliac artery
    E) popliteal artery
    Answer: E

Pages: 648
68. Blood flows directly into the superior vena cava from the:
    A) inferior vena cava
    B) brachiocephalic veins
    C) coronary sinus
    D) internal jugular veins
    E) axillary veins
    Answer: B

Pages: 651
69. Blood flows from the cephalic vein into the:
    A) basilic vein
    B) median cubital vein
    C) brachiocephalic vein
    D) axillary vein
    E) internal jugular vein
    Answer: D

70. The only postnatal arteries that carry deoxygenated blood are the:
    A) coronary arteries
    B) bronchial arteries
    C) pulmonary arteries
    D) common carotid arteries
    E) dorsalis pedis arteries
    Answer: C

71. Compared with systemic capillaries, pulmonary capillaries have:
    A) large spaces between their endothelial cells
    B) a much lower blood colloid osmotic pressure
    C) thicker walls
    D) a much higher net filtration pressure
    E) a much lower blood hydrostatic pressure
    Answer: E

72. The fetal vessel carrying the blood with the highest oxygen levels is the:
    A) umbilical artery
    B) umbilical vein
    C) ductus arteriosus
    D) ductus venosus
    E) foramen ovale
    Answer: B

73. The ligamentum teres (round ligament) of the liver forms from the fetal:
    A) umbilical artery
    B) umbilical vein
    C) ductus arteriosus
    D) ductus venosus
    E) foramen ovale
    Answer: B

74. The minimum diastolic blood pressure indicating Stage I hypertension is:
    A) 150 mm Hg
    B) 125 mm Hg
    C) 100 mm Hg
    D) 90 mm Hg
    E) 75 mm Hg
    Answer: D

75. Primary hypertension is hypertension which is:
   A) not attributable to any identifiable cause
   B) due to kidney disease
   C) caused by chronically high levels of aldosterone
   D) caused by a stroke in the cardiovascular center
   E) caused by failure of the fight-or-flight mechanisms to shut down
   Answer: A

True-False

Write T if the statement is true and F if the statement is false.

1. The tunica interna of an artery is composed of smooth muscle.
   Answer: False

2. Increased sympathetic stimulation to blood vessels with alpha adrenergic receptors causes vasodilation.
   Answer: False

3. The common carotid and subclavian arteries are examples of elastic arteries.
   Answer: True

4. Vessels classified as muscular arteries are so-called because their primary function is distribution of blood to skeletal muscles.
   Answer: False

5. Blood flow through capillaries is regulated primarily by changing the diameter of the veins.
   Answer: False

6. The sinuses of the brain normally contain venous blood.
   Answer: True

7. Most solutes are exchanged between plasma and interstitial fluid via simple diffusion.
   Answer: True

8. Blood hydrostatic pressure is due primarily to plasma proteins.
   Answer: False

9. Interstitial fluid osmotic pressure tends to move fluid out of capillaries and into interstitial fluid.
   Answer: True

10. The term "blood flow" refers to the velocity at which blood moves through vessels.
    Answer: False

11. If stroke volume remains constant, parasympathetic stimulation to the heart results in a decreased cardiac output.
    Answer: True

12. Increased levels of aldosterone raise systemic blood pressure by increasing the viscosity of blood.
    Answer: False

13. During severe hemorrhage, ADH helps maintain systemic blood pressure by causing vasoconstriction.
    Answer: True

14. As the cross-sectional area of a blood vessel increases, the velocity of blood flow decreases.
    Answer: True

15. As the diameter of a blood vessel increases, the resistance decreases.
    Answer: True

Short Answer

Write the word or phrase that best completes each statement or answers the question.

1. The study of the forces involved in circulating blood throughout the body is called _____.
   Answer: hemodynamics

2. Blood flows through the _____ of a blood vessel.
   Answer: lumen

3. The tissue that forms a continuous layer of cells lining the inner surface of the heart and blood vessels is _____.
   Answer: endothelium

4. Increased sympathetic stimulation to blood vessels with beta adrenergic receptors results in

   _____.
   Answer: vasodilation

5. An abnormal increase in interstitial fluid volume is called _____.
   Answer: edema

6. The cardiovascular center is located in the _____.
   Answer: medulla oblongata

7. Sympathetic fibers that transmit impulses to smooth muscle in blood vessel walls are known as
   _____ nerves.
   Answer: vasomotor

8. Baroreceptors in the wall of the aortic arch send sensory information to the brain via sensory
   fibers of the _____ nerve.
   Answer: vagus

9. Hypercapnia and severe hypoxia stimulate _____ located in the walls of the carotid sinus
   and aortic arch.
   Answer: chemoreceptors

10. The term _____ refers to a rapid resting heart/pulse rate; the term _____ refers to a
    slow resting heart/pulse rate.
    Answer: tachycardia; bradycardia

11. The various sounds that are heard while taking blood pressure readings are called _____
    sounds.
    Answer: Korotkoff

12. Blood pressure is usually measured in the left _____ artery using a device called a(n)

    _____.
    Answer: brachial; sphygmomanometer

13. The difference between systolic and diastolic blood pressures is called _____.
    Answer: pulse pressure

14. The first vessel to branch off the aorta after the coronary arteries is the _____.
    Answer: brachiocephalic trunk

Pages: 632
15. The common hepatic, left gastric, and splenic arteries all branch from the _____ artery.
Answer: celiac

Pages: 639
16. The left and right vertebral arteries merge to form the _____ artery.
Answer: basilar

Pages: 648
17. Blood flows into the superior vena cava from the left and right _____ veins.
Answer: brachiocephalic

Pages: 648
18. Blood flows into the inferior vena cava from the left and right _____ veins.
Answer: common iliac

Pages: 658
19. The _____ vein is formed by the union of the superior mesenteric and splenic veins.
Answer: hepatic portal

Pages: 662
20. Blood bypasses the fetal liver through a vessel called the _____.
Answer: ductus venosus

Pages: 666
21. A weakened section of the wall of an artery or vein that bulges outward is called a(n)
_____.
Answer: aneurysm

Pages: 627
22. Both vasoconstriction and increased secretion of aldosterone are stimulated by the hormone
_____.
Answer: angiotensin II

Pages: 628
23. The effect of nitric oxide on blood vessels is to cause _____.
Answer: vasodilation

Pages: 614
24. Microscopic vessels through which substances are exchanged between blood and body tissues are
the _____. Blood flow into these vessels is regulated in part by changes in the diameters
of larger vessels called _____.
Answer: capillaries; arterioles

Pages: 615
25. Capillaries, such as those in the kidneys and choroid plexuses, that have pores in the plasma
membranes of their endothelial cells are said to be _____.
Answer: fenestrated

26. Capillaries, such as those in the liver, that are wider in diameter than other capillaries and that have spaces between the endothelial cells in their walls are called _____.
Answer: sinusoids

27. The forces that tend to move fluid from blood into interstitial fluid are _____ and _____.
Answer: blood hydrostatic pressure; interstitial fluid osmotic pressure

28. The forces that tend to move fluid from interstitial fluid into blood are _____ and _____.
Answer: blood colloid osmotic pressure; interstitial fluid hydrostatic pressure

29. If the number of red blood cells increases relative to plasma volume, then resistance _____ due to the increase in _____.
Answer: increases; viscosity

30. Weak venous valves cause pressure overloads in veins and loss of vascular tone, which results in a condition known as _____.
Answer: varicose veins

Matching

Choose the item from Column 2 that best matches each item in Column 1.

Match the following:

1. artery

   vessels with relatively thick walls; blood under highest pressure

2. vein

   vessels with valves; blood under lowest pressure

3. fenestrated capillary

   special capillary in which plasma membranes of endothelial cells contain pores

4. arteriole

   nearly microscopic vessel that plays key role in regulating flow into capillaries

5. venule   formed by the merging
                                   of capillaries

6. thoroughfare channel   a low resistance
                                   pathway bypassing a
                                   capillary bed; the
                                   distal end of a
                                   metarteriole

7. vasa vasorum   vessels providing
                                   blood flow to the
                                   cells in the walls of
                                   large vessels

8. sinusoid   special capillary
                                   with spaces between
                                   endothelial cells and
                                   lacking a complete
                                   basement membrane

9. vascular sinus   a vein whose wall
                                   lacks smooth muscle
                                   and is supported
                                   instead by dense
                                   connective tissue

10. anastomosis   the union of vessels
                                   serving the same body
                                   region providing
                                   collateral
                                   circulation

Match the following...

11. right subclavian artery   brachiocephalic trunk

12. left subclavian artery   aorta

13. brachial artery   axillary artery

14. basilar artery   vertebral artery

15. common hepatic artery   celiac artery

16. brachiocephalic vein   internal jugular vein

17. azygos vein   superior vena cava

18. femoral vein   great saphenous vein

19. posterior tibial vein   peroneal vein

20. axillary vein   basilic vein

Essay

Write your answer in the space provided or on a separate sheet of paper.

Pages: 612
1. Explain the mechanism by which elastic arteries function as a pressure reservoir.
   Answer: Walls of elastic arteries stretch to accommodate the surge of blood accompanying each
            heart beat. This stretching of elastic fibers stores energy, which is converted to
            kinetic energy during recoil to assist in moving blood forward.

Pages: 612
2. Contrast the structure and functioning of elastic vs. muscular arteries.
   Answer: · Elastic arteries - walls thin compared to diameter; more elastic fibers and less
            smooth muscle in tunica media; act as pressure reservoirs to assist in maintaining
            blood flow
            · Muscular arteries - walls relatively thick; tunica media has more smooth muscle and
            fewer elastic fibers; capable of greater vasoconstriction and vasodilation to regulate
            blood flow to suit needs of structure supplied.

Pages: 620
3. Describe the mechanisms by which edema can develop.
   Answer: Blood hydrostatic pressure may increase due to high blood volume created by excretion
            problems or due to increased venous pressure created by heart failure or intravascular
            clotting. Blood colloid osmotic pressure may be low if plasma protein levels are
            reduced by malnutrition, burns, or liver/kidney disease. Interstitial fluid osmotic
            pressure may be high if plasma proteins leak out of vessels and water follows in
            condition of inflammation or if lymph flow is blocked preventing return of proteins
            and fluid to blood.

Pages: 622
4. Identify and discuss the factors that contribute to systemic vascular resistance.
   Answer: · Blood viscosity - ratio of formed elements and proteins to plasma; increasing
            viscosity via increasing formed elements or decreasing plasma volume increases
            resistance
            · Total blood vessel length - directly proportional to resistance; increasing length
            of circuit (by adding new blood vessels to serve added tissue) increases resistance
            · Diameter of blood vessels - has major effect on resistance; increased diameter
            decreases resistance, thus increasing flow; controlled by ANS

Pages: 618
5. Describe the basic mechanisms by which substances enter and leave capillaries.
    Answer: 1) Diffusion - mostly simple diffusion for gases, amino acids, glucose, and hormones; lipid-soluble materials pass directly through phospholipid membranes; water-soluble substances pass through fenestrations or intercellular clefts
    2) Vesicular transport - for large lipid-insoluble substances, such as maternal antibodies passing to fetus; enter endothelial cells by endocytosis on plasma side, exit by exocytosis on opposite side
    3) Bulk flow - for regulation of fluid volumes; filtration and reabsorption; passive movement in response to pressure gradients

Pages: 626
6. Explain the mechanisms by which the autonomic nervous system regulates flow of blood to particular tissues in any given circumstance.
    Answer: Control is principally via sympathetic vasomotor nerves. Small arteries and arterioles have alpha adrenergic receptors for NE, and are vasoconstricted when stimulated. Most vessels in skeletal muscle and heart have beta adrenergic receptors, and are vasodilated when stimulated. Some sympathetic fibers to skeletal muscles are cholinergic, and dilate in response to ACh. Tissues with dilated arterioles receive greater blood flow.

Pages: 663
7. Describe the structural differences between fetal and postnatal circulation. Explain the functions of the special structures described.
    Answer: Exchange of materials between fetal and maternal circulation occurs via diffusion across the placenta. Deoxygenated blood passes from the fetus to the placenta via two umbilical veins. Oxygenated blood returns to the fetus via the umbilical vein. Most blood bypasses the liver via the ductus venosus. Most blood bypasses the lungs via the foramen ovale in the interatrial septum and the ductus arteriosus between the pulmonary trunk and the aorta.

Pages: 629
8. List five of the signs and symptoms of circulatory shock. Explain the physiological basis for the development of each.
    Answer: Increased epinephrine and norepinephrine cause 1) clammy, cool skin due to vasoconstriction in skin vessels, 2) tachycardia, 3) sweating, 4) nausea. Decreased cardiac output leads to 1) hypotension and weak, rapid pulse 2) altered mental state. Decreased urine formation results from increased ADH and aldosterone and from hypotension. Thirst is triggered by loss of ECF. Acidosis results from an increase in production of lactic acid from anaerobic respiration.

Pages: 627
9. Name four of the hormones involved in the control of systemic blood pressure. Describe the mechanism by which each contributes to the regulation of blood pressure.
    Answer: Hormones include: 1) Angiotensin II; 2) Aldosterone; 3) Epinephrine and norepinephrine; 4) ADH; 5) ANP; 6) PTH; 7) calcitriol; 8) Histamines and kinins. For functions, see page 627.

Pages: 631

10. The bacteria that cause cholera produce a toxin that alters intestinal permeability, causing severe diarrhea. Without treatment, as many as 70% of victims die from shock. Explain in detail why fatal shock develops in these cases.

    Answer: Decreased fluid volume decreases total plasma volume, which decreases venous return and stroke volume. Decreased cardiac output to heart reduces force of contraction. Decreased cardiac output to brain reduces vasomotor tone, thus increasing net vasodilation. Decreased cardiac output and increased vasodilation decrease blood pressure, further decreasing venous return and cardiac ouput.

Pages: 630

11. The bacteria that cause cholera produce a toxin that alters intestinal permeability, causing severe diarrhea. Describe two negative feedback loops - one involving the nervous system, and one involving the endocrine system - that would be activated to combat the fluid loss to prevent development of shock.

    Answer: Nervous system loops start with baroreceptors sensing drop in blood pressure due to drop in blood volume. Information is transmitted to the cardiovascular center in the medulla, which increases sympathetic stimulation to increase heart rate, force of contraction (stroke volume), and net vasoconstriction. All will increase cardiac output. Endocrine loops include renin-angiotensin-aldosterone loop, ADH loop, epinephrine loop, all of which either increase fluid volume or enhance vasoconstriction (or both). See text for details.

Multiple-Choice

Choose the one alternative that best completes the statement or answers the question.

Pages: 680
1. Mucus and cilia are mechanisms of non-specific resistance found in the:
   A) blood
   B) urinary tract
   C) upper respiratory tract
   D) lower intestinal tract
   E) spleen
   Answer: C

Pages: 680
2. Lysozyme is:
   A) an enzyme found in body fluids that flow over epithelial surfaces that destroys certain bacteria
   B) a type of antibody that makes something more recognizable to a phagocyte
   C) a lymphokine produced by helper T cells
   D) one of the self-antigens on the surface of antigen-presenting cells
   E) an antihistamine released by eosinophils
   Answer: A

Pages: 691
3. Antibodies are:
   A) plasma cells
   B) B lymphocytes
   C) T lymphocytes
   D) gamma globulin proteins
   E) cytokines released by macrophages
   Answer: D

Pages: 685
4. An important benefit of fever is that it causes:
   A) increased activity of phagocytes
   B) death of interfering normal flora
   C) opening of active sites on antibody molecules
   D) modification of antigenic determinant sites to better activate the immune system
   E) reduced blood flow to the site of infection
   Answer: A

Pages: 686
5. In cell-mediated immunity, the antigenic cell/molecule is destroyed by:
    A) killer T cells
    B) mast cells
    C) opsonizing antibodies
    D) complement
    E) plasma cells
    Answer: A

Pages: 680
6. Keratin provides protection by:
    A) providing a mechanical barrier in skin
    B) acting as a chemical attractant for macrophages
    C) causing direct lysis of bacteria
    D) increasing the rate at which antibodies are produced
    E) activating killer T cells
    Answer: A

Pages: 682
7. Wandering macrophages are the same thing as:
    A) neutrophils
    B) mast cells
    C) monocytes
    D) antibodies
    E) complement
    Answer: C

Pages: 695
8. Antibodies are produced by:
    A) macrophages
    B) killer T cells
    C) neutrophils
    D) mast cells
    E) plasma cells
    Answer: E

Pages: 691
9. The most common structural class of antibody molecules is:
    A) IgA
    B) IgM
    C) IgG
    D) IgD
    E) IgE
    Answer: C

10. "Teaching" lymphocytes to recognize self from non-self antigens is the function of the:
    A) plasma cell
    B) spleen
    C) thymus
    D) macrophage
    E) liver
    Answer: C

11. The spleen is located:
    A) on the right side, just inferior to the liver
    B) on the left side, between the fundus of the stomach and the diaphragm
    C) underneath the sternum
    D) on the right side, behind the appendix
    E) on the left side, just inferior to the left kidney
    Answer: B

12. Immunoglobulins that circulate in the bloodstream attached to mast cells are classed as:
    A) IgA
    B) IgM
    C) IgG
    D) IgD
    E) IgE
    Answer: E

13. Plasma cells are a form of:
    A) helper T cell
    B) B cell
    C) killer T cell
    D) macrophage
    E) complement
    Answer: B

14. The thoracic duct empties lymph into the:
    A) right lymphatic duct
    B) cysterna chyli
    C) junction of the left subclavian and internal jugular veins
    D) ventricles of the brain
    E) right atrium of the heart
    Answer: C

Pages: 674
15. The cysterna chyli receives lymph from the:
    A) thoracic duct
    B) lumbar and intestinal trunks
    C) bronchomediastinal and subclavian trunks
    D) spleen
    E) axillary lymph nodes
    Answer: B

Pages: 694
16. The role of suppressor T cells is:
    A) to hold the antigen against a blood vessel wall while the killer T cell does its job
    B) to shut down the immune response after the pathogen has been destroyed
    C) to keep you from having an immune response to your own tissues
    D) to reset the thermoregulatory cells in the hypothalamus to restore normal body temperature
       after fever has developed
    E) there is no known function
    Answer: B

Pages: 692
17. During specific immunity, competent T cells are activated by:
    A) plasma cells
    B) complement
    C) antibodies
    D) interleukin-1
    E) histamine
    Answer: D

Pages: 689
18. During specific immunity, the first contact with an antigen is usually made by a(n):
    A) interleukin
    B) plasma cell
    C) killer T cell
    D) suppressor T cell
    E) macrophage
    Answer: E

Pages: 671
19. Lymph capillaries are different from blood capillaries because they:
    A) are not made of endothelium
    B) are not microscopic
    C) are more permeable
    D) do not transport any formed elements
    E) are more likely to develop cholesterol plaques
    Answer: C

Pages: 703
20. Possibly fatal constriction of the bronchioles and a rapid drop in blood pressure are typical of:
    A) anaphylactic hypersensitivity
    B) phagocytosis when it occurs too rapidly
    C) overproduction of memory B cells
    D) delayed hypersensitivity
    E) immune complex hypersensitivity
    Answer: A

Pages: 671
21. T cell and B cells are:
    A) phagocytes
    B) antibodies
    C) lymphocytes
    D) complement proteins
    E) both phagocytes and lymphocytes
    Answer: C

Pages: 683
22. The process of coating an antigenic microbe with antibodies to make it more susceptible to phagocytosis is called:
    A) chemotaxis
    B) opsonization
    C) cloning
    D) anergy
    E) inflammation
    Answer: B

Pages: 689
23. Dendritic cells are:
    A) forms of lymphocytes that provide non-specific immunity
    B) T cells that are destroyed during negative selection
    C) T cells that have not been activated
    D) antigen-presenting cells found in skin and mucous membranes
    E) phagocytic cells found in splenic cords
    Answer: D

Pages: 697
24. Giving someone an antiserum (immune globulin) would:
    A) protect him from a specific disease by giving him passively acquired immunity
    B) cause him to produce his own antibodies to the pathogen causing the disease
    C) protect him for several years
    D) trigger formation of memory B cells that can make antibodies to protect him from this disease in the future
    E) all of the above are correct
    Answer: A

Pages: 689
25. In order for an antigen to activate or sensitize a T cell, the antigen must first be:
    A) coated with opsonizing antibodies
    B) displayed on the surface of a macrophage with self antigens
    C) displayed on the surface of another T cell with IgD antibodies
    D) producing lymphokines
    E) partly digested by a natural killer cell
    Answer: B

Pages: 692
26. The immunoglobulin important for providing passively acquired immunity to the fetus in utero is:
    A) IgA
    B) IgG
    C) IgM
    D) IgD
    E) IgE
    Answer: B

Pages: 703
27. Which of the following occurs in delayed hypersensitivity?
    A) The complement system is activated by killer T cells.
    B) Memory B cells produce antibodies within hours of a second exposure to an antigen.
    C) Sensitized T cells migrate to the antigen site within 48-72 hours.
    D) IgE antibodies cause histamine to be released from mast cells when they combine with an antigen.
    E) Immune complexes precipitate into joints and kidney tubules.
    Answer: C

Pages: 697
28. Receiving an immunization with an altered form of the tetanus toxin results in:
    A) naturally acquired active immunity
    B) naturally acquired passive immunity
    C) artificially acquired active immunity
    D) artificially acquired passive immunity
    E) no response, because altered toxins cannot act as antigens
    Answer: C

Pages: 698
29. The term **immunological tolerance** refers to:
    A) inability of the immune system to respond to a particular antigen
    B) lack of reactivity to peptide fragments from one's own proteins
    C) the maximum dosage of an antigen to which one can be exposed without initiating an immune reponse
    D) the actual number microbes necessary to cause signs and symptoms of a disease
    E) the ability of memory cells to recognize an antigen from prior exposure
    Answer: B

Pages: 698
30. The term **anergy** refers to:
    A) immobilization of a bacterium by specific antibodies
    B) making a microbe more susceptible to phagocytosis by coating it with antibodies
    C) attraction of phagocytes to an area of tissue damage by chemicals released from damaged cells
    D) the lack of reactivity to peptide fragments from one's own proteins
    E) failure of a self-reactive lymphocyte to respond to antigenic stimulation
    Answer: E

Pages: 671
31. Lymphatic tissue is a specialized form of:
    A) adipose tissue
    B) dense connective tissue
    C) simple squamous epithelium
    D) reticular connective tissue
    E) smooth muscle
    Answer: D

Pages: 671
32. You would expect anchoring filaments to open spaces between endothelial cells in lymph capillaries if:
    A) blood hydrostatic pressure is low
    B) blood colloid osmotic pressure is high
    C) interstitial fluid hydrostatic pressure is high
    D) interstitial fluid osmotic pressure is low
    E) sympathetic stimulation to the filaments increases
    Answer: C

Pages: 673
33. Chyle is:
    A) lymph with a high fat content
    B) the blood in the red pulp of the spleen
    C) a chemical released into phagosomes to destroy foreign materials
    D) one of the chemical mediators of inflammation
    E) hormone-rich lymph in the thymus gland that stimulates stem cell differentiation
    Answer: A

Pages: 671
34. There are no lymph capillaries in the:
    A) subcutaneous layer of skin
    B) brain
    C) kidneys
    D) bones
    E) liver
    Answer: B

Pages: 673
35. People who are confined to bed for long periods of time often develop edema because:
    A) their blood pressure becomes elevated, forcing more fluid into interstitial spaces as blood hydrostatic pressure rises
    B) lack of motor activity leads to reduced sympathetic stimulation to lymphatic vessels, so lymph tends to pool
    C) without skeletal muscle contraction to force lymph through lymphatic vessels, fluid tends to accumulate in interstitial spaces
    D) heart rate and force of contraction are reduced, so the pressure gradient is insufficient to maintain lymph flow
    E) reduced vasomotor tone allows proteins to leak from plasma, and water follows the osmotic gradient
    Answer: C

Pages: 675
36. Which of the following is considered to be a primary lymphatic organ?
    A) red bone marrow
    B) spleen
    C) any lymph node
    D) pharyngeal tonsil
    E) liver
    Answer: A

Pages: 676
37. Which person most likely has the largest thymus gland?
    A) third trimester fetus
    B) two-year-old
    C) 12-year-old
    D) 25-year-old
    E) 65-year-old
    Answer: C

Pages: 676
38. Which of the following is normally found in the germinal centers of a lymph node?
    A) pre-T cells
    B) B cells
    C) mature T cells
    D) mast cells
    E) fibroblasts
    Answer: B

Pages: 681
39. A chemical that is produced by virus-infected cells and released to provide non-specific anti-viral protection to neighboring cells is:
    A) transferrin
    B) interleukin-1
    C) histamine
    D) interleukin-2
    E) interferon
    Answer: E

Pages: 681
40. Complement protein C3 enhances phagocytosis by:
    A) anergy
    B) opsonization
    C) cytolysis
    D) T cell activation
    E) triggering production of more macrophages
    Answer: B

Pages: 681
41. Which of the following is a non-specific mechanism of resistance?
    A) activation of the complement system via the alternative pathway
    B) binding of an allergen to IgE molecules on mast cells
    C) a delayed hypersensitivity response to poison ivy
    D) cloning of B cells in response to a measles vaccine
    E) a transfusion reaction between incompatible blood types
    Answer: A

Pages: 685
42. Which of the following is an example of a specific immune response?
    A) release of histamine from damaged cells
    B) adherence of a macrophage to a microbe
    C) release of interferon from virus-infected cells
    D) opsonization of an antigen by IgG molecules
    E) cytolysis of microbes by complement proteins
    Answer: D

Pages: 684
43. The swelling associated with inflammation is caused by:
    A) the large numbers of phagocytes attracted to the area
    B) blockage of blood flow in capillaries by infecting bacteria
    C) movement of fluid out of capillaries due to increased capillary permeability
    D) the accumulation of intracellular fluid released by damaged cells
    E) a larger volume of blood in vasodilated vessels
    Answer: C

Pages: 681
44. **ALL** of the following are functions of complement proteins **EXCEPT**:
    A) opsonization of microbes
    B) cytolysis of microbes
    C) activation of B cell differentiation
    D) activation of inflammation
    E) attraction of phagocytes
    Answer: C

45. Chemicals that intensify and prolong the pain associated with inflammation are:
A) prostaglandins
B) complement proteins
C) interferons
D) interleukins
E) defensins
Answer: A

46. The term **reactivity**, when applied to an antigen, refers to the antigen's ability to:
A) provoke an immune response
B) react specifically with complement proteins to activate the complement system
C) be fragmented by antigen-presenting cells
D) react with receptors on macrophages
E) react specifically with antibodies or cells produced in an immune response
Answer: E

47. Haptens are:
A) chemicals that destroy materials within phagocytic vesicles
B) cytokines released by activated T cells that activate other T cells
C) antigenic substances that have reactivity but lack immunogenicity
D) types of activated complement proteins
E) the H chains of an antibody molecule
Answer: C

48. MHC-II antigens are located on:
A) antigen-presenting cells
B) red blood cells
C) bacteria and viruses
D) immunoglobulins
E) embryonic tissues only
Answer: A

49. Heat as a sign of inflammation results primarily from:
A) alteration of the thermoregulatory centers in the hypothalamus
B) vasoconstriction in the skin, preventing heat loss
C) acceleration of local anaerobic metabolic reactions
D) increased blood flow, which brings more heat to the area
E) chemical reactions between cytokines and microbes
Answer: D

Pages: 695

50. Which of the following is **TRUE** about antibodies?
    A) An IgM antibody molecule is able to react with ten different types of antigenic determinants.
    B) Some types of antibodies provide non-specific resistance to disease.
    C) The five different immunoglobulin classes are determined by the function of the molecules.
    D) All antibodies secreted by a particular plasma cell combine specifically with only one type of antigenic determinant.
    E) People who receive an antiserum will develop active immunity once the injected antibodies begin to reproduce.
    Answer: D

Pages: 691

51. The immunoglobulin class of an antibody molecule is determined by the:
    A) structure of the L chains
    B) structure of the variable region
    C) structure of the constant region of the H chains
    D) function of the molecule
    E) type of antigen that stimulates production of the antibodies
    Answer: C

Pages: 691

52. The antigen-binding site of an antibody molecule is contained in the:
    A) hinge region
    B) disulfide bonds
    C) constant region of the L chains
    D) constant region of the H chains
    E) variable regions of the H and L chains
    Answer: E

Pages: 692

53. In cell-mediated immunity, the next step after antigen recognition by a T cell receptor is:
    A) physical interaction with a plasma cell
    B) costimulation via cytokines or via interaction with molecules on antigen-presenting cells
    C) formation of a phagocytic vesicle
    D) chemical inactivation of the antigen by a cytotoxic T cell
    E) somatic mutation of the T cell's genes
    Answer: B

Pages: 694

54. Killer T cells destroy antigens by:
    A) releasing lethal oxidants into a phagocytic vesicle containing the antigen
    B) releasing cytokines that activate macrophages that do the actual "killing"
    C) secreting antibodies that cause cytolysis
    D) activating cytolytic complement proteins
    E) releasing cytolytic chemicals and toxins that activate DNA fragmentation in the target cell
    Answer: E

Pages: 697
55. **ALL** of the following are characteristic of the secondary antibody response **EXCEPT**:
    A) proliferation and differentiation of memory cells
    B) development of a higher antibody titer than in the primary response
    C) production of antibodies with a higher affinity for the antigen than those in the primary response
    D) predominant production of IgM antibodies
    E) response occurring within hours of exposure
    Answer: D

Pages: 697
56. You had a case of chickenpox when you were six years old. When you were ten, your playmates developed chickenpox, but you did not. This was most likely due to:
    A) naturally acquired active immunity
    B) naturally acquired passive immunity
    C) artificially acquired active immunity
    D) artificially acquired passive immunity
    E) changes in the surface antigen of the chickenpox virus
    Answer: A

Pages: 699
57. The term **immunological surveillance** refers to:
    A) identification and destruction of cells displaying tumor antigens
    B) the mechanical protection provided by intact skin and mucous membranes
    C) the deletion of B cells that recognize self antigens during development in the bone marrow
    D) fragmentation of an antigen by antigen-presenting cells to find the most immunogenic portion
    E) filtration of lymph through lymph nodes
    Answer: A

Pages: 684
58. **ALL** of the following are considered "cardinal" signs (i.e., occurring in all cases) of inflammation **EXCEPT**:
    A) heat
    B) redness
    C) edema
    D) pus
    E) pain
    Answer: D

Pages: 699
59. The human immunodeficiency virus primarily infects:
    A) plasma cells
    B) helper T cells
    C) killer T cells
    D) red blood cells
    E) epithelial cells lining the genitourinary tract
    Answer: B

Pages: 699
60. The function of the enzyme reverse transcriptase in HIV infection is to:
    A) produce new viral RNA from the host cell's DNA
    B) attach the virus to the host cell's plasma membranes
    C) produce DNA that can be incorporated into the host cell's DNA from viral RNA
    D) block the ability of antigen-presenting cells to fragment viral antigens
    E) convert host RNA into viral RNA
    Answer: C

Pages: 703
61. A patient entering the emergency room has been stung by a bee. The patient is wheezing and is exhibiting hypotension. Which of the following best explains this situation?
    A) The bee venom attached to tissues in the bronchioles and blood vessels, causing constriction of the bronchioles and increased capillary permeability.
    B) The patient is pathologically afraid of bees and developed a severe parasympathetic response.
    C) The bee venom antigens were bound by IgE antibodies on mast cells, causing a massive release of histamine.
    D) The bee venom antigens mimicked the action of IgE antibodies, causing a massive release of histamine.
    E) The bee venom antigens combines with IgG antibodies, and the resulting immune complex activated the complement system.
    Answer: C

Pages: 703
62. Which of the following best describes the events of Type III hypersensitivity reactions?
    A) Antigens combine with IgE antibodies on mast cells and basophils, causing release of histamine from these cells.
    B) Sensitized T cells migrate to the site of the antigen, and release cytokines that cause local inflammation.
    C) Antigens directly activate the complement system.
    D) Microbes whose antigens are similar to self antigens evade antigen-presenting cells and cause damage to cells of the immune system.
    E) Foreign antigens combine with antibodies and precipitate in basement membranes under the endothelium of blood vessels, activating the complement system, and triggering inflammation.
    Answer: E

Pages: 704
63. A transplant of tissue between individuals of the same species, but with different genetic backgrounds is called a(n):
    A) autograft
    B) isograft
    C) allograft
    D) xenograft
    E) homeograft
    Answer: C

64. The role of interleukin-1 in cell-mediated immunity is to:
    A) provide non-specific protection against viral invasion
    B) act as a costimulator in T cell activation
    C) stimulate B cell differentiation
    D) activate the complement system
    E) initiate the release of histamine from mast cells
    Answer: B

65. Which of the following events occurs first in inflammation?
    A) vasodilation
    B) phagocyte migration
    C) tissue repair
    D) antigen fragmentation
    E) complement activation
    Answer: A

True-False

Write T if the statement is true and F if the statement is false.

1. Interstitial fluid and lymph have basically the same composition.
    Answer: True

2. Lymph capillaries are present in all body tissues.
    Answer: False

3. Flow of lymph through lymphatic vessels is dependent on the pressure gradient set up by the pumping action of the heart.
    Answer: False

4. Lymph is returned to the blood at the junctions of the internal jugular and subclavian veins.
    Answer: True

5. The cisterna chyli is a chain of lymph nodes located in the lumbar region.
    Answer: False

6. The spleen is considered a primary lymphatic organ.
    Answer: False

7. Primary lymphatic organs are so-called because they are the lymphatic organs that produce B cells and T cells.
Answer: True

8. The thymus gland is located in the neck, anterior to the trachea.
Answer: False

9. The thymus gland continues to grow larger throughout one's life.
Answer: False

10. The white pulp of the spleen is lymphatic tissue, while the red pulp consists of venous sinuses.
Answer: True

11. Interferons provide non-specific resistance to disease.
Answer: True

12. Natural killer cells are a type of phagocytic cell.
Answer: False

13. Plasma cells are involved in the development of cell-mediated immunity.
Answer: False

14. In order for a T cell to respond to an antigen, an antigen-presenting cell must present an appropriate antigen fragment together with class II MHC antigens.
Answer: True

15. Antigen-presenting cells include B cells, dendritic cells, and macrophages.
Answer: True

Short Answer

Write the word or phrase that best completes each statement or answers the question.

1. Disease-producing organisms are called _____.
Answer: pathogens

Pages: 671
2. Defense mechanisms that provide general protection against a wide range of disease-producing organisms are collectively called _____.
   Answer: non-specific resistance

Pages: 671
3. Lymph capillaries in the villi of the small intestine that are specialized for fat transport are called _____.
   Answer: lacteals

Pages: 674
4. The main collecting duct of the lymphatic system is the _____.
   Answer: thoracic (left lymphatic) duct

Pages: 675
5. The primary lymphatic organs are the _____ and the _____.
   Answer: red bone marrow; thymus gland

Pages: 675
6. The major secondary lymphatic organs are the _____ and the _____.
   Answer: lymph nodes; spleen

Pages: 677
7. Efferent lymphatic vessels and blood vessels penetrate a lymph node at a small depression called the _____.
   Answer: hilus

Pages: 679
8. Lymphatic nodules located in the mucous membranes of the mouth and throat are the _____.
   Answer: tonsils

Pages: 680
9. The first line of defense against disease-producing organisms is intact _____ and _____.
   Answer: skin; mucous membranes

Pages: 680
10. A group of about 20 normally inactive proteins in the blood and on cell membranes that enhance phagocytosis and inflammation is called the _____.
    Answer: complement system

Pages: 680
11. The type of lymphocyte that provides non-specific resistance is the _____.
    Answer: natural killer cell

Pages: 683
12. The attraction of phagocytes to a particular area by chemicals released from damaged cells or by pathogens is called _____.
    Answer: chemotaxis

Pages: 686
13. Substances that are recognized as foreign by the immune system and that provoke immune reponses are called _____.
Answer: antigens

Pages: 690
14. A monokine that induces fever is _____.
Answer: interleukin-1 (others possibly acceptable)

Pages: 686
15. Cytokines secreted by virus-infected cells that inhibit viral replication in uninfected cells are _____.
Answer: (alpha and beta) interferons

Pages: 691
16. The antigen binding site is located in the _____ region of an immunoglobulin molecule.
Answer: variable

Pages: 691
17. The structural class of an immunoglobulin molecule is determined by the structure of the constant region of the _____ of the molecule.
Answer: H chain

Pages: 692
18. Most antibodies belong to the _____ class of immunoglobulins.
Answer: IgG

Pages: 692
19. Antibodies that are located on mast cells and basophils belong to the _____ class of immunoglobulins.
Answer: IgE

Pages: 695
20. Activated B cells clone and differentiate into antibody-secreting cells called _____ or into _____ for future protection from the same antigen.
Answer: plasma cells; memory B cells

Pages: 696
21. The process by which antibodies coat a microorganism to make it more susceptible to phagocytosis is called _____.
Answer: opsonization

Pages: 694
22. T cells exhibiting CD8 proteins on their plasma membranes are known as _____ T cells.
Answer: cytotoxic (killer)

Pages: 694
23. T cells that exhibit CD4 proteins on their plasma membranes are known as _____ T cells.
Answer: helper

Pages: 698
24. Lymphocytes' lack of reactivity to peptide fragments from your own proteins is called

_____.

Answer: immunological tolerance

Pages: 703
25. Interactions of allergens with IgE antibodies on the surface of mast cells and basophils leads to a reaction called _____.
Answer: anaphylaxis (Type I hypersensitivity)

Pages: 688
26. Partial antigens that have reactivity but lack immunogenicity are called _____.
Answer: haptens

Pages: 688
27. The specific portion of an antigen that triggers the immune response is called the _____.
Answer: antigenic determinant (epitope)

Pages: 692
28. The first antibody secreted by plasma cells after intitial exposure to any antigen belong to the _____ immunoglobulin class.
Answer: IgM

Pages: 697
29. The measured amount of antibody in serum is called the antibody _____.
Answer: titer

Pages: 697
30. Acquired immunity in which an individual produces antibodies in reponse to direct antigen exposure is referred to as _____ immunity; acquired immunity in which a person is protected by antibodies made by other people is referred to as _____ immunity.
Answer: active; passive

Matching

Choose the item from Column 2 that best matches each item in Column 1.

1. helper T cells   display CD4 proteins;
                               secrete cytokines to
                               enhance lymphocyte
                               proliferation

2. cytotoxic T cells   display CD8
                               molecules; destroy
                               antigens in
                               cell-mediated
                               immunity

3. natural killer cells   lymphocytes that provide non-specific immunity

4. B cells   may act as antigen-presenting cells; differentiate into antibody-producing cells or memory cells

5. plasma cells   descendants of B cells that produce antibodies

6. dendritic cells   process and present antigens to T and B cells; found in skin and mucous membranes

7. macrophages   wandering or fixed phagocytic cells; process and present antigens to T cells

8. suppressor T cells   down-regulate immune response by chemically inhibiting lymphocyte proliferation

9. mast cells   release histamine in Type I hypersensitivity

10. memory T cells   remain in lymphatic tissue ready to respond to antigen possibly years after original exposure to that antigen

Essay

Write your answer in the space provided or on a separate sheet of paper.

Pages: 671

1. Identify the components of the lymphatic system, and describe the functions of the lymphatic system.

    Answer: Components include lymph, lymphatic vessels, lymphatic tissue (nodes, spleen, thymus, red bone marrow, MALT). Functions include returning lost fluid and proteins to blood plasma to maintain BHP and BCOP, transport of dietary fats and lipid-soluble vitamins from the GI tract to the blood, and protection via specific and non-specific immunity.

Pages: 671

2. Describe how lymph capillaries differ structurally from blood capillaries, and explain how these structural differences relate to the function of these vessels.

    Answer: Lymph capillaries have a larger diameter and overlapping endothelial cells. Cells separate when IHP is above lymphatic pressure so that fluid flows into vessels. During edema, swelling pulls on anchoring filaments to make openings larger.

Pages: 678

3. Describe the location, structure, and functions of the spleen.

    Answer: The spleen is about 5" long, located on the left side between the stomach and the diaphragm. The organ has a capsule and a hilus. The parenchyma consists of white pulp (lymphatic tissue - lymphocytes around arteries) and red pulp (venous sinuses). Splenic cords in red pulp consist of RBCs, macrophages, plasma cells, lymphocytes, and granulocytes. The spleen is a site of B cell proliferation and phagocytosis and is an important blood reservoir. In the fetus it participates in RBC formation.

Pages: 696

4. List and briefly describe the mechanisms by which antibodies work to destroy antigens.

    Answer: Antibodies block action of microbial toxins, prevent viral attachment to cells, block microbial motility, cause agglutination or precipitation of antigens, opsonize antigens, or activate complement. The latter three mechanisms enhance phagocytosis.

Pages: 683

5. Describe the process of phagocytosis, and identify the cell types involved.

    Answer: Macrophages and neutrophils are the major phagocytes. Process involves: 1) chemotaxis - attraction of phagocytes to chemicals released at site of damage; 2) adherence - attachment of phagocyte to foreign material; 3) ingestion - formation of phagocytic vesicle around foreign material; 4) digestion - lysosomes merge with phagocytic vesicle; enzymes, defensins, and lethal oxidants released into vesicle to destroy contents

Pages: 684

6. List the four cardinal signs of inflammation, and describe how each develops. What benefit is derived from the development of each sign?

   Answer: 1) Pain - kinins and PGs stimulate pain receptors; makes person aware of damage; 2) Swelling - histamine increases capillary permeability; plasma proteins, cells, and water leave blood vessels; protective cells and proteins reach site; 3) Redness - local blood flow increases as histamine dilates vessels; increased influx of oxygen & nutrients and protective cells and proteins; 4) Heat - due to increased blood flow (see above); increased local temperature increases rate of local chemical reactions

Pages: 691

7. Describe the similarities and differences between cell-mediated immunity and antibody-mediated immunity.

   Answer: Both are forms of specific immunity. Both require presentation of processed antigen with MHC markers to activate T cells. In CMI, antigen is destroyed chemically by action of cytotoxic T cells. In AMI, antigen is destroyed by the action of antibody molecules (several mechanisms - see question above). CMI is directed against intracellular pathogens, some cancer cells, and foreign tissues. Antibody-mediated immunity is directed against extracellular pathogens. Both may be involved in fighting a particular pathogen.

Pages: 689

8. Describe the process by which antigen-presenting cells process and present an exogenous antigen.

   Answer: Steps include: 1) phagocytosis or endocytosis of antigen; 2) digestion of antigen into fragments within vesicle; 3) vesicles containing fragments and MHC-II molecules merge and fuse; 4) antigen fragments bind to MHC-II molecules; 5) vesicle undergoes exocytosis, and Ag-MHC-II complexes are inserted into membrane of APC.

Pages: 692

9. Name the five immunoglobulin classes, and describe the functions of each.

   Answer: · IgG - most abundant; enhance phagocytosis, neutralize toxins, activate complement; pass through placenta
   · IgM - first Ab produced in primary immune response; activate complement, cause agglutination and lysis of microbes
   · IgA - present particularly in fluids secreted onto skin or mucous membranes
   · IgD - antigen receptors on B cell surfaces; involved in B cell activation
   · IgE - located on surfaces of mast cells and basophils; involved in hypersensitivity reactions; provides protection against parasitic worms

Pages: 698

10. Every cell in the body possesses molecules that could act as antigens in others. How does your own immune system "learn" not to attack your own antigens?

    Answer: Some immature T cells in thymus undergo positive selection when they become able to recognize self-MHC molecules. Those that cannot recognize MHC molecules die. Those that survive undergo negative selection, in which those cells that recognize fragments of self antigens are eliminated (deletion) or inactivated (anergy) by failure of costimulation. Similar selection of B cells occurs in red marrow and peripheral tissues.

Chapter 23   The Respiratory System

Multiple-Choice

Choose the one alternative that best completes the statement or answers the question.

Pages: 708
1. The internal nares are the openings between the:
    A) nasal cavity and the nasopharynx
    B) nasal cavity and the oropharynx
    C) nasal cavity and the middle ear
    D) nasopharynx and the oropharynx
    E) nasopharynx and the middle ear
    Answer: A

Pages: 712
2. Which of the following lists the structures in the correct order of air flow?
    A) trachea, laryngopharynx, nasopharynx, oropharynx, larynx
    B) nasopharynx, oropharynx, laryngopharynx, trachea, larynx
    C) nasopharynx, oropharynx, laryngopharynx, larynx, trachea
    D) oropharynx, nasopharynx, laryngopharynx, larynx, trachea
    E) nasopharynx, laryngopharynx, oropharynx, larynx, trachea
    Answer: C

Pages: 714
3. The epithelial portion of the mucous membrane the larynx is:
    A) stratified squamous epithelium
    B) simple squamous epithelium
    C) smooth muscle
    D) transitional epithelium
    E) pseudostratified columnar epithelium
    Answer: E

Pages: 715
4. The trachea extends from the:
    A) larynx to vertebra T5
    B) soft palate to the hyoid bone
    C) atlas to vertebra C7
    D) epiglottis to the thyroid cartilage
    E) foramen magnum to vertebra C5
    Answer: A

Pages: 714
5. The vocal folds are part of the:
    A) nasal cavity
    B) laryngopharynx
    C) trachea
    D) larynx
    E) lungs
    Answer: D

Pages: 715
6. If tension on the vocal folds increases, then the:
   A) loudness of the sound decreases
   B) pitch of the sound increases
   C) loudness of the sound increases
   D) pitch of the sound decreases
   E) epiglottis closes
   Answer: B

Pages: 714
7. The function of the epiglottis is to:
   A) hold the pharynx open during speech
   B) produce surfactant
   C) close off the nasal cavity during swallowing
   D) close off the larynx during swallowing
   E) vibrate to produce sound as air passes over it
   Answer: D

Pages: 717
8. The carina is an important medical marker that is located at the:
   A) junction of the nasopharynx and the oropharynx
   B) junction of the laryngopharynx and the larynx
   C) hilus of each lung
   D) point at which the trachea divides into the right and left bronchi
   E) part of the oropharynx visible when a patient opens her mouth
   Answer: D

Pages: 720
9. The cardiac notch is located on the:
   A) lateral surface of the left lung
   B) lateral surface of the right lung
   C) medial surface of the left lung
   D) medial surface of the right lung
   E) inferior surface of the right lung
   Answer: C

Pages: 719
10. What is normally found between the visceral and parietal layers of the pleura?
    A) the lungs
    B) venous blood
    C) serous fluid
    D) air
    E) lymph
    Answer: C

Pages: 720
11. Each bronchopulmonary segment is a segment of lung tissue served by a:
   A) primary bronchus
   B) secondary bronchus
   C) tertiary bronchus
   D) respiratory bronchiole
   E) terminal bronchiole
   Answer: C

Pages: 727
12. The function of Type II (septal) cells is to:
   A) help control what passes between squamous epithelial cells of the alveoli and the endothelial cells of the capillaries
   B) produce surfactant
   C) act as phagocytes
   D) produce mucus in the upper respiratory tract
   E) store oxygen until it can be transported into the blood
   Answer: B

Pages: 727
13. Airway resistance is affected primarily by the:
   A) amount of surfactant
   B) thickness of the cartilage in the bronchial wall
   C) amount of elastic tissue in the lungs
   D) diameter of the bronchioles
   E) partial pressure of each type of gas in inspired air
   Answer: D

Pages: 730
14. Dalton's Law states that:
   A) at a constant temperature, the volume of a gas varies inversely with the pressure
   B) at a constant pressure, the volume of a gas is directly proportional to the temperature
   C) the rate of diffusion is directly proportional to the surface area of the membrane
   D) in a mixture of gases each gas exerts its own partial pressure
   E) at a constant temperature, the volume of a gas is directly proportional to the pressure
   Answer: D

Pages: 723
15. Boyle's Law states that:
   A) at a constant temperature, the volume of a gas varies inversely with the pressure
   B) at a constant pressure, the volume of a gas is directly proportional to the temperature
   C) the rate of diffusion is directly proportional to the surface area of the membrane
   D) in a mixture of gases, each gas exerts its own partial pressure
   E) at a constant temperature, the volume of a gas is directly proportional to the pressure
   Answer: A

Pages: 731
16. During external respiration, gases are exchanged between the:
    A) outside air and the alveoli
    B) alveoli and the blood
    C) blood and cells
    D) outside air and blood in the dermis
    E) cytosol and mitochondria
    Answer: B

Pages: 727
17. Compliance is affected primarily by the amount of elastic tissue in the lungs and the:
    A) amount of surfactant
    B) thickness of the cartilage in the bronchial wall
    C) partial pressure of oxygen in inspired air
    D) diameter of the bronchioles
    E) temperature of inspired air
    Answer: A

Pages: 728
18. The anatomic dead space is so-called because:
    A) it is the amount of air left in the lungs after death
    B) there is no exchange of gases between the air in these spaces and the blood
    C) the air in these spaces cannot be moved either into the lungs or into the outside air
    D) it results from pathological changes in the respiratory tract
    E) it describes areas into which air can only flow after death
    Answer: B

Pages: 732
19. During internal respiration, oxygen moves:
    A) into cells by primary active transport
    B) out of cells by primary active transport
    C) into cells by diffusion
    D) out of cells by diffusion
    E) into cells by secondary active transport
    Answer: C

Pages: 732
20. Most oxygen is transported in blood by:
    A) the heme portion of hemoglobin
    B) the globin portion of hemoglobin
    C) simply dissolving in plasma
    D) conversion to bicarbonate ion
    E) any type of plasma protein
    Answer: A

Pages: 736
21. Carbamino compounds are formed when:
    A) oxygen binds to hemoglobin
    B) bicarbonate ions leave red blood cells
    C) carbon dioxide binds to proteins
    D) carbon dioxide enters the cerebrospinal fluid
    E) carbon monoxide binds to the oxygen binding sites on hemoglobin
    Answer: C

Pages: 736
22. If the partial pressure of carbon dioxide rises within homeostatic range, then:
    A) more oxygen can attach to hemoglobin
    B) the pH of blood increases
    C) more bicarbonate ions are produced from carbonic acid
    D) respiratory rate decreases
    E) chemoreceptors in the walls of the carotid sinus and aortic arch fire fewer action potentials
    Answer: C

Pages: 734
23. If the pH of blood and interstitial fluid rises within the homeostatic range, then:
    A) more oxygen can combine with hemoglobin
    B) less oxygen can stay attached to hemoglobin
    C) the level of hydrogen ions in these fluids has increased
    D) the increase was caused by an elevated pCO_2
    E) respiratory rate will increase to compensate
    Answer: A

Pages: 736
24. Carbonic acid is produced when:
    A) O_2 combines with bicarbonate ion
    B) CO_2 combines with bicarbonate ion
    C) CO_2 combines with water
    D) CO_2 combines with O_2
    E) CO_2 combines with hemoglobin
    Answer: C

Pages: 736
25. Most carbon dioxide is transported in blood by:
    A) the heme portion of hemoglobin
    B) the globin portion of hemoglobin
    C) simply dissolving in plasma
    D) conversion to bicarbonate ion
    E) any plasma protein
    Answer: D

Pages: 739
26. Peripheral chemoreceptors are located in the:
    A) medulla oblongata
    B) pons
    C) walls of the aorta and carotid sinus
    D) alveoli
    E) walls of the secondary bronchi
    Answer: C

Pages: 733
27. If the partial pressure of oxygen increases, then:
    A) more oxygen can attach to hemoglobin
    B) less oxygen can stay attached to hemoglobin
    C) more bicarbonate ions are produced from carbonic acid
    D) the pH of blood decreases
    E) respiratory rate increases
    Answer: A

Pages: 739
28. The function of the apneustic area is to:
    A) set the basic pattern of breathing
    B) prevent overinflation of the lungs
    C) prolong inspiration
    D) produce surfactant
    E) alter blood pressure in response to changes in respiratory rate
    Answer: C

Pages: 739
29. Chemoreceptors increase their rate of firing action potentials when the:
    A) partial pressure of oxygen rises
    B) partial pressure of carbon dioxide rises
    C) pH of cerebrospinal fluid rises
    D) blood pressure rises
    E) either pCO_2 or pH rise
    Answer: B

Pages: 723
30. Which of the following best describes the process of inspiration?
    A) The diaphragm contracts, thus decreasing the depth of the thoracic cavity. As the size of the thoracic cavity decreases, intrapulmonic pressure decreases below atmospheric pressure, and air comes in along its own pressure gradient.
    B) The diaphragm relaxes, thus increasing the depth of the thoracic cavity. As the size of the thoracic cavity increases, intrapulmonic pressure decreases below atmospheric pressure, and air comes in along its own pressure gradient.
    C) The diaphragm contracts, thus increasing the depth of the thoracic cavity. As the size of the thoracic cavity increases, intrapulmonic pressure increases above atmospheric pressure, and air goes out along its own pressure gradient.
    D) The diaphragm contracts, thus increasing the depth of the thoracic cavity. As the size of the thoracic cavity increases, intrapulmonic pressure decreases below atmospheric pressure, and air comes in along its own pressure gradient.
    E) The diaphragm relaxes, thus decreasing the depth of the thoracic cavity. As the size of the thoracic cavity decreases, intrapulmonic pressure decreases below atmospheric pressure, and air comes in along its own pressure gradient.
    Answer: D

Pages: 719
31. The pleura is:
    A) the serous membrane surrounding the lungs
    B) the membrane gases must cross during external respiration
    C) the lining of the alveoli and respiratory passages
    D) the lung tissue itself
    E) both A and B are correct
    Answer: A

Pages: 734
32. The chloride shift occurs when:
    A) partial pressure of oxygen is too high, and oxygen starts binding with chloride ions in plasma
    B) so much carbon dioxide is being produced that the hydrogen ions that form as a result start combining with chloride to make hydrochloric acid
    C) carbon dioxide needs to be actively transported across the alveolar-capillary membrane
    D) bicarbonate ions cross the cell membrane of an erythrocyte, and to maintain electrical balance within the cell, chloride ions are exchanged for the bicarbonate ions
    E) low pH in cerebrospinal fluid causes chemically-gated chloride channels to open in central chemoreceptors
    Answer: D

Pages: 722
33. The walls of the alveoli are made mostly of:
    A) simple squamous epithelium
    B) smooth muscle
    C) pseudostratified columnar epithelium
    D) hyaline cartilage rings
    E) transitional epithelium
    Answer: A

Pages: 730
34. Based on your knowledge of the gas laws and molecular activity, which of the following would you expect to result from an increase in temperature?
    A) More of a particular gas can be dissolved in a liquid.
    B) A particular gas will diffuse across membranes at a faster rate.
    C) The volume of a particular gas will decrease.
    D) The solubility coefficient of a particular gas will increase.
    E) The partial pressure of the gas will increase.
    Answer: B

Pages: 730
35. Where would you expect to find the highest partial pressure of carbon dioxide?
    A) in the atmosphere
    B) in pulmonary arteries
    C) in the pulmonary veins
    D) in alveolar air
    E) in the intracellular fluid
    Answer: E

Pages: 737
36. The basic pattern of breathing is set by nuclei of neurons located in the:
    A) pons
    B) diaphragm
    C) medulla oblongata
    D) lungs
    E) thoracic region of the spinal cord
    Answer: C

Pages: 739
37. The apneustic and pneumotaxic areas are located in the:
    A) pons
    B) diaphragm
    C) medulla oblongata
    D) lungs
    E) thoracic region of the spinal cord
    Answer: A

Pages: 733
38. The hemoglobin dissociation curve compares the:
    A) partial pressures of oxygen and carbon dioxide
    B) partial pressure of oxygen and the amount of oxygen attached to each hemoglobin molecule
    C) partial pressure of carbon dioxide and pH
    D) partial pressure of carbon dioxide and the amount of hemoglobin within red blood cells
    E) amount of oxyhemoglobin and the amount of carbaminohemoglobin
    Answer: B

Pages: 734
39. In metabolically active tissues you would expect:
    A) the percent saturation of hemoglobin will be less than it is near the lungs
    B) the partial pressure of oxygen will be higher than in the alveoli
    C) the pH will be slightly higher than it is in the fluid close to the lungs
    D) the partial pressure of carbon dioxide will be at its lowest point
    E) all of these are correct
    Answer: A

Pages: 734
40. Hemoglobin will tend to bind more oxygen at a given pO_2 if:
    A) the pCO_2 is increased
    B) the temperature is increased
    C) the pH is increased
    D) BPG concentrations increase
    E) the concentration of hydrogen ions increases
    Answer: C

Pages: 730
41. The major gases in alveolar air are oxygen, carbon dioxide, nitrogen, and:
    A) carbon monoxide
    B) water vapor
    C) sulfur dioxide
    D) ozone
    E) hydrogen
    Answer: B

Pages: 727
42. During anaphylactic hypersensitivity, the primary effect of histamine on the respiratory system is to:
    A) decrease respiratory rate and depth
    B) decrease compliance by reducing surfactant production
    C) force fluid from the pulmonary vessels into the lungs
    D) increase airway resistance
    E) decrease compliance by reducing elasticity of membranes
    Answer: D

Pages: 708
43. **ALL** of the following are functions of the nose **EXCEPT**:
    A) warming of incoming air
    B) filtering incoming air
    C) external respiration
    D) resonance for speech
    E) moistening incoming air
    Answer: C

44. Tha cartilage that forms the inferior wall of the larynx and that serves as a landmark for performing a tracheostomy is the:
    A) arytenoid cartilage
    B) thyroid cartilage
    C) cricoid cartilage
    D) corniculate cartilage
    E) epiglottis
    Answer: C

45. Contraction of the external intercostal muscles results in:
    A) an increase in the volume of the thoracic cavity by moving the ribs and sternum
    B) a decrease in the volume of the thoracic cavity by moving the ribs and sternum
    C) an increase in the volume of the thoracic cavity by increasing the strength of contraction of the diaphragm
    D) an increase in the pitch of sound produced during speech
    E) closing off of the larynx to create greater intrathoracic pressure
    Answer: A

46. Low levels of surfactant in premature babies results most directly in:
    A) high airway resistance
    B) low airway resistance
    C) greater friction between the parietal pleura and visceral pleura
    D) high compliance
    E) low compliance
    Answer: E

47. The residual volume is the amount of air:
    A) remaining in the lungs after the lungs collapse
    B) that can be inhaled above tidal volume
    C) remaining in the lungs after forced expiration
    D) contained in air spaces above the alveoli
    E) that can be exhaled above tidal volume
    Answer: C

48. The tidal volume is the:
    A) volume of air the lungs can hold when maximally inflated
    B) volume of air moved in and out of the lungs in a single quiet breath
    C) percentage of alveolar air that is water vapor
    D) sum of the inspiratory and expiratory reserve volumes
    E) volume of air left in the lungs after a forced expiration
    Answer: B

Pages: 730
49. The use of hyperbaric oxygen therapy, in which a greater than normal volume of oxygen can be dissolved in blood by using pressure, is a clinical application of:
   A) Henry's law
   B) Boyle's law
   C) the Bohr effect
   D) the Haldane effect
   E) Dalton's law
   Answer: A

Pages: 730
50. You would expect pO_2 to be highest in the:
   A) pulmonary arteries
   B) pulmonary veins
   C) hepatic portal vein
   D) intracellular fluid
   E) interstitial fluid
   Answer: B

Pages: 732
51. Several small alveoli merge to form one single, larger air space. This results in a(n):
   A) increased rate of gas exchange due to an increased volume of air within the alveolus
   B) increased rate of gas exchange due to increased pO_2 and decreased pCO_2 within the alveolus
   C) decreased rate of gas exchange due to decreased pO_2 and increased pCO_2 within the alveolus
   D) decreased rate of gas exchange due to a decrease in surface area
   E) decreased rate of gas exchange due to an increase in the thickness of the alveolar-capillary membrane
   Answer: D

Pages: 734
52. The Bohr effect refers to the:
   A) tendency of hemoglobin to become more saturated as the partial pressure of oxygen increases
   B) decreased affinity of hemoglobin for oxygen when the temperature increases
   C) decreased affinity of hemoglobin for oxygen when pH decreases
   D) movement of chloride ions into erythrocytes in exchange for bicarbonate ions
   E) the fact that fetal hemoglobin has a higher affinity for oxygen than adult hemoglobin has
   Answer: C

Pages: 734
53. The function of carbonic anhydrase is to:
   A) transport chloride ions across erythrocyte membranes during the chloride shift
   B) catalyze the conversion of carbon dioxide and water to carbonic acid
   C) buffer hydrogen ions produced as the partial pressure of carbon dioxide rises
   D) attach carbon dioxide to the globin portion of hemoglobin
   E) catalyze the formation of BPG inside red blood cells
   Answer: B

Pages: 735
54. Narrowing of blood vessels serving a particular tissue results in what form of hypoxia in that tissue?
    A) hypoxic
    B) anemic
    C) stagnant
    D) histotoxic
    E) hypercapnic
    Answer: C

Pages: 736
55. Destruction of erythrocytes by the malaria parasite could lead to what form of hypoxia?
    A) hypoxic
    B) anemic
    C) stagnant
    D) histotoxic
    E) hypercapnic
    Answer: B

Pages: 736
56. The Haldane effect refers to the decreased:
    A) affinity of hemoglobin for oxygen in the presence of high pCO_2
    B) affinity of hemoglobin for oxygen as pH falls
    C) affinity of hemoglobin for carbon dioxide as temperature increases
    D) affinity of hemoglobin for carbon dioxide as pO_2 increases
    E) rate of external respiration as surface area decreases
    Answer: D

Pages: 736
57. If the partial pressure of carbon dioxide is decreasing, then:
    A) the partial pressure of oxygen must be increasing
    B) the pH will also be decreasing
    C) the affinity of hemoglobin for oxygen is decreasing
    D) there is an increase in the rate of the reaction converting carbonic acid into water and carbon dioxide
    E) there is an increase in the rate of the reaction converting carbonic acid into hydrogen ion and bicarbonate ion
    Answer: D

Pages: 737
58. The diaphragm is stimulated to contract by the:
    A) phrenic nerve
    B) vagus nerve
    C) axillary nerve
    D) glossopharyngeal nerve
    E) intercostal nerves
    Answer: A

59. The function of the pneumotaxic area is to:
    A) inhibit the inspiratory area
    B) stimulate the inspiratory area
    C) set the basic pattern of breathing
    D) stimulate the diaphragm to contract
    E) inhibit the expiratory area
    Answer: A

60. Peripheral chemoreceptors are stimulated if the:
    A) partial pressure of carbon dioxide falls below 40 mm Hg
    B) partial pressure of oxygen falls below 50 mm Hg
    C) concentration of hydrogen ions increases in the cerebrospinal fluid
    D) concentration of hydrogen ions decreases in arterial blood
    E) concentration of bicarbonate ions increases in arterial blood
    Answer: B

61. If a person is hypoventilating, then:
    A) the partial pressure of carbon dioxide is decreasing
    B) the rate at which carbonic acid is dissociating into hydrogen ions and bicarbonate ions is increasing
    C) the rate at which carbonic acid is dissociating into hydrogen ions and bicarbonate ions is decreasing
    D) the pH of cerebrospinal fluid is increasing
    E) more oxygen will be able to bind to hemoglobin
    Answer: B

62. The receptors in the inflation (Hering-Breuer) reflex are the:
    A) central chemoreceptors
    B) peripheral chemoreceptors
    C) aortic baroreceptors
    D) muscle spindles in the diaphragm
    E) stretch receptors in the walls of the bronchi and bronchioles
    Answer: E

63. The output in the inflation (Hering-Breuer) reflex is:
    A) prolonged inspiration
    B) increased respiratory depth
    C) expiration
    D) increased blood pressure
    E) bronchodilation
    Answer: C

Pages: 742
64. The oxygen-diffusing capacity increases during exercise primarily due to:
    A) an increase in the solubility coefficient for oxygen
    B) a decrease in the thickness of the alveolar-capillary membrane as the lungs are stretched
    C) an increase in the partial pressure of oxygen in alveolar air
    D) opening of gated oxygen channels in the alveolar-capillary membrane
    E) an increase in surface area for diffusion as perfusion of pulmonary capillaries increases
    Answer: E

Pages: 745
65. Oxygen-diffusing capacity is decreased during pneumonia primarily because:
    A) invading bacteria consume the oxygen before it can diffuse into the blood
    B) pulmonary capillaries are blocked by bacteria
    C) blood pressure falls too low to maintain pulmonary perfusion
    D) the alveolar-capillary membrane is thicker due to inflammation and fluid buildup
    E) damaged alveoli merge, thus decreasing surface area for diffusion
    Answer: D

Pages: 717
66. C-shaped cartilage rings support the:
    A) laryngopharynx
    B) larynx
    C) trachea
    D) tertiary bronchi
    E) all of the above
    Answer: C

Pages: 730
67. The partial pressure of oxygen in alveoli is:
    A) 40 mm Hg
    B) 45 mm Hg
    C) 100 mm Hg
    D) 105 mm Hg
    E) 160 mm Hg
    Answer: D

Pages: 730
68. The partial pressure of carbon dioxide in blood in the pulmonary veins is:
    A) 40 mm Hg
    B) 45 mm Hg
    C) 100 mm Hg
    D) 105 mm Hg
    E) 160 mm Hg
    Answer: A

Pages: 730
69. The partial pressure of oxygen in blood in the pulmonary veins is:
    A) 40 mm Hg
    B) 45 mm Hg
    C) 100 mm Hg
    D) 105 mm Hg
    E) 160 mm Hg
    Answer: C

Pages: 730
70. The partial pressure of carbon dioxide in atmospheric air at sea level is:
    A) 0.04 mm Hg
    B) 0.3 mm Hg
    C) 40 mm Hg
    D) 45 mm Hg
    E) 160 mm Hg
    Answer: B

Pages: 730
71. Which of the following contains the highest percentage of carbon dioxide?
    A) inspired air
    B) alveolar air
    C) expired air
    D) atmospheric air
    E) all of the above contain the same percentage of carbon dioxide
    Answer: B

Pages: 723
72. Atmospheric pressure at sea level is:
    A) 100 mm Hg
    B) 160 mm Hg
    C) 500 mm Hg
    D) 760 mm Hg
    E) 1000 mm Hg
    Answer: D

Pages: 729
73. The normal adult value for residual volume is:
    A) 350 ml
    B) 500 ml
    C) 1200 ml
    D) 3600 ml
    E) 4800 ml
    Answer: C

74. The partial pressure of oxygen in blood in the pulmonary arteries is:
    A) 40 mm Hg
    B) 45 mm Hg
    C) 100 mm Hg
    D) 105 mm Hg
    E) 160 mm Hg
    Answer: A

75. The alveolar ventilation rate for someone whose tidal volume = 450 ml, whose dead space air = 150 ml, and whose respiratory rate is 15 respirations per minute is:
    A) 2250 ml/min
    B) 4500 ml/min
    C) 6750 ml/min
    D) 9000 ml/min
    E) There is not enough information to calculate the alveolar ventilation rate.
    Answer: B

True-False

Write T if the statement is true and F if the statement is false.

1. The term **external respiration** refers to movement of air between the atmosphere and the lungs.
   Answer: False

2. The trachea and bronchi are considered to be part of the lower respiratory tract.
   Answer: True

3. The fauces is the opening from the mouth to the oropharynx.
   Answer: True

4. The function of the cricoid cartilage is to prevent swallowed materials from entering the larynx.
   Answer: False

5. Aspirated objects are more likely to lodge in the left primary bronchus than the right primary bronchus due to anatomical differences.
   Answer: False

6. The pleural cavity is normally filled with air.
   Answer: False

7. The left lung has two lobes; the right lung has three lobes.
Answer: True

Pages: 723
8. At a constant temperature, the pressure of a gas in a closed container is inversely proportional to the volume of the container.
Answer: True

Pages: 723
9. When the diaphragm contracts, the volume of the thoracic cavity decreases.
Answer: False

Pages: 726
10. Under normal conditions, intrapulmonic pressure is always subatmospheric.
Answer: False

Pages: 727
11. If surface tension of alveolar fluid decreases, then lung compliance increases.
Answer: True

Pages: 730
12. The partial pressure of oxygen in alveolar air is normally higher than the partial pressure of carbon dioxide.
Answer: True

Pages: 730
13. The partial pressure of carbon dioxide is higher in alveolar air than it is in the atmosphere.
Answer: True

Pages: 734
14. If the partial pressure of carbon dioxide increases in blood, then the percent saturation of hemoglobin increases.
Answer: False

Pages: 736
15. If the partial pressure of carbon dioxide increases in extracellular fluid, then the pH increases.
Answer: False

Short Answer

Write the word or phrase that best completes each statement or answers the question.

Pages: 734
1. The conversion of carbon dioxide and water to carbonic acid is catalyzed by the enzyme

   _____.

   Answer: carbonic anhydrase

Pages: 736
2. When carbon dioxide combines with hemoglobin, the resulting compound is called _____.
   Answer: carbaminohemoglobin

Pages: 736
3. The greatest percentage of carbon dioxide is transported in plasma as _____.
   Answer: bicarbonate ion

Pages: 737
4. The respiratory center of the brain consists of the _____ area in the medulla oblongata and the _____ and _____ areas in the pons.
   Answer: medullary rhythmicity; pneumotaxic; apneustic

Pages: 739
5. Central chemoreceptors are located in the _____; peripheral chemoreceptors are located in the _____ and _____.
   Answer: medulla oblongata; carotid bodies; aortic bodies

Pages: 739
6. Central chemoreceptors respond to changes in pH or partial pressure of carbon dioxide in _____.
   Answer: cerebrospinal fluid

Pages: 740
7. In a person who is hyperventilating, the partial pressure of carbon dioxide in arterial blood _____.
   Answer: decreases

Pages: 740
8. In a person who is hypoventilating, the concentration of hydrogen ions _____, causing extracellular pH to _____.
   Answer: increases; decrease

Pages: 739
9. If partial pressure of carbon dioxide in arterial blood exceeds 40 mm Hg, the condition is called _____.
   Answer: hypercapnia

Pages: 748
10. The medical term for a nosebleed is _____.
    Answer: epistaxis

Pages: 748
11. Spitting of blood from the respiratory tract is called _____.
    Answer: hemoptysis

Pages: 708
12. Exchange of gases between blood in systemic capillaries and tissue cells is known as _____.
    Answer: internal respiration

Pages: 708
13. The openings between the nasal cavity and the nasopharynx are the _____.
Answer: internal nares

Pages: 712
14. The anatomical term for the Adam's apple is the _____, which is part of the _____.
Answer: thyroid cartilage; larynx

Pages: 720
15. When the pleural cavity fills with air, the condition is known as _____.
Answer: pneumothorax

Pages: 720
16. The section of a lung supplied by a tertiary bronchus is called a(n) _____.
Answer: bronchopulmonary segment

Pages: 723
17. The process by which gases are exchanged between the atmosphere and the lung alveoli is called _____.
Answer: pulmonary ventilation

Pages: 723
18. Boyle's law states that at a constant temperature, the pressure of a gas _____ as the volume of the container increases.
Answer: decreases

Pages: 726
19. Normal quiet breathing is called _____; temporary cessation of breathing is called _____; painful, labored breathing is called _____.
Answer: eupnea; apnea; dyspnea

Pages: 727
20. The detergent-like substance that reduces the surface tension of alveolar fluid is called _____.
Answer: surfactant

Pages: 727
21. The ease with which the lungs and thoracic wall can be expanded is called _____.
Answer: compliance

Pages: 727
22. Greater sympathetic stimulation to bronchioles causes _____, which causes airway resistance to _____.
Answer: bronchodilation; decrease

Pages: 728
23. The volume of air that moves in and out of the airways with each inspiration and expiration during normal quiet breathing is called the _____.
Answer: tidal volume

Pages: 729
24. Air that remains in the lungs even after the expiratory reserve volume is expelled is called the _____ .
Answer: residual volume

Pages: 730
25. _____ states that each gas in a mixture of gases exerts its own pressure as if all the other gases were not present.
Answer: Dalton's law

Pages: 730
26. Henry's law states that at a constant temperature, the quantity of a gas that will dissolve in a liquid is proportional to the _____ of the gas and its _____ .
Answer: partial pressure; solubility coefficient

Pages: 733
27. As the partial pressure of oxygen in blood increases, the percent saturation of hemoglobin _____ .
Answer: increases

Pages: 734
28. As extracellular pH decreases, percent saturation of hemoglobin _____ .
Answer: decreases

Pages: 735
29. As temperature of blood increases, percent saturation of hemoglobin _____ .
Answer: decreases

Pages: 740
30. As respiratory rate decreases, partial pressure of carbon dioxide in arterial blood _____ , resulting in a(n) _____ in extracellular pH.
Answer: increases; decrease

Matching

Choose the item from Column 2 that best matches each item in Column 1.

Match the following:

1. tidal volume   volume of air moved
                                     in and out of the
                                     lungs during normal
                                     quiet breathing

2. inspiratory reserve volume   volume of air the can
                                     be inhaled beyond
                                     tidal volume

3. expiratory reserve volume   volume of air that
can be exhaled beyond
tidal volume

4. minimal volume   volume of air
remaining in the
lungs after the lungs
collapse

5. residual volume   volume of air   dead space air
remaining in the
lungs after forced
expiration

6. inspiratory capacity   tidal volume +
inspiratory reserve
volume

7. functional residual
capacity   residual volume +
expiratory reserve
volume

8. vital capacity   tidal volume +   total lung capacity
inspiratory reserve
volume + expiratory
reserve volume

9. alveolar ventilation rate   (tidal volume - dead
space air) X
respirations per
minute

10. minute ventilation   tidal volume X
respirations per
minute

Match the following...

11. total lung capacity   6000 ml

12. vital capacity   4800 ml

13. expiratory reserve volume   1200 ml

14. inspiratory reserve volume   3100 ml

15. inspiratory capacity   3600 ml

16. functional residual
capacity   2400 ml

Essay

Write your answer in the space provided or on a separate sheet of paper.

Pages: 717

1. Describe the structural changes that are seen as the bronchial tree branches.
   Answer: Epithelium changes from pseudostratified ciliated columnar to nonciliated simple columnar. Walls change from being supported by partial rings of cartilage to plates of cartilage to no cartilage. As the amount of cartilage in walls decreases, the amount of smooth muscle increases. Diameter of lumen steadily decreases as branching increases.

Pages: 722

2. Describe the structure and function of the alveolar-capillary membrane.
   Answer: The membrane consists of the alveolar wall (type I and II alveolar cells and alveolar macrophages), the epithelial basement membrane, the capillary basement membrane, and the endothelium of the capillary wall. The large surface area and thinness (0.5 μ m) make it specialized for rapid diffusion of gases.

Pages: 723

3. State Boyle's law, and explain how it relates to the process of pulmonary ventilation:
   Answer: Boyle's law says that at a constant temperature, the pressure of a gas in closed container is inversely proportional to the volume of the container. During inspiration, as the diaphragm contracts and flattens, the volume of the thoracic cavity is increased, thus decreasing the air pressure in the lungs. As the pressure falls below atmospheric pressure, air can flow into the lungs until equilibrium is reached. During expiration, the diaphragm relaxes, reducing the size of the thoracic cavity, increasing the intrapulmonic pressure and forcing air out of the lungs.

Pages: 731

4. Identify and briefly discuss that factors that affect the rate of external respiration.
   Answer: 1) Partial pressure of gases - as gradients across the alveolar-capillary membrane change, rate changes proportionally
   2) Total surface area for gas exchange - rate is directly proportional to surface area
   3) Diffusion distance - rate is inversely proportional to thickness of alveolar-capillary membrane
   4) Solubility combined with molecular weight of gases - rate is directly proportional to solubility, but inversely proportional to molecular weight

Pages: 708

5. Identify the structures included in the respiratory system, and state the functions of the respiratory system.
   Answer: Structures include the nose, pharynx, larynx, trachea, bronchi, bronchioles, and lungs. The functions include gas exchange, a role in the sense of smell, protection via filtering air, production of sound, a role in acid-base balance and water balance, and elimination of wastes.

Pages: 736

6. Write out the chemical formula that describes the conversion of carbon dioxide to bicarbonate ion, and use the formula to explain why hyperventilation and hypoventilation affect acid-base balance.

    Answer: $CO_2 + H_2O$ <-----> H_2CO_3 <-----> $H^+ + HCO_3^-$
    During hyperventilation, pCO_2 decreases, driving reaction to left, thus reducing H^+ and raising pH. During hypoventilation, pCO_2 increases, driving reaction to the right, increasing H^+ and lowering pH.

Pages: 733

7. Define the term percent saturation of hemoglobin, and identify the factors that affect the percent saturation of hemoglobin. What is the significance of these factors in the delivery of oxygen to tissues?

    Answer: Percent saturation of hemoglobin refers to the ratio of oxyhemoglobin to total hemoglobin. As the partial pressure of oxygen increases, so does percent saturation. Conditions present in and around metabolically active tissues, such as increased partial pressure of carbon dioxide, lower pH, higher temperature, and higher BPG levels, all decrease affinity of hemoglobin for oxygen, so that the oxygen is released where it is most needed. When these conditions are reversed, as they would be in the pulmonary capillaries, more oxygen can be picked up.

Pages: 731

8. In chronic emphysema, some alveoli merge together and some are replaced with fibrous connective tissue. In addition, the bronchioles are often inflamed, and expiratory volume is reduced. Using proper respiratory system terminology, explain at least four reasons why affected individuals will have problems with ventilation and external respiration.

    Answer: Answers could include: reduced compliance (reduces ability to increase thoracic volume); increased airway resistance (decreases tidal volume); decreased diffusion due to increased diffusion distance, due to decreased surface area, and due to changes in partial pressures of gases. Other answers may be acceptable.

Pages: 727

9. A patient entering the emergency room has been stung by a bee, and is exhibiting anaphylaxis. Explain the immunological basis for this reaction, and explain the respiratory symptoms you would expect and the reasons for these symptoms.

    Answer: [note: This question is intended to cover multiple chapters.] Bee venom antigens combine with IgE antibodies on mast cells and basophils, causing them to release large quantities of histamine and other mediators in inflammation. Among other effects, histamine causes bronchconstriction, which increases airway resistance, making ventilation difficult, and reducing pO_2 in alveolar air, leading to hypoxic hypoxia.

Pages: 736

10. You have just moved from your home on the Atlantic coast to a small town high in the Rocky Mountains. Explain the effects of this move on your respiratory system. What are the short-term and long-term compensations you might expect your body to make to this new environment?

    Answer: [note: This question is intended to cover multiple chapters.]
    Reduced partial pressure of oxygen is sensed by peripheral chemoreceptors, which trigger an increase in respiratory rate and depth. Blood pressure may be increased to maintain oxygen delivery. Over time one might see an increase in the red blood cell count to optimize pick-up of available oxygen. (Other answers may be correct.)

Chapter 24   The Digestive System

Multiple-Choice

Choose the one alternative that best completes the statement or answers the question.

Pages: 754
1. The type of chemical reaction catalyzed by the digestive enzymes in the digestive juices of the alimentary canal is:
   A) oxidation
   B) reduction
   C) hydrolysis
   D) dehydration
   E) phosphorylation
   Answer: C

Pages: 754
2. In areas of the gastrointestinal tract specialized for absorption of nutrients, the type of epithelium seen in the mucosa is:
   A) simple squamous
   B) stratified squamous
   C) transitional
   D) simple columnar
   E) pseudostratified ciliated columnar
   Answer: D

Pages: 768
3. Rugae are:
   A) folds in the mucosa of the walls of hollow organs
   B) bacteria that are always present in the large intestine
   C) bumps on the surface of the tongue
   D) folds of the peritoneum that hold the liver in place
   E) bands of smooth muscle running longitudinally along the large intestine
   Answer: A

Pages: 757
4. The liver is separated from the stomach and duodenum by the:
   A) diaphragm
   B) lamina propria
   C) lesser omentum
   D) pancreas
   E) greater omentum
   Answer: C

Pages: 756
5. The small intestine is attached to the posterior abdominal wall by a fold of the peritoneum called the:
A) mesocolon
B) mesentery
C) falciform ligament
D) taeniae coli
E) greater omentum
Answer: B

Pages: 759
6. The major digestive enzyme in saliva is:
A) amylase
B) pepsin
C) carboxypeptidase
D) lipase
E) maltase
Answer: A

Pages: 759
7. Saliva is produced by the:
A) pancreas
B) parotid glands
C) liver
D) taste buds
E) parietal cells
Answer: B

Pages: 763
8. Which of the following is an example of mechanical digestion?
A) glycolysis
B) defecation
C) oxidation-reduction
D) mastication
E) hydrolysis
Answer: D

Pages: 768
9. Parietal cells of the gastric mucosa secrete:
A) pepsinogen
B) trypsinogen
C) hydrochloric acid
D) saliva
E) cholecystokinin
Answer: C

10. The pyloric sphincter is located at the junction of the:
    A) esophagus and stomach
    B) stomach and duodenum
    C) ileum and cecum
    D) esophagus and larynx
    E) sigmoid colon and rectum
    Answer: B

11. The major chemical digestion that occurs in the adult stomach is:
    A) hydrolysis of fats by gastric lipase
    B) formation of chylomicrons
    C) conversion of ammonia to urea
    D) hydrolysis of sucrose by sucrase
    E) hydrolysis of proteins by pepsin
    Answer: E

12. The acini are:
    A) the exocrine portion of the pancreas
    B) enteroendocrine cells in the duodenum
    C) cells that secrete gastric juice
    D) liver cells that produce bile
    E) a transport form of triglycerides
    Answer: A

13. Increased activity of the sympathetic nervous system will:
    A) increase production of all hydrolytic enzymes by abdominal organs
    B) increase only production of those digestive juices rich in buffers
    C) have no effect on the digestive system
    D) decrease production of digestive juices
    E) increase movement of food through the alimentary canal
    Answer: D

14. The functions of the gallbladder include:
    A) production of bile
    B) storage and concentration of bile
    C) formation of urea
    D) secretion of cholecystokinin
    E) both A and B are correct
    Answer: B

Pages: 784

15. The most proximal portion of the small intestine is the:
    A) ileum
    B) jejunum
    C) cecum
    D) ascending colon
    E) duodenum
    Answer: E

Pages: 784

16. Folds in the mucosa of the small intestine that increase the surface area for diffusion are called:
    A) microvilli
    B) villi
    C) rugae
    D) taeniae coli
    E) haustra
    Answer: B

Pages: 780

17. The common bile duct is formed by the union of the:
    A) right and left hepatic ducts
    B) cystic and pancreatic ducts
    C) common hepatic and cystic ducts
    D) all bile capillaries
    E) pancreatic and accessory ducts
    Answer: C

Pages: 788

18. Most absorption of nutrients occurs in the:
    A) mouth
    B) transverse colon
    C) stomach
    D) small intestine
    E) rectum
    Answer: D

Pages: 780

19. The function of bile is to:
    A) emulsify fats
    B) transport fats through the blood
    C) hydrolyze fats
    D) actively transport fats through epithelial cell membranes
    E) all of these are correct
    Answer: A

Pages: 783
20. Production of bile is increased primarily by the hormone:
    A) trypsin
    B) GIP
    C) CCK
    D) secretin
    E) gastrin
    Answer: D

Pages: 789
21. The digestive juice containing the most types of hydrolytic enzymes is produced by the:
    A) liver
    B) stomach
    C) large intestine
    D) parotid glands
    E) pancreas
    Answer: E

Pages: 789
22. Specific disaccharides are hydrolyzed by enzymes found in:
    A) gastric juice
    B) intestinal juice
    C) saliva
    D) pancreatic juice
    E) bile
    Answer: B

Pages: 789
23. The hydrolytic reactions catalyzed by trypsin and chymotrypsin would result in the production of:
    A) fatty acids and glycerol
    B) monosaccharides
    C) peptides
    D) nucleotides
    E) dextrin
    Answer: C

Pages: 754
24. Which of the following would be considered an accessory organ of the digestive system?
    A) pancreas
    B) stomach
    C) esophagus
    D) large intestine
    E) small intestine
    Answer: A

Pages: 759
25. The largest of the salivary glands are located:
    A) at the base of the tongue
    B) along the alveolar processes of the rami of the mandible
    C) in the gingivae
    D) embedded in the walls of the fauces
    E) just anterior to the ears, near the junction of the mandible and the temporal bones
    Answer: E

Pages: 764
26. Which of the following lists the tubing in the correct order of food movement?
    A) nasopharynx, oropharynx, laryngopharynx, larynx, esophagus
    B) oropharynx, laryngopharynx, esophagus, stomach, pyloric valve
    C) laryngopharynx, oropharynx, esophagus, stomach, pyloric valve
    D) oropharynx, laryngopharynx, esophagus, pyloric valve, stomach
    E) nasopharynx, oropharynx, larynx, esophagus, stomach
    Answer: B

Pages: 794
27. Which of the following lists the tubing in the correct order of food movement?
    A) descending colon, splenic flexure, transverse colon, hepatic flexure, ascending colon, sigmoid colon
    B) ascending colon, hepatic flexure, transverse colon, splenic flexure, descending colon, sigmoid colon
    C) sigmoid colon, ascending colon, hepatic flexure, transverse colon, splenic flexure, descending colon
    D) ascending colon, splenic flexure, transverse colon, hepatic flexure, descending colon, sigmoid colon
    E) sigmoid colon, descending colon, splenic flexure, transverse colon, hepatic flexure, ascending colon
    Answer: B

Pages: 780
28. The function of stellate reticuloendothelial cells of the liver is:
    A) to produce bile
    B) to produce hydrochloric acid
    C) to produce red blood cells
    D) to store fat
    E) phagocytosis
    Answer: E

Pages: 784
29. Which of the following lists the tubing in the correct order of food movement?
    A) pyloric valve, duodenum, ileum, jejunum, ileocecal valve
    B) pyloric valve, jejunum, duodenum, ileum, ileocecal valve
    C) ileocecal valve, ileum, jejunum, duodenum, pyloric valve
    D) ileocecal valve, duodenum, jejunum, ileum, pyloric valve
    E) pyloric valve, duodenum, jejunum, ileum, ileocecal valve
    Answer: E

30. Haustra are:
   A) bands of smooth muscle that run along the large intestine
   B) pouches formed by taeniae coli in the large intestine
   C) cells that produce enzymes in the duodenum
   D) valves that prevent backflow of chyme along the intestinal tract
   E) pads of adipose tissue hanging from the large intestine
   Answer: B

31. During swallowing, the nasal cavity is closed off by the soft palate and the:
   A) epiglottis
   B) uvula
   C) palatine tonsils
   D) fauces
   E) tongue
   Answer: B

32. Which of the following is an accurate description of the location of the pancreas?
   A) It lies just under the diaphragm and superior to the stomach.
   B) It is suspended from the cecum.
   C) It lies posterior to the greater curvature of the stomach.
   D) It lies between the right and quadrate lobes on the posterior surface of the liver.
   E) It extends inferiorly and anteriorly over the transverse colon and small intestine.
   Answer: C

33. The longest portion of the small intestine is the:
   A) duodenum
   B) jejunum
   C) ileum
   D) pylorus
   E) cecum
   Answer: C

34. Which of the following best describes the location of the esophagus?
   A) from the fauces to the stomach
   B) posterior to the vertebral column from the laryngopharynx to the stomach
   C) retroperitoneal
   D) posterior to the trachea, anterior to the vertebral column from laryngopharynx to stomach
   E) anterior to the trachea and heart, just under the sternum
   Answer: D

Pages: 754
35. The contractions of the muscularis that result in waves of movement of food through the gastrointestinal tract are known as:
A) rhythmic segmentation
B) haustal churning
C) peristalsis
D) pendular movements
E) plicae circularis
Answer: C

Pages: 780
36. The greenish color of bile is the result of the presence of breakdown products of:
A) hemoglobin
B) urea
C) starch
D) the B vitamins
E) fats
Answer: A

Pages: 755
37. Skeletal muscles constitute the muscularis layer of the wall of the gastrointestinal tract in the:
A) oropharynx
B) stomach
C) duodenum
D) ascending colon
E) rectum
Answer: A

Pages: 796
38. The large intestine absorbs mostly:
A) amino acids
B) monosaccharides
C) bile pigments
D) water
E) triglycerides
Answer: D

Pages: 783
39. Gastric emptying is stimulated by **ALL** of the following **EXCEPT**:
A) distension of the stomach
B) gastrin
C) CCK
D) partially digested proteins
E) the vagus nerve
Answer: C

Pages: 774
40. Partially digested food is usually passed from the stomach to the small intestine about how long after consumption?
    A) an hour or less
    B) 2-4 hours
    C) 6-8 hours
    D) 10-12 hours
    E) 24 hours
    Answer: B

Pages: 783
41. The difference between the effects of secretin and the effects of CCK on the pancreas is that:
    A) secretin stimulates secretion of pancreatic juice, while CCK inhibits secretion of pancreatic juice
    B) secretin stimulates the acini of the pancreas, while CCK stimulates the pancreatic islets
    C) secretin stimulates alpha cells and CCK stimulates beta cells
    D) secretin causes dilation of the pancreatic duct, while CCK causes constriction of the duct
    E) secretin stimulates secretion of pancreatic juice rich in bicarbonate, while CCK stimulates secretion of pancreatic juice rich in digestive enzymes
    Answer: E

Pages: 780
42. A portal triad consists of:
    A) one branch each from the hepatic portal vein, the hepatic artery, and a bile duct
    B) the three branches of the hepatic portal vein within the liver
    C) the three smallest lobes of the liver
    D) one branch each from the hepatic portal vein, the hepatic vein, and a central vein
    E) all of the hepatocytes surrounding a branch of the hepatic portal vein
    Answer: A

Pages: 781
43. The liver performs the process of gluconeogenesis, which is conversion of:
    A) glucose to glycogen
    B) glucose to triglycerides
    C) glycogen to glucose
    D) non-carbohydrates to glucose
    E) glucose to amino acids
    Answer: D

Pages: 782
44. The liver produces urea to:
    A) detoxify ammonia produced via deamination of proteins
    B) keep bile in an inactive form until it reaches the small intestine
    C) convert into glucose when blood glucose is low
    D) store iron
    E) bind to ingested poisons to detoxify them
    Answer: A

45. The function of apoferritin in hepatocytes is to:
    A) detoxify ammonia produced via deamination of proteins
    B) keep bile in an inactive form until it reaches the small intestine
    C) convert glycogen into glucose when blood glucose is low
    D) store iron
    E) bind to ingested poisons to detoxify them
    Answer: D

46. The digestive hormone that stimulates secretion of insulin is:
    A) secretin
    B) CCK
    C) GIP
    D) gastrin
    E) pepsin
    Answer: C

47. The function of Paneth cells within intestinal glands is to:
    A) produce mucus
    B) secrete gut hormones
    C) provide protection from bacteria
    D) secrete intestinal acids
    E) produce digestive enzymes
    Answer: C

48. The term **segmentation** refers to the:
    A) anatomical characteristics of the large intestine
    B) action of bacteria on undigested food in the large intestine
    C) forward movement of food through the gastrointestinal tract
    D) movement of intestinal contents between haustra in the large intestine
    E) localized mechanical digestion of chyme in the small intestine
    Answer: E

49. Maltose is a product of the hydrolysis of:
    A) fatty acids
    B) sucrose
    C) glycerol
    D) starch
    E) protein
    Answer: D

Pages: 789
50. Glucose is transported into epithelial cells of villi via:
    A) secondary active transport coupled to active transport of sodium ions
    B) secondary active transport coupled to active transport of galactose
    C) facilitated diffusion
    D) primary active transport
    E) pinocytosis
    Answer: A

Pages: 790
51. The role of micelles in absorption of triglycerides is to:
    A) transport triglycerides through the lymph
    B) make triglycerides more soluble in the water of intestinal fluid
    C) actively transport triglycerides into the intestinal cells
    D) hydrolyze triglycerides to fatty acids and glycerol
    E) protect triglycerides from premature hydrolysis in the stomach
    Answer: B

Pages: 796
52. The normal color of feces is due primarily to the:
    A) pigments in the bacteria present
    B) breakdown products of hemoglobin
    C) pigments in foods consumed
    D) pigments in epithelial cells sloughed off from the gastrointestinal tract wall
    E) chemical interactions of undigested foods
    Answer: B

Pages: 796
53. The receptors in the defecation reflex are:
    A) stretch receptors in the ileocecal sphincter
    B) baroreceptors in the rectal veins
    C) osmoreceptors in the sigmoid colon
    D) chemoreceptors in the rectal wall
    E) stretch receptors in the rectal wall
    Answer: E

Pages: 798
54. The embryonic stomodeum develops into the:
    A) stomach
    B) anus
    C) oral cavity
    D) pancreas
    E) liver
    Answer: C

Pages: 800
55. The bacterium, <u>Helicobacter</u> <u>pylori</u>, is thought to be an important cause of:
    A) dental caries
    B) hepatitis
    C) colon cancer
    D) gastric ulcers
    E) bulimia
    Answer: D

Pages: 800
56. Hepatitis B is transmitted by:
    A) direct contact with body fluids
    B) the fecal-oral route
    C) inhalation of the virus
    D) biting insects
    E) contact with infected pets
    Answer: A

Pages: 763
57. The process of mastication results in:
    A) passage of food from the oral cavity into the esophagus
    B) removal of pathogens from partially digested food by MALT tissues
    C) mechanical mixing of food with saliva and shaping of food into a bolus
    D) sudden movement of colonic contents into the rectum
    E) passage of feces from the anus
    Answer: C

Pages: 770
58. The muscularis of most of the organs of the gastrointestinal tract consists of two layers of smooth muscle except in the:
    A) duodenum
    B) ileum
    C) sigmoid colon
    D) stomach
    E) ascending colon
    Answer: D

Pages: 770
59. Which of the following has the lowest pH?
    A) saliva
    B) gastric juice
    C) pancreatic juice
    D) bile
    E) intestinal juice
    Answer: B

60. The role of bicarbonate ions in pancreatic juice is to:
    A) stimulate opening of the sphincter of Oddi
    B) emulsify lipids to facilitate absorption
    C) buffer acidic mucus in intestinal juice
    D) activate trypsin
    E) buffer HCl in gastric juice
    Answer: E

61. The pulp cavity of a tooth is normally filled with:
    A) air
    B) blood vessels
    C) cementum
    D) dentin
    E) enamel
    Answer: B

62. The structure of incisors makes them specialized for:
    A) cutting food
    B) tearing and shredding food
    C) crushing and grinding food
    D) mixing food with saliva
    E) gustatory reception
    Answer: A

63. Which of the following occurs during the cephalic phase of gastric digestion?
    A) Chemoreceptors detect a change in pH of gastric juice.
    B) Stretch receptors detect distension of the stomach.
    C) Chemoreceptors detect fatty acids in the duodenum.
    D) Sight, smell, thought, or taste of food triggers parasympathetic impulses.
    E) CCK is secreted by enteroendocrine cells.
    Answer: D

64. In the neural negative feedback loop controlling secretion of gastric juice, the output in response to entry of food into the stomach would include:
    A) decreased parasympathetic impulses
    B) increased secretory activity by parietal cells
    C) increased distension of the stomach wall
    D) increased sympathetic impulses
    E) decreased secretory activity of chief cells
    Answer: B

Pages: 780
65. Blood in liver sinusoids passes next into the:
     A) hepatic artery
     B) hepatic portal vein
     C) central vein
     D) inferior vena cava
     E) right and left hepatic ducts
     Answer: C

True-False

Write T if the statement is true and F if the statement is false.

Pages: 753
1. The liver is a part of the gastrointestinal tract.
    Answer: False

Pages: 754
2. Chemical digestion is a series of hydrolysis reactions.
    Answer: True

Pages: 755
3. Gastrointestinal motility is controlled by the myenteric plexus (of Auerbach).
    Answer: True

Pages: 756
4. The stomach and duodenum are suspended from the liver by the mesentery.
    Answer: False

Pages: 759
5. The function of the uvula is to prevent swallowed food from entering the nasopharynx and the nasal cavity.
    Answer: True

Pages: 759
6. Greater sympathetic stimulation to the salivary glands increases production of saliva.
    Answer: False

Pages: 759
7. Salivary amylase initiates the hydrolysis of triglycerides.
    Answer: False

Pages: 768
8. Chief cells and parietal cells in the stomach are examples of enteroendocrine cells.
    Answer: False

Pages: 768
9. A function of hydrochloric acid in gastric juice is to convert pepsinogen to pepsin.
    Answer: True

Pages: 783
10. Secretin is a hormone that decreases gastric secretions.
Answer: True

Pages: 776
11. The principal triglyceride-digesting enzyme in adults is produced by the pancreas.
Answer: True

Pages: 780
12. The function of bile is to catalyze the hydrolysis of lipids.
Answer: False

Pages: 780
13. The gallbladder secretes bile.
Answer: False

Pages: 788
14. Most absorption of nutrients occurs in the small intestine.
Answer: True

Pages: 791
15. Chylomicrons are protein-coated lipid globules that are transported via the lymphatic vessels to the blood.
Answer: True

Short Answer

Write the word or phrase that best completes each statement or answers the question.

Pages: 784
1. Projections of the mucosa of the small intestine that increase surface area for absorption and digestion are called _____.
Answer: villi

Pages: 776
2. The exocrine cells of the pancreas are arranged in clusters called _____; the endocrine cells of the pancreas are organized into clusters called _____.
Answer: acini; pancreatic islets (of Langerhans)

Pages: 780
3. The common bile duct is formed by the union of the _____ and the _____.
Answer: common hepatic duct; cystic duct

Pages: 780
4. Bile is produced by the _____.
Answer: liver (hepatocytes)

Pages: 780
5. The function of the stellate reticuloendothelial cells of the liver is _____.
Answer: phagocytosis

Pages: 782
6. The hormone that stimulates ejection of bile from the gallbladder is _____.
   Answer: cholecystokinin

Pages: 784
7. Partially digested food passes from the _____ region of the stomach into the _____ region of the small intestine.
   Answer: pylorus; duodenum

Pages: 780
8. The process by which bile salts break triglycerides into one millimeter droplets is called _____.
   Answer: emulsification

Pages: 789
9. Trypsin and chymotrypsin are enzymes produced by the _____.
   Answer: pancreas

Pages: 789
10. The substrate for salivary amylase is _____.
    Answer: starch

Pages: 789
11. The substrate for trypsin is _____.
    Answer: protein

Pages: 789
12. The substrate for sucrase is _____, which is hydrolyzed into _____ and _____.
    Answer: sucrose; glucose; fructose

Pages: 794
13. The last 20 cm of the large intestine is called the _____.
    Answer: rectum

Pages: 794
14. The prominent bands of smooth muscle running longitudinally along the large intestine are the _____.
    Answer: taeniae coli

Pages: 796
15. The contents of the colon are moved quickly from the transverse colon into the rectum by a movement called _____.
    Answer: mass peristalsis

Pages: 754
16. The layer of areolar connective tissue underlying the epithelium of the mucosa of the gastrointestinal tract is called the _____.
    Answer: lamina propria

Pages: 757
17. Accumulation of serous fluid in the peritoneal cavity is called _____.
Answer: ascites

Pages: 757
18. The largest of the peritoneal folds that drapes over the transverse colon and small intestine is the _____.
Answer: greater omentum

Pages: 756
19. The salivary glands located inferior and anterior to the ears between the skin and masseter muscle are the _____.
Answer: parotid glands

Pages: 761
20. Teeth are composed primarily of a calcified connective tissue called _____.
Answer: dentin

Pages: 764
21. The act of swallowing is also called _____.
Answer: deglutition

Pages: 764
22. The condition in which the stomach protrudes above the diaphragm through the esophageal hiatus is known as a(n) _____.
Answer: hiatal hernia

Pages: 768
23. Chief cells of the stomach produce _____.
Answer: pepsinogen

Pages: 770
24. G cells of the stomach secrete _____.
Answer: gastrin

Pages: 768
25. Folds in the gastric mucosa are called _____.
Answer: rugae

Pages: 776
26. The pancreatic duct and the common bile duct unite to form the _____.
Answer: hepatopancreatic ampulla

Pages: 776
27. Trypsinogen is converted to trypsin by the action of the enzyme _____.
Answer: enterokinase

Pages: 780
28. The right and left hepatic ducts are formed by the merging of smaller ducts called _____.
Answer: bile canaliculi

29. The microscopic, finger-like projections of the apical membranes of absorptive cells in the small intestine are called _____.
    Answer: microvilli

30. A yellowish coloration of the sclerae and mucous membranes due to the buildup of bilirubin is called _____.
    Answer: jaundice

Matching

Choose the item from Column 2 that best matches each item in Column 1.

1. stomach	fundus
2. small intestine	jejunum
3. large intestine	cecum
4. liver	caudate lobe
5. gallbladder	cystic duct
6. pancreas	acini
7. peritoneum	mesentery
8. tongue	fungiform papillae
9. teeth	cementum
10. parotid gland	Stensen's duct

Essay

Write your answer in the space provided or on a separate sheet of paper.

1. Identify and briefly define the six basic digestive processes.
   Answer: Ingestion = eating; secretion = addition of juices into the GI tract by endocrine and exocrine cells; mixing and propulsion via action of muscle tissue moves food through GI tract; digestion = breakdown - either mechanical or chemical (via enzymes); absorption = taking materials from the lumen of the GI tract and transporting them into mucosal epithelium to blood or lymph; defecation = elimination of indigestible substances and bacteria through the anus

Pages: 754
2. Identify the layers of the wall of the gastrointestinal tract, and briefly describe their structures and functions.

   Answer: The mucosa is the inner layer, consisting of epithelium (stratified squamous or simple columnar), the lamina propria (areolar connective tissue), and a layer of smooth muscle. There may be lymphatic tissue here. The submucosa is loose connective tissue, is highly vascular, and contains the myenteric plexus. The muscularis is either skeletal muscle for chewing and swallowing, or smooth muscle (2 or 3 layers). The serosa is the outer layer in digestive organs inferior to the diaphragm. It is the visceral peritoneum.

Pages: 770
3. Describe the digestive activities occurring in the stomach.

   Answer: Mechanical digestion occurs via mixing waves and peristalsis. Chief cells produce pepsinogen, which is activated by HCl from parietal cells to pepsin for protein digestion. Chief cells also secrete gastric lipase for fat digestion in infants. Parietal cells also secrete intrinsic factor for absorption of vitamin B_{12}. Mucous cells make mucus for protection. G cells secrete gastrin to control gastric digestive activity.

Pages: 775
4. Describe the location, gross anatomy, and microscopic anatomy of the pancreas. What are the functions of the different types of cells in the pancreas?

   Answer: The pancreas lies posterior to the greater curvature of the stomach, and is connected by the pancreatic and accessory ducts to the duodenum. It is retroperitoneal and about 6" long and 1" thick. The head fits into the duodenum; the body and tail extend away from the head. Acini produce digestive enzymes, and make up about 99% of pancreatic cells. The pancreatic islets produce glucagon (from alpha cells), insulin (from beta cells), and somatostatin (from delta cells).

Pages: 789
5. Identify the protein-hydrolyzing enzymes in the digestive tract, and name their sources. Why are these enzymes released in an inactive form?

   Answer: Pepsin from stomach; trypsin, chymotrypsin, carboxypeptidase from pancreas; aminopeptidase and dipeptidase from small intestine. The enzymes are not activated until in the lumen of the stomach or small intestine because they would otherwise digest the proteins in the cells that produce them.

Pages: 780
6. Describe the flow of blood into, through, and out of the liver. What is the functional significance of this arrangement?

   Answer: Oxygenated blood comes from the hepatic artery. Deoxygenated blood containing newly absorbed nutrients comes from the hepatic portal vein. Blood from each mixes in sinusoids, where oxygen, nutrients, and poisons are extracted, and other nutrients are added. Blood then passes into the central vein, then to the hepatic vein. Nutrients newly absorbed are processed quickly, without having to make the full vascular circuit.

Pages: 781
7. Describe the role of the liver in carbohydrate metabolism.
    Answer: When blood glucose is low, the liver performs glycogenolysis (glycogen to glucose) and
             gluconeogenesis (amino acids, etc., to glucose) to add glucose to blood. When blood
             glucose is high, the liver performs glycogenesis (glucose to glycogen) and lipogenesis
             (glucose to triglycerides) for storage.

Pages: 782
8. Describe the role of the liver in disposing of worn-out red blood cells.
    Answer: Bilirubin, derived from heme, is absorbed by the liver and secreted in the bile. Iron
             is stored in the form of ferritin. Stellate reticuloendothelial cells phagocytize old
             red blood cells.

Pages: 796
9. A patient is passing feces that are chalky white in color. What problems might this indicate?
    Explain your answer.
    Answer: Lack of brown coloration indicates lack of bile pigment derivatives in the feces.
             Since these come from the action of the liver, the problem could be with the liver
             itself or with the route the bile takes to get to the small intestine.

Pages: 789
10. This morning you drank your coffee black with one teaspoon of sugar and you had a donut with
    powdered sugar on it. Describe the fates of the table sugar, grain sugar, and fats from this
    meal in the digestive tract.
    Answer: Sucrose (table sugar) is broken down by sucrase in the small intestine into glucose
             and fructose. Maltose (grain sugar) is broken down by maltase in the small intestine
             into two molecules of glucose. The sugars are absorbed by secondary active transport
             (glucose) or facilitated diffusion (fructose). The fats are digested by lipases
             (lingual, pancreatic), emulsified by bile, and absorbed either directly into
             epithelial cells, to blood, or into the lymph.

Chapter 25   Metabolism

Multiple-Choice

Choose the one alternative that best completes the statement or answers the question.

Pages: 818
1. The reactions of the Krebs cycle and the electron transport chain occur in the:
   A) cytosol
   B) mitochondria
   C) nucleus
   D) interstitial fluid
   E) ribosomes
   Answer: B

Pages: 818
2. Lactic acid is produced as a result of the chemical reduction of:
   A) acetyl CoA
   B) oxaloacetic acid
   C) pyruvic acid
   D) cytochromes
   E) NAD
   Answer: C

Pages: 825
3. The end-product of glycogenolysis in skeletal muscle is:
   A) glycogen
   B) glucose 1-phosphate
   C) carbon dioxide
   D) pyruvic acid
   E) acetyl CoA
   Answer: B

Pages: 824
4. The end-products of the complete aerobic oxidation of glucose are:
   A) fatty acids and glycerol
   B) ATP and oxygen
   C) amino acids
   D) carbon dioxide and water
   E) pyruvic acid and lactic acid
   Answer: D

Pages: 818
5. The function of decarboxylases is to:
   A) convert ADP to ATP
   B) convert ammonia into urea
   C) carry 2-carbon units into the Krebs cycle
   D) remove carbon dioxide from compounds
   E) transport hydrogen ions into mitochondria
   Answer: D

Pages: 827
6. Beta oxidation is the process by which:
   A) hydrogen ions are removed from compounds in the Krebs cycle
   B) carbon dioxide is removed from compounds in the Krebs cycle
   C) amine groups are removed from proteins
   D) fatty acids are broken down for use in the Krebs cycle
   E) ADP is converted to ATP
   Answer: D

Pages: 828
7. Urea is produced in the process of detoxifying:
   A) ammonia
   B) lactic acid
   C) carbon dioxide
   D) pyruvic acid
   E) ketone bodies
   Answer: A

Pages: 828
8. The compound that is converted into urea by the liver is formed from the:
   A) acetyl units formed during lipolysis
   B) amine groups removed during deamination
   C) reactions of the electron transport chain
   D) reactions of glycogenolysis
   E) lactic acid formed during anaerobic respiration
   Answer: B

Pages: 826
9. The conversion of glycerol into glyceraldehyde 3-phosphate for use in glycolysis is an example of:
   A) glycogenolysis
   B) deamination
   C) beta oxidation
   D) oxidative phosphorylation
   E) gluconeogenesis
   Answer: E

Pages: 809
10. The body's thermostat is located in the:
    A) liver
    B) skin
    C) hypothalamus
    D) rectum
    E) kidneys
    Answer: C

11. The exchange of heat via the movement of fluids is called:
    A) radiation
    B) conduction
    C) convection
    D) evaporation
    E) transduction
    Answer: C

12. The complete hydrolysis of proteins yields:
    A) amino acids
    B) fatty acids and glycerol
    C) nucleic acids
    D) monosaccharides
    E) carbon dioxide and water
    Answer: A

13. The complete hydrolysis of carbohydrates yields:
    A) amino acids
    B) fatty acids and glycerol
    C) nucleic acids
    D) monosaccharides
    E) carbon dioxide and water
    Answer: D

14. **ALL** of the following hormones raise blood glucose **EXCEPT**:
    A) human growth hormone
    B) glucagon
    C) cortisol
    D) insulin
    E) epinephrine
    Answer: D

15. The processes of lipogenesis, protein synthesis, and glycogenesis, are all promoted by the hormone:
    A) human growth hormone
    B) glucagon
    C) cortisol
    D) insulin
    E) epinephrine
    Answer: D

Pages: 807
16. The function of the satiety center is to:
    A) regulate the rate of lipogenesis
    B) stimulate consumption of food
    C) regulate the release of insulin
    D) stimulate cessation of feeding
    E) regulate body temperature
    Answer: D

Pages: 810
17. The hormones primarily responsible for daily regulation of production of body heat are produced by the:
    A) pancreas
    B) thyroid gland
    C) adrenal cortex
    D) adrenal medulla
    E) hypothalamus
    Answer: B

Pages: 813
18. Most biological oxidations are:
    A) dehydrogenation reactions
    B) dehydration reactions
    C) hydrolysis reactions
    D) phosphorylation reactions
    E) decarboxylation reactions
    Answer: A

Pages: 813
19. The function of NAD^+ and FAD is to:
    A) carry acetyl units into the Krebs cycle
    B) increase the rate of glycolysis
    C) increase the rate of facilitated diffusion of glucose into cells
    D) transfer a high-energy phosphate group from glucose 6-phosphate onto ADP
    E) carry hydrogen atoms between compounds
    Answer: E

Pages: 813
20. Conversion of NAD^+ to NADH and H^+ is an example of:
    A) oxidation
    B) reduction
    C) phosphorylation
    D) hydrolysis
    E) dehydration
    Answer: B

21. The role of insulin in the body's utilization of glucose is to:
    A) catalyze the conversion of glucose 6-phosphate into glycogen
    B) carry acetyl units into the Krebs cycle
    C) increase the rate of facilitated diffusion of glucose into cells
    D) transport hydrogen ions between compounds of the Krebs cycle and compounds of the electron transport chain
    E) increase the rate of glycolysis
    Answer: C

22. Glycolysis is said to be an anaerobic process because it does not require:
    A) glucose
    B) NAD
    C) oxygen
    D) ATP
    E) enzymes
    Answer: C

23. The function of coenzyme A in glucose metabolism is to:
    A) reduce pyruvic acid to lactic acid
    B) convert glucose 6-phosphate into glycogen
    C) transport glucose from the blood across cell membranes
    D) carry hydrogen ions between compounds in the Krebs cycle and compounds in the electron transport chain
    E) carry two-carbon units into the Krebs cycle
    Answer: E

24. The primary significance of the Krebs cycle in terms of ATP production is:
    A) production of large amounts of GTP that can be converted to ATP
    B) transfer of energy into NADH and $FADH_2$
    C) generation of carbon dioxide for use in the electron transport chain
    D) transfer of energy into ATP between each step of the cycle
    E) production of acetyl units
    Answer: B

25. The enzyme found in the hydrogen ion channels between the inner and outer mitochondrial membranes is:
    A) cytochrome oxidase
    B) ATP synthase
    C) NADH dehydrogenase
    D) citric synthetase
    E) succinyl kinase
    Answer: B

Pages: 818
26. Upon entry into the Krebs cycle, acetyl CoA combines with:
    A) acetoacetic acid
    B) pyruvic acid
    C) citric acid
    D) oxaloacetic acid
    E) NAD
    Answer: D

Pages: 829
27. A positive nitrogen balance probably exists in:
    A) a starving person
    B) an uncontrolled diabetic
    C) someone recovering from surgery
    D) someone on a low-calorie, low-fat diet
    E) a patient with muscular dystrophy
    Answer: C

Pages: 828
28. Which of the following compounds is a ketone body?
    A) acetoacetic acid
    B) pyruvic acid
    C) cholesterol
    D) oxaloacetic acid
    E) glycerol
    Answer: A

Pages: 828
29. A person who is excreting large amounts of ketone bodies is probably:
    A) using mostly fatty acids for energy production
    B) producing large amounts of new protein
    C) just starting to exercise after a long period of inactivity
    D) not taking in enough oxygen
    E) in renal failure
    Answer: A

Pages: 818
30. Most of the lactic acid produced as a result of anaerobic metabolic activity is:
    A) converted into acetyl units for use in the Krebs cycle
    B) converted back into pyruvic acid
    C) deaminated
    D) converted into ketone bodies
    E) detoxified by binding to oxygen
    Answer: B

Pages: 827
31. Beta oxidation of fatty acids yields:
    A) glycerol
    B) cytochromes
    C) cholesterol
    D) acetyl units
    E) glucose 6-phosphate
    Answer: D

Pages: 825
32. For glycerol to be used in carbohydrate metabolism, it is first converted into:
    A) fatty acids
    B) oxaloacetic acid
    C) glyceraldehyde 3-phosphate
    D) glucose 6-phosphate
    E) citric acid
    Answer: C

Pages: 825
33. For glycogen to be used for energy production, it must first be converted into:
    A) glycerol
    B) glucose 6-phosphate
    C) carbon dioxide
    D) lactic acid
    E) an acetyl unit
    Answer: B

Pages: 840
34. Coenzymes FAD and FMN are derivatives of the vitamin:
    A) folic acid
    B) thiamine
    C) riboflavin
    D) ascorbic acid
    E) pantothenic acid
    Answer: C

Pages: 840
35. The fat soluble vitamin needed for proper formation of prothrombin by the liver is:
    A) vitamin A
    B) pantothenic acid
    C) vitamin C
    D) vitamin K
    E) folic acid
    Answer: D

Pages: 818

36. In the absence of adequate oxygen to enter the Krebs cycle, pyruvic acid will be converted into:
    A) lactic acid
    B) acetone
    C) oxaloacetic acid
    D) urea
    E) ketone bodies
    Answer: A

Pages: 841

37. The B vitamin that is used for production of coenzyme A is:
    A) pantothenic acid
    B) folic acid
    C) biotin
    D) cyanocobalamin
    E) riboflavin
    Answer: A

Pages: 841

38. The only vitamin not found in fruits and vegetables is:
    A) ascorbic acid
    B) vitamin K
    C) riboflavin
    D) cyanocobalamin
    E) folic acid
    Answer: D

Pages: 838

39. The mineral that is needed for proper functioning of carbonic anhydrase is:
    A) zinc
    B) sodium
    C) manganese
    D) iron
    E) calcium
    Answer: A

Pages: 811

40. Vasoconstriction is considered to be a heat-saving mechanism because:
    A) the energy released by vascular smooth muscle contraction warms you up
    B) sweat glands don't receive enough blood supply to produce sweat
    C) less heat can be conducted to the surface and radiated away
    D) more heat is generated by friction as a larger volume of blood is forced through a smaller space
    E) all of these are correct
    Answer: C

Pages: 836

41. According to the USDA Food Guide Pyramid, the largest number of food servings per day should come from the:
    A) fruit group
    B) vegetable group
    C) bread, cereal, rice, and pasta group
    D) milk, yogurt, and cheese group
    E) meat, poultry, fish, beans, eggs, and nuts group
    Answer: C

Pages: 828

42. Before amino acids can enter the Krebs cycle, they must be:
    A) oxidized
    B) reduced
    C) decarboxylated
    D) deaminated
    E) dehydrated
    Answer: D

Pages: 828

43. A compound that is a product of ketogenesis is:
    A) oxaloacetic acid
    B) cholesterol
    C) pyruvic acid
    D) glucose 6-phosphate
    E) beta-hydroxybutyric acid
    Answer: E

Pages: 825

44. Glycogen is synthesized by liver cells and:
    A) adipocytes
    B) skeletal muscle fibers
    C) neurons in the brain
    D) endocrine cells in the pancreas
    E) columnar epithelial cells in the small intestine
    Answer: B

Pages: 823

45. In chemiosmosis, ATP is produced when:
    A) a high energy phosphate group is passed from glucose 6-phosphate to ADP
    B) hydrogen ions are bound to NAD and FAD
    C) pyruvic acid is converted to an acetyl unit
    D) hydrogen ions diffuse into the mitochondrial matrix
    E) glucose is transported across the cell membrane
    Answer: D

46. Glucose moves into most cells by:
    A) primary active transport
    B) secondary active transport
    C) facilitated diffusion
    D) pinocytosis
    E) simple diffusion
    Answer: C

47. Each molecule of acetyl CoA that enters the Krebs cycle produces how many molecules of carbon dioxide?
    A) one
    B) two
    C) four
    D) six
    E) 36
    Answer: B

48. During the complete oxidation of one glucose molecule, the electron transport chain yields how many molecules of water?
    A) one
    B) two
    C) four
    D) six
    E) 32
    Answer: D

49. Most ATP generated by the complete oxidation of glucose results from the reactions of:
    A) glycogenolysis
    B) glycolysis
    C) the Krebs cycle
    D) the electron transport chain
    E) gluconeogenesis
    Answer: D

50. GTP is generated by the reactions of:
    A) glycogenolysis
    B) glycolysis
    C) the Krebs cycle
    D) the electron transport chain
    E) gluconeogenesis
    Answer: C

Pages: 828
51. Acidosis may result from buildup of:
   A) ketone bodies
   B) glucose
   C) glycogen
   D) fat-soluble vitamins
   E) glycerol
   Answer: A

Pages: 841
52. Scurvy results from a(n):
   A) excessive intake of vitamin A
   B) deficiency of folic acid
   C) deficiency of vitamin D
   D) excessive intake of vitamin D
   E) deficiency of vitamin C
   Answer: E

Pages: 841
53. Pernicious anemia results from a deficiency of:
   A) niacin
   B) cyanocobalamin
   C) folic acid
   D) vitamin D
   E) ascorbic acid
   Answer: B

Pages: 838
54. A mineral that acts as an antioxidant is:
   A) zinc
   B) chlorine
   C) selenium
   D) sulfur
   E) chromium
   Answer: C

Pages: 838
55. The mineral needed for synthesis of hemoglobin and found as part of an enzyme needed for melanin synthesis is:
   A) iron
   B) magnesium
   C) zinc
   D) calcium
   E) copper
   Answer: E

Pages: 836
56. According to nutrition experts, complex carbohydrates should make up what percentage of caloric intake?
A) less than 10%
B) at least 35-45%
C) 75-80%
D) 25-30%
E) 12-15%
Answer: B

Pages: 835
57. Gluconeogenesis occurs primarily in hepatocytes and:
A) adipocytes
B) neurons
C) skeletal muscle fibers
D) kidney cortex cells
E) cardiac muscle cells
Answer: D

Pages: 835
58. During a prolonged fast or in starvation, most tissues will:
A) oxidize glucose at a faster rate
B) break down the coenzymes NAD and FAD to make ATP
C) switch from glucose to fatty acids as the primary energy source
D) increase the storage of glycogen in skeletal muscle fibers
E) convert more glucose to triglycerides
Answer: C

Pages: 832
59. Which of the following is **NOT** characteristic of the absorptive state?
A) production of ATP by oxidizing glucose to carbon dioxide and water by most body cells
B) storage of dietary lipids in adipose tissue
C) gluconeogenesis using amino acids in the liver
D) packaging of fatty acids and triglycerides into VLDLs
E) transport of amino acids into body cells
Answer: C

Pages: 833
60. Which of the following is **NOT** characteristic of the postabsorptive state?
A) glycogenolysis of liver glycogen
B) gluconeogenesis using lactic acid in skeletal muscle fibers
C) beta oxidation of fatty acids by most body cells
D) oxidation of ketone bodies by heart and kidneys
E) lipogenesis by hepatocytes
Answer: E

Pages: 833
61. The primary hormone regulating the metabolic reactions and membrane transport activities of the absorptive state is:
   A) glucagon
   B) thyroxine
   C) epinephrine
   D) insulin
   E) cortisol
   Answer: D

Pages: 830
62. Which of the following is a nonessential amino acid?
   A) tyrosine
   B) valine
   C) leucine
   D) methionine
   E) phenylalanine
   Answer: A

Pages: 830
63. Nonessential amino acids are produced from essential amino acids by the process of:
   A) oxidation
   B) reduction
   C) deamination
   D) dehydrogenation
   E) transamination
   Answer: E

Pages: 820
64. Decarboxylation reactions occur in:
   A) the electron transport chain
   B) the Krebs cycle
   C) chemiosmosis
   D) glycogenolysis
   E) ketogenesis
   Answer: B

Pages: 823
65. The function of oxygen in aerobic respiration is to:
   A) carry hydrogen ions from the Krebs cycle to the electron transport chain
   B) bind electrons and hydrogen ions at the end of the electron transport chain
   C) attach a high energy phosphate group to ADP
   D) form carbon dioxide in the Krebs cycle
   E) donate electrons to NADH to make it release a hydrogen ion
   Answer: B

True-False

Write T if the statement is true and F if the statement is false.

Pages: 807

1. The lipostatic theory states that most cells in the body will use fatty acids as their primary energy source when caloric intake is low.
Answer: False

Pages: 809

2. The thermoregulatory centers of the hypothalamus are located in the preoptic area.
Answer: True

Pages: 810

3. If body temperature decreases, the hypothalamus secretes more TRH.
Answer: True

Pages: 813

4. FAD is the reduced form of the coenzyme, and $FADH_2$ is the oxidized form.
Answer: False

Pages: 814

5. Kinases are enzymes that catalyze phosphorylation reactions.
Answer: True

Pages: 814

6. Synthesis of triglycerides is known as lipogenesis.
Answer: True

Pages: 815

7. The reactions of the Krebs cycle are part of anaerobic cellular respiration.
Answer: False

Pages: 819

8. When acetyl CoA enters the Krebs cycle, the acetyl unit combines with citric acid to form isocitric acid.
Answer: False

Pages: 819

9. During the Krebs cycle, a molecule of GTP is formed as succinyl CoA is converted to succinic acid.
Answer: True

Pages: 822

10. In the electron transport chain, cytochrome a_3 passes its electrons to cytochrome c.
Answer: False

11. A gram of triglyceride produces a little more than twice the kilocalories as a gram of carbohydrate or protein.
   Answer: True

12. Conversion of amino acids into glucose is an example of gluconeogenesis.
   Answer: True

13. Ribose 5-phosphate is synthesized principally from beta-hydroxybutyric acid.
   Answer: False

14. The predominant hormone regulating the postabsorptive state is insulin.
   Answer: False

15. During starvation, nervous tissue uses glucose and ketone bodies for ATP production.
   Answer: True

Short Answer

Write the word or phrase that best completes each statement or answers the question.

1. The feeding and satiety centers are located in the _____.
   Answer: hypothalamus

2. The term that refers to all the chemical reactions of the body is _____.
   Answer: metabolism

3. Those chemical reactions that break down molecules and release energy are collectively known as _____. Those that synthesize larger molecules from smaller ones are collectively known as _____.
   Answer: catabolism; anabolism

4. The rate at which the resting, fasting body breaks down nutrients to liberate energy is called the _____.
   Answer: basal metabolic rate

5. Transfer of heat between objects without physical contact is called _____.
   Answer: radiation

Pages: 810
6. Transfer of heat between solid objects in contact with each other is called _____.
   Answer: conduction

Pages: 812
7. A fever-producing substance is called a _____.
   Answer: pyrogen

Pages: 810
8. The effect of ingested food on metabolic rate is called _____.
   Answer: specific dynamic action

Pages: 812
9. The lowering of body temperature below thirty-five degrees Centigrade is called _____.
   Answer: hypothermia

Pages: 814
10. Formation of ATP by transfer of a high-energy phosphate group from an intermediate phosphorylated compound to ADP is called _____ phosphorylation.
    Answer: substrate-level

Pages: 814
11. Formation of ATP via energy released during the reactions of the electron transport chain is called _____ phosphorylation.
    Answer: oxidative

Pages: 815
12. The oxidation of glucose by cells is also called _____.
    Answer: cellular respiration

Pages: 816
13. Glycolysis is the oxidation of glucose to _____.
    Answer: pyruvic acid

Pages: 818
14. Loss of a molecule of carbon dioxide by a substance is called _____.
    Answer: decarboxylation

Pages: 818
15. When acetyl CoA enters the Krebs cycle, the acetyl unit combines with _____ to form _____.
    Answer: oxaloacetic acid; citric acid

Pages: 819
16. During the Krebs cycle, two molecules of _____ are generated by substrate-level phosphorylation.
    Answer: GTP

Pages: 821
17. The linking of ATP generation with the pumping of hydrogen ions across the inner and outer mitochondrial membranes is called _____.
Answer: chemiosmosis

Pages: 821
18. Iron-containing proteins involved in the reactions of the electron transport chain are called

_____.
Answer: cytochromes

Pages: 824
19. The complete aerobic oxidation of glucose yields _____ and _____.
Answer: carbon dioxide; water

Pages: 824
20. The theoretical maximum number of ATP molecules produced as a result of the aerobic respiration of one glucose molecule is _____.
Answer: 38

Pages: 825
21. Glycogen is stored in the _____ and the _____.
Answer: liver; skeletal muscle tissue

Pages: 825
22. The process by which new glucose is formed from non-carbohydrate sources is called _____.
Answer: gluconeogenesis

Pages: 827
23. The process by which fatty acids are catabolized to two-carbon fragments is called _____.
Answer: beta oxidation

Pages: 828
24. Acetone, acetoacetic acid, and beta-hydroxybutyric acid are collectively known as _____.
Answer: ketone bodies

Pages: 828
25. Lipogenesis is stimulated by the hormone _____.
Answer: insulin

Pages: 828
26. During the deamination of proteins, an amino group is removed and converted to _____, which is then converted to _____ by the liver.
Answer: ammonia; urea

Pages: 833
27. During the postabsorptive state, blood glucose levels are normally maintained at _____ per 100 ml.
Answer: 70-110 mg

28. The breakdown of proteins, particularly in skeletal muscle fibers, is stimulated primarily by the hormone _____.
Answer: cortisol

29. The most abundant cation in the body is _____.
Answer: calcium ion

30. The vitamin that is essential for the formation of photopigments is _____.
Answer: vitamin A

Matching

Choose the item from Column 2 that best matches each item in Column 1.

1. vitamin A   essential for formation of photopigments

2. vitamin D   essential for the absorption and utilization of calcium from the gastrointestinal tract

3. vitamin E   antioxidant thought to inhibit catabolism of fatty acids in cell membranes

4. vitamin K   essential for synthesis of prothrombin by the liver

5. vitamin C   antioxidant important for collagen production and wound healing

6. vitamin B_1 important for the synthesis of acetylcholine

7. vitamin B_2 component of coenzymes FAD and FMN

8. niacin   component of coenzymes NAD and NADP

9. vitamin B_6 coenzyme for amino acid metabolism and production of antibodies

10. vitamin B_{12} necessary for red blood cell formation and synthesis of methionine

11. pantothenic acid   part of coenzyme A

12. folic acid   necessary for purine and pyrimidine synthesis and blood cell production

Essay

Write your answer in the space provided or on a separate sheet of paper.

Pages: 807
1. What functions define a molecule as a nutrient?
   Answer: A molecule is a nutrient if it is used to 1) supply energy for a life sustaining process, 2) synthesize structural or functional molecules, or 3) is stored for future use.

Pages: 812
2. Your lab partner has the flu, but came to school anyway. Now it's two days later, and you feel chills. Explain this reaction using your knowledge of non-specific resistance and thermoregulatory mechanisms.
   Answer: [note: This question is designed to integrate material from multiple chapters.] A macrophage phagocytizes the pathogen, and secretes IL-1, which acts as a pyrogen on the hypothalamus. The hypothalamus secretes PGEs to reset the hypothalamic thermosta to a higher level. Heat promoting responses, including vasoconstriction, increased metabolism, and shivering, help raise the body temperature to the new level. The increased temperature increases T cell production, macrophage activity, heart rate, etc.

Pages: 814
3. Briefly outline the possible fates of glucose in the body.
 Answer: 1) immediate oxidation for ATP production
 2) synthesis of amino acids for protein synthesis
 3) synthesis of glycogen for storage in liver and skeletal muscle
 4) formation of triglycerides via lipogenesis for long-term storage after glycogen stores are full
 5) excretion in urine if blood glucose very high

Pages: 831
4. What are the possible fates of pyruvic acid in the body? What is the primary determinant of the fate of pyruvic acid? What is the fate of the compounds to which pyruvic acid may be converted?
 Answer: Pyruvic acid in the presence of low oxygen levels is reduced to lactic acid, which is converted to either glycogen or carbon dioxide. In the presence of high oxygen levels, pyruvic acid is converted to an acetyl unit, which may be carried into the Krebs cycle by coenzyme A, or converted into fatty acids, ketone bodies, or cholesterol.

Pages: 830
5. What are the possible fates of glucose 6-phosphate within human cells?
 Answer: G6P may be dephosphorylated to glucose, which can enter the bloodstream. It may be converted into glycogen and stored. It can be oxidized to pyruvic acid for ATP production, or converted into ribose 5-phosphate for use in synthesis of nucleic acids.

Pages: 828
6. Describe the process of ketogenesis. Under what circumstances would you expect to see elevated levels of the end-products of this process?
 Answer: The liver converts the acetyl units of acetyl CoA to acetoacetic acid, which is converted to acetone and beta-hydroxybutyric acid. These reactions are increased during periods of fasting and starvation and also in cases of uncontrolled diabetes mellitus. In these situations, lipolysis reactions are increased, which leads to ketogenesis.

Pages: 824
7. Summarize the mechanisms by which ATP is produced during the aerobic respiration of one glucose molecule.
 Answer: Glycolysis yields 2 ATP via substrate-level phosphorylation and 4 or 6 via oxidative phosphorylation in the electron transport chain from $2NADH + 2H^+$. During the formation of acetyl CoA, 6 ATP are made from $2NADH + 2H^+$ entering the electron transport chain. The Krebs cycle generates 2 GTP. $6NADH + 6H^+ + 2FADH_2$ yields 22 ATP via the electron transport chain.

Pages: 818
8. Describe the chemiosmotic mechanism of ATP generation.
 Answer: The proton pump in the inner mitochondrial membrane expels hydrogen ions from the mitochondrial matrix to create an electrochemical gradient of hydrogen ions. The inner mitochondrial membrane is nearly impermeable to hydrogen ions, which results in a gradient that creates the proton motive force. In regions where specific channels exist, hydrogen ions diffuse back across the membrane, triggering ATP production via action of ATP synthase in the channels.

Pages: 818
9. Summarize the Krebs cycle. What defines this process as a "cycle?"

Answer: Acetyl units (2C) enter via CoA to combine with oxaloacetic acid (4C) to form citric acid (6C). Two decarboxylations remove two carbon dioxides, returning the process to the 4C compound (hence the "cycle"). 2GTP are formed by substrate-level phosphorylation. 6NADH, 6H$^+$, and 2FADH$_2$ are formed by the redox reactions that occur in the cycle, all of which enter the electron transport chain to yield 22 ATP. All reactions occur in the matrix of the mitochondria.

Pages: 831
10. What are the possible sources and fates of acetyl CoA?

Answer: Acetyl units are derived from pyruvic acid, beta oxidation of fatty acids, and from certain amino acids. They may be used to form fatty acids (which are used to form triglycerides), ketone bodies, or cholesterol. The acetyl units may also be carried into the Krebs cycle for use in ATP production.

Chapter 26 The Urinary System

Multiple-Choice

Choose the one alternative that best completes the statement or answers the question.

Pages: 857
1. The part of a juxtamedullary nephron that is in the renal medulla is the:
   A) glomerulus only
   B) glomerular (Bowman's) capsule only
   C) renal corpsucle
   D) loop of Henle
   E) entire nephron
   Answer: D

Pages: 854
2. Which of the following lists the vessels in the correct order of blood flow?
   A) efferent arteriole, glomerulus, afferent arteriole, peritubular capillaries
   B) peritubular capillaries, efferent arteriole, glomerulus, afferent arteriole
   C) afferent arteriole, efferent arteriole, peritubular capillaries, glomerulus
   D) afferent arteriole, glomerulus, efferent arteriole, peritubular capillaries
   E) efferent arteriole, afferent arteriole, glomerulus, peritubular capillaries
   Answer: D

Pages: 857
3. Which of the following lists the nephron regions in the correct order of fluid flow?
   A) glomerular capsule, distal convoluted tubule, loop of Henle, proximal convoluted tubule
   B) proximal convoluted tubule, loop of Henle, distal convoluted tubule, glomerular capsule
   C) glomerular capsule, proximal convoluted tubule, loop of Henle, distal convoluted tubule
   D) loop of Henle, glomerular capsule, proximal convoluted tubule, distal convoluted tubule
   E) distal convoluted tubule, loop of Henle, proximal convoluted tubule, glomerular capsule
   Answer: C

Pages: 864
4. The most important function of the juxtaglomerular (JG) apparatus is to:
   A) secrete water and sodium into the tubular fluid
   B) release renin in response to a drop in renal blood pressure or blood flow
   C) make sure that the diameter of the efferent arteriole is kept larger than that of the afferent arteriole
   D) produce antidiuretic hormone in response to increased glomerular filtration rate (GFR)
   E) produce chemicals that change the diameter of the loop of Henle
   Answer: B

Pages: 863
5. If there were an obstruction in the renal artery, one might expect to see:
    A) a decrease in glomerular filtration rate (GFR)
    B) an increase in the release of renin
    C) an increase in glomerular filtration rate (GFR)
    D) both B and C are correct
    E) both A and B are correct
    Answer: E

Pages: 864
6. Atrial natriuretic peptide works to:
    A) counteract the effects of aldosterone
    B) raise systemic blood pressure
    C) enhance the effects of antidiuretic hormone
    D) increase blood flow to the kidney
    E) both B and C are correct
    Answer: A

Pages: 866
7. Substances reabsorbed in the proximal convoluted tubule (PCT) by facilitated diffusion are:
    A) moving only if the cells of the PCT are using ATP to facilitate the movement
    B) moving down a concentration gradient with the help of a carrier
    C) moving up a concentration gradient
    D) always proteins
    E) both A and B are correct
    Answer: B

Pages: 868
8. The **transport maximum** is the:
    A) highest the glomerular filtration rate can increase without inhibiting kidney function
    B) greatest percentage of plasma entering the glomerulus that can become filtrate
    C) upper limit of reabsorption due to saturation of carrier systems
    D) steepest any concentration gradient can become
    E) fastest rate at which fluid can flow through the renal tubules
    Answer: C

Pages: 870
9. Most water is reabsorbed in the proximal convoluted tubule by **obligatory reabsorption**, which means that:
    A) water is moving up its own gradient
    B) water is "following" sodium and other ions/molecules to maintain osmotic balance
    C) the carrier that transports sodium cannot do so without binding water first
    D) the proximal convoluted tubule cannot physically hold the volume of water that enters from the glomerular capsule, so water is reabsorbed because of hydrostatic pressure
    E) the rate of water reabsorption never changes, regardless of water intake
    Answer: B

10. If the level of aldosterone in the blood increases, then:
    A) more potassium is excreted in the urine
    B) more sodium is excreted in the urine
    C) blood pressure will drop
    D) glomerular filtration rate will drop
    E) first B, then C, then D
    Answer: A

11. The significance of secretion of ammonium (NH_4^+) ions by the tubule cells is:
    A) it triggers the release of renin
    B) it results from generation of new bicarbonate ions that can be reabsorbed to help maintain pH
    C) it keeps the ascending limb of Henle's loop from reabsorbing water
    D) it carries urea across the endothelium of the vasa recta
    E) there is no apparent function for this type of secretion
    Answer: B

12. Renal pyramids are part of the:
    A) renal cortex
    B) renal medulla
    C) renal capsule
    D) renal pelvis
    E) glomerulus
    Answer: B

13. Podocytes are cells specialized for filtration that are found in the:
    A) walls of the vasa recta
    B) ascending limb of the loop of Henle
    C) urinary bladder
    D) visceral layer of the glomerular capsule
    E) collecting duct
    Answer: D

14. The function of the macula densa cells is to:
    A) prevent water reabsorption in the ascending limb of the loop of Henle
    B) prevent over-distension of the urinary bladder
    C) add bicarbonate ions to the tubular fluid in the proximal convoluted tubule
    D) monitor NaCl concentration in the tubular fluid
    E) produce the carrier molecules used to actively transport ions into the peritubular space
    Answer: D

Pages: 856

15. In comparison with capillaries elsewhere in the body, glomerular capillaries:
    A) have a lower blood hydrostatic pressure
    B) are more permeable
    C) have a thicker basement membrane
    D) have no fenestrations
    E) reabsorb almost as much fluid as they filter
    Answer: B

Pages: 862

16. If the diameter of the efferent arteriole is smaller than the diameter of the afferent arteriole, then:
    A) blood pressure in the glomerulus stays low
    B) blood pressure in the glomerulus stays high
    C) there must be an abnormal blockage in the peritubular capillaries
    D) the endothelial-capsular membrane filters less blood than normal
    E) capsular hydrostatic pressure increases to levels higher than glomerular blood hydrostatic pressure
    Answer: B

Pages: 849

17. Around each kidney is a renal capsule made up of:
    A) loose connective tissue
    B) simple cuboidal epithelium
    C) adipose tissue
    D) smooth muscle
    E) dense fibrous connective tissue
    Answer: E

Pages: 862

18. Glomerular filtrate contains:
    A) everything in blood
    B) everything in blood except cells and proteins
    C) water and electrolytes only
    D) water and wastes only
    E) water only
    Answer: B

Pages: 855

19. Glomeruli are located in the renal:
    A) cortex
    B) pelvis
    C) medulla
    D) sinus
    E) calyx
    Answer: A

Pages: 862
20. An obstruction in the proximal convoluted tubule decreases glomerular filtration rate because:
    A) blood hydrostatic pressure in the glomerulus decreases when blood can't flow through the tubule
    B) osmotic pressure in the glomerular capsule increases due to leakage of more proteins into the filtrate
    C) hydrostatic pressure in the glomerular capsule increases, which decreases net filtration pressure
    D) hydrostatic pressure in the glomerular capsule decreases due to leakage of more filtrate into the peritubular space
    E) release of renin decreases as fluid flow to the macula densa decreases
    Answer: C

Pages: 866
21. As substances are reabsorbed in the proximal convoluted tubules of the kidneys, they move from:
    A) filtered fluid to epithelial cells, to interstitial fluid, to peritubular capillaries
    B) filtered fluid to interstitial fluid, to epithelial cells, to peritubular capillaries
    C) peritubular capillaries to interstitial fluid, to epithelial cells, to filtered fluid
    D) vasa recta to epithelial cells, to interstitial fluid, to filtered fluid
    E) peritubular capillaries to epithelial cells, to interstitial fluid, to filtered fluid
    Answer: A

Pages: 866
22. Normally most filtered bicarbonate ions are reabsorbed in the:
    A) proximal convoluted tubule
    B) loop of Henle
    C) distal convoluted tubule
    D) collecting ducts
    E) glomerulus
    Answer: A

Pages: 877
23. The renal clearance of a large protein such as albumin would be closest to which of the following values?
    A) the rate of renal blood flow
    B) the total blood volume entering both kidneys each minute
    C) the average glomerular filtration rate
    D) the transport maximum for glucose
    E) zero
    Answer: E

Pages: 870
24. The permeability of the collecting ducts to water is regulated by:
    A) aldosterone
    B) renin
    C) antidiuretic hormone
    D) atrial natriuretic peptide
    E) angiotensinogen
    Answer: C

Pages: 870

25. In a patient who has lost fluid due to vomiting and diarrhea, which is likely to be higher than normal in a blood sample?
    A) antidiuretic hormone (ADH) only
    B) aldosterone only
    C) atrial natriuretic peptide (ANP) only
    D) both ADH and aldosterone
    E) both aldosterone and ANP
    Answer: D

Pages: 874

26. The countercurrent mechanism in the loop of Henle builds and maintains an osmotic gradient in the renal medulla. Which of the following is **NOT** a contributing factor?
    A) Fluid flows in opposite directions in the ascending and descending limbs of the loop of Henle.
    B) Chloride ions passively diffuse from the interstitial fluid into the thick portion of the ascending limb.
    C) The thick portion of the ascending limb is impermeable to water.
    D) The descending limb is permeable to water.
    E) Fluid in the descending limb is in osmotic equilibrium with the surrounding interstitial fluid.
    Answer: B

Pages: 883

27. The most common cause of glucosuria (glycosuria) is untreated diabetes mellitus, in which the:
    A) rate at which glucose enters the glomerular filtrate is higher than the rate at which the distal convoluted tubules and collecting ducts can reabsorb it
    B) Na^+-glucose symports in the distal convoluted tubules are defective
    C) Na^+-glucose symports in the proximal convoluted tubule are defective
    D) transport maximum for glucose reabsorption in the proximal convoluted tubule is exceeded
    E) countercurrent multiplier fails to keep up with glucose reabsorption
    Answer: D

Pages: 870

28. Cells that have receptors for aldosterone include:
    A) podocytes
    B) intercalated cells in the collecting ducts
    C) cells in the thick ascending limb of the loop of Henle
    D) cells in the proximal convoluted tubules
    E) principal cells in the collecting ducts
    Answer: E

Pages: 853

29. What percentage of the resting cardiac output flows to the kidneys?
    A) 1%
    B) 5%
    C) 20%
    D) 40%
    E) 75%
    Answer: C

30. An increase in blood pressure in the afferent arterioles of the kidney will result in **ALL** of the following **EXCEPT**:
    A) a decrease in the release of atrial natriuretic peptide
    B) an increase in glomerular filtration rate
    C) an increase in glomerular hydrostatic pressure
    D) a somewhat greater output of urine
    E) a decrease in release of renin by juxtaglomerular cells
    Answer: A

31. In the kidneys, the largest volume of water reabsorption occurs in the:
    A) glomerular capillaries
    B) collecting ducts
    C) proximal convoluted tubules
    D) distal convoluted tubules
    E) ascending limbs of the loops of Henle
    Answer: C

32. In the proximal convoluted tubule, symporters that use the sodium ion concentration difference are responsible for reabsorption of:
    A) glucose only
    B) amino acids only
    C) bicarbonate ions only
    D) both glucose and bicarbonate ions
    E) both glucose and amino acids
    Answer: E

33. Which of the following would pass through the glomerular filters in the kidneys most easily?
    A) albumin
    B) antibodies
    C) glucose
    D) platelets
    E) angiotensinogen
    Answer: C

34. Filtration of blood in the glomeruli is promoted by:
    A) blood colloid osmotic pressure
    B) blood hydrostatic pressure
    C) capsular hydrostatic pressure
    D) both blood hydrostatic pressure and capsular hydrostatic pressure
    E) both blood colloid osmotic pressure and capsular hydrosatic pressure
    Answer: B

Pages: 878
35. Which of the following lists the structures in the correct order of urine flow?
    A) minor calyces, major calyces, renal pelvis, urethra, urinary bladder, ureter
    B) major calyces, minor calyces, renal pelvis, urethra, urinary bladder, ureter
    C) renal pelvis, major calyces, minor calyces, ureter, urinary bladder, urethra
    D) minor calyces, major calyces, renal pelvis, ureter, urinary bladder, urethra
    E) renal pelvis, ureter, major calyces, minor calyces, urinary bladder, urethra
    Answer: D

Pages: 853
36. Which of the following lists the arteries of the kidneys in the correct order of blood flow?
    A) interlobular, arcuate, interlobar, segmental, renal
    B) renal, segmental, interlobar, arcuate, interlobular
    C) segmental, arcuate, interlobar, interlobular, renal
    D) renal, interlobar, interlobular, segmental, arcuate
    E) segmental, interlobar, arcuate, interlobular, renal
    Answer: B

Pages: 854
37. Sympathetic nerves from the renal plexus are distributed to the:
    A) renal blood vessels
    B) convoluted tubules
    C) renal pyramids
    D) collecting ducts
    E) both renal blood vessels and convoluted tubules
    Answer: A

Pages: 860
38. The cells making up the proximal and distal convoluted tubules are:
    A) stratified squamous epithelial cells
    B) simple squamous epithelial cells
    C) simple cuboidal epithelial cells
    D) transitional epithelial cells
    E) smooth muscle cells
    Answer: C

Pages: 855
39. The renal corpuscle consists of:
    A) the proximal and distal convoluted tubules
    B) the glomerulus and the glomerular (Bowman's) capsule
    C) the descending and ascending limbs of the loop of Henle
    D) all the renal pyramids
    E) the glomerulus and the vasa recta
    Answer: B

Pages: 862
40. Which of the following pressures is highest in the renal corpuscle under normal circumstances?
   A) blood colloid osmotic pressure
   B) capsular hydrostatic pressure
   C) capsular colloid osmotic pressure
   D) glomerular blood hydrostatic pressure
   E) None is higher than the others; all pressures are equal under normal circumstances.
   Answer: D

Pages: 864
41. Angiotensinogen is produced by the:
   A) kidneys
   B) liver
   C) lungs
   D) adrenal cortex
   E) hypothalamus
   Answer: B

Pages: 877
42. If glomerular filtration rate = 100 ml/min, urine flow rate = 1.1 ml/min, plasma concentration of urea = 20 mg/100 ml, and urine concentration of urea = 1000 mg/100 ml, the clearance of urea is about:
   A) 1.1 ml/min
   B) 5.0 ml/min
   C) 25.0 ml/min
   D) 35.0 ml/min
   E) 55.0 ml/min
   Answer: E

Pages: 864
43. Although angiotensin II has several target tissues, all of its actions are directed toward:
   A) decreasing glomerular filtration rate
   B) increasing glomerular filtration rate
   C) increasing the surface area of the nephron
   D) decreasing the concentration of urine
   E) decreasing the secretion of aldosterone
   Answer: B

Pages: 864
44. An increase in blood volume would result in an increase in secretion of:
   A) aldosterone only
   B) antidiuretic hormone only
   C) renin only
   D) atrial natriuretic peptide only
   E) renin, aldosterone, and antidiuretic hormone
   Answer: D

Pages: 866
45. During a fight-or-flight response, sympathetic stimulation of renal blood vessels causes:
   A) increased glomerular filtration rate due to generalized vasodilation of renal vessels
   B) increased glomerular filtration rate due to stimulation of vasoconstriction in the efferent arterioles only
   C) decreased glomerular filtration rate due to vasoconstriction predominantly in the afferent arterioles
   D) decreased glomerular filtration rate due to reduced secretion of renin
   E) no change in glomerular filtration rate
   Answer: C

Pages: 866
46. Small proteins and peptides that pass the glomerular filter are usually reabsorbed by:
   A) osmosis
   B) simple diffusion
   C) facilitated diffusion
   D) pinocytosis
   E) phagocytosis
   Answer: D

Pages: 868
47. The passive diffusion of sodium ions from the tubular fluid of the proximal convoluted tubules into the tubule cells occurs because:
   A) the sodium ions are following water, which is actively transported, so that osmotic balance is maintained
   B) the sodium ions are moving down an electrochemical gradient, which is maintained by the action of the sodium pumps
   C) the sodium ions are following potassium ions, which are being actively transported into the tubule cells, to maintain electrical balance within the cells
   D) the sodium ions are consumed during the reaction that produces new bicarbonate ions, so there is a chemical gradient
   E) chloride ions are being actively transported into the tubular fluid by the tubule cells, so sodium ions are exchanged with them to maintain electrical balance within the cells
   Answer: B

Pages: 868
48. If a substance has exceeded its transport maximum in the kidney tubules, it is likely that:
   A) urine volume will increase
   B) urine volume will decrease
   C) the concentration of the substance in the blood is unusually high
   D) the concentration of the substance in the blood cannot increase further
   E) both A and C are correct
   Answer: E

Pages: 870
49. The effect of aldosterone on the principal cells of the distal convoluted tubule is to:
    A) increase the synthesis of sodium pumps
    B) increase the cells' permeability to water
    C) increase retention of potassium ions
    D) increase the cells' secretion of antidiuretic hormone
    E) trigger the release of renin
    Answer: A

Pages: 871
50. The enzyme carbonic anhydrase is necessary for:
    A) secretion of renin
    B) secretion of potassium ions from the cells of the collecting duct
    C) production of carbonic acid by the tubule cells
    D) response of principal cells to antidiuretic hormone
    E) vasoconstriction of the efferent arterioles
    Answer: C

Pages: 872
51. Blood in the renal vein may have a higher concentration of bicarbonate ions than blood in the renal artery because:
    A) bicarbonate ions cannot pass the glomerular filter, so blood becomes more concentrated as water enters the nephrons and bicarbonate ions stay behind
    B) the rate of cellular respiration is higher in the endothelium of the renal venules than in the endothelium of the renal arterioles
    C) newly produced bicarbonate ions (as opposed to filtered bicarbonate ions) are added to the blood by the intercalated cells of the collecting ducts as they secrete hydrogen ions into the tubular fluid
    D) hydrogen ions must be added to venous blood to maintain pH balance, so bicarbonate ions are reabsorbed at the same time to maintain electrical balance
    E) large ions like bicarbonate ions are necessary in venous blood to help maintain blood colloid osmotic pressure
    Answer: C

Pages: 873
52. Urine that is hypotonic to blood plasma is produced when:
    A) levels of aldosterone are high
    B) levels of antidiuretic hormone are high
    C) levels of antidiuretic hormone are low
    D) plasma concentration of sodium ions is high
    E) levels of both aldosterone and antidiuretic hormone are high
    Answer: C

Pages: 874
53. The concentration of solutes in tubular fluid is greatest in the:
    A) glomerular (Bowman's) capsule
    B) proximal convoluted tubule
    C) hairpin turn of the loop of Henle
    D) ascending limb of the loop of Henle
    E) distal convoluted tubule
    Answer: C

Pages: 880
54. In the female, the urinary bladder is located:
    A) posterior to the pubic symphysis, superior to the uterus, and anterior to the vagina
    B) anterior to the pubic symphysis, inferior to the uterus, and anterior to the vagina
    C) posterior to the pubic symphysis, inferior to the uterus, and posterior to the vagina
    D) posterior to the pubic symphysis, inferior to the uterus, and anterior to the vagina
    E) anterior to the pubic symphysis, superior to the uterus, and posterior to the vagina
    Answer: D

Pages: 880
55. The mucosa of the urinary bladder is made up of the lamina propria and:
    A) transitional epithelium
    B) simple cuboidal epithelium
    C) stratified squamous epithelium
    D) smooth muscle
    E) adipose tissue
    Answer: A

Pages: 882
56. The normal daily volume of urine produced is:
    A) under 200 ml
    B) 200-400 ml
    C) 1000-2000 ml
    D) 3 liters
    E) 180 liters
    Answer: C

Pages: 882
57. **ALL** of the following are normally found in urine **EXCEPT**:
    A) urea
    B) uric acid
    C) creatinine
    D) glucose
    E) sodium and chloride ions
    Answer: D

Pages: 877
58. The reason inulin is used to determine glomerular filtration rate is because it is:
    A) both filtered and efficiently secreted by nephrons
    B) both filtered and completely reabsorbed by nephrons
    C) reabsorbed and secreted, but not filtered by nephrons
    D) only filtered, and neither reabsorbed nor secreted by the nephrons
    E) only secreted, and neither filtered nor reabsorbed by nephrons
    Answer: D

Pages: 874
59. **Urea recycling** in the renal medulla refers to the:
    A) conversion of urea to ammonia by the tubule cells
    B) conversion of ammonia to ammonium ions by the tubule cells
    C) conversion of urea into amino acids in the vasa recta
    D) mechanism by which urea leaves the collecting duct and re-enters the loop of Henle, thus helping to maintain the hypertonic conditions of the interstitial spaces
    E) mechanism by which urea leaves the collecting ducts and enters the vasa recta, thus helping to maintain the correct blood volume in the vasa recta
    Answer: D

Pages: 862
60. The forces that oppose filtration of substances from the glomerular capillaries into the glomerular (Bowman's) space include:
    A) capsular colloid osmotic pressure
    B) blood colloid osmotic pressure
    C) glomerular blood hydrostatic pressure
    D) capsular hydrostatic pressure
    E) both blood colloid osmotic and colloid hydrostatic pressure
    Answer: E

Pages: 882
61. The "sweet" odor of diabetic urine is due to the presence of:
    A) sucrose
    B) glucose
    C) ketone bodies
    D) urobilinogen
    E) albumin
    Answer: C

Pages: 877
62. So-called "loop diuretics" work as diuretics by:
    A) increasing secretion of atrial natriuretic peptide, which causes water to be excreted with sodium
    B) decreasing secretion of ADH, which inhibits facultative reabsorption
    C) decreasing secretion of aldosterone, which causes water to be excreted with sodium
    D) blocking the action of carbonic anhydrase
    E) inhibiting the Na^+-K^+-$2Cl^-$ symporters in the thick ascending limb of the loop of Henle
    Answer: E

Pages: 866
63. Sympathetic nerves to the kidney regulate the:
    A) diameter of arterioles within the kidney
    B) permeability of the endothelial-capsular membrane
    C) effect of ADH on tubule cells
    D) colloid osmotic pressure in the glomerular capsule
    E) diameter of the convoluted tubules
    Answer: A

Pages: 870

64. Principal cells in the distal convoluted tubules:
    A) secrete renin
    B) monitor sodium and chloride ion concentrations in tubular fluid
    C) secrete hydrogen ions when pH in the extracellular fluid is low
    D) filter large proteins
    E) respond to ADH and aldosterone
    Answer: E

Pages: 882

65. Which of the following would be in the highest concentration in normal urine?
    A) albumin
    B) bilirubin
    C) creatinine
    D) acetoacetic acid
    E) urobilinogen
    Answer: C

True-False

Write T if the statement is true and F if the statement is false.

Pages: 870
1. Antidiuretic hormone causes the collecting ducts to become less permeable to water.
   Answer: False

Pages: 869
2. The ascending limb of the loop of Henle is relatively permeable to water and relatively impermeable to solutes.
   Answer: False

Pages: 878
3. The principal function of the ureters is to transport urine from the renal pelvis to the urinary bladder.
   Answer: True

Pages: 861
4. The macula densa is part of the renal corpuscle.
   Answer: False

Pages: 854
5. Blood enters the glomerulus through the afferent arteriole.
   Answer: True

Pages: 854
6. The nerve supply to the kidneys is part of the sympathetic division of the autonomic nervous system.
   Answer: True

Pages: 862
7. The primary force that promotes filtration in the kidney is glomerular blood hydrostatic pressure.
    Answer: True

Pages: 862
8. The reason that blood hydrostatic pressure is so high in the glomerulus is that the diameter of the efferent arteriole is larger than the diameter of the afferent arteriole.
    Answer: False

Pages: 864
9. When delivery of fluid and NaCl to the macula densa is decreased, less vasoconstrictor is released from the JG cells, and the afferent arterioles dilate.
    Answer: True

Pages: 864
10. An increase in level of angiotensin II will cause vasoconstriction in the efferent arterioles.
    Answer: True

Pages: 853
11. The kidneys receive 20-25% of resting cardiac output.
    Answer: True

Pages: 862
12. The diameter of the afferent arteriole is normally smaller than the diameter of the efferent arteriole.
    Answer: False

Pages: 872
13. Intercalated cells of the distal convoluted tubule secrete hydrogen ions into the tubular fluid.
    Answer: True

Pages: 869
14. Cells in the thick ascending limb of the loop of Henle have symporters that simultaneously reabsorb one sodium ion, one potassium ion, and two chloride ions.
    Answer: True

Pages: 880
15. The reflex arc for the micturition reflex is a parasympathetic reflex arc.
    Answer: True

Short Answer

Write the word or phrase that best completes each statement or answers the question.

Pages: 878
1. The 10"-12" tubes carrying urine from the kidneys to the urinary bladder are the _____
    Answer: ureters

2. The mucosa of the urinary bladder includes _____ epithelium.
   Answer: transitional

3. The smooth muscle layers surrounding the mucosa of the urinary bladder are collectively known as the _____.
   Answer: detrusor muscle

4. The normal component of urine that is derived from the detoxification of ammonia produced as a result of deamination of proteins is _____.
   Answer: urea

5. The hormone that increases the water permeability of the principal cells of the distal convoluted tubule is _____.
   Answer: antidiuretic hormone

6. The plasma concentration at which a substance begins to spill into the urine because its transport maximum has been surpassed is called the _____.
   Answer: renal threshold

7. The enzyme secreted by the juxtaglomerular cells in response to impulses from renal sympathetic nerves is _____.
   Answer: renin

8. The substrate for the enzyme secreted by the juxtaglomerular cells is _____.
   Answer: angiotensinogen

9. Blood colloid osmotic pressure is due to the presence of _____ in blood plasma.
   Answer: proteins

10. The blood vessels surrounding the loop of Henle that help maintain the hypertonic conditions in the peritubular spaces of the renal medulla are called the _____.
    Answer: vasa recta

11. The apex of a renal pyramid is called a renal _____.
    Answer: papilla

12. Urine drains into the renal pelvis from cup-like structures called _____.
    Answer: major calyces

Pages: 855
13. A glomerulus and its associated glomerular capsule make up a _____.
Answer: renal corpuscle

Pages: 856
14. Specialized epithelial cells with pedicels covering glomerular capillaries are called

_____.
Answer: podocytes

Pages: 861
15. The percentage of plasma in afferent arterioles that becomes glomerular filtrate is called the

_____.
Answer: filtration fraction

Pages: 862
16. In the formula for calculating net filtration pressure, those forces that oppose glomerular filtration are _____ and _____.
Answer: capsular hydrostatic pressure; blood colloid osmotic pressure

Pages: 886
17. When daily urine output is less than 50 ml per day, the condition is known as _____.
Answer: anuria

Pages: 868
18. Membrane proteins that perform secondary active transport are called _____.
Answer: symporters

Pages: 871
19. The enzyme that catalyzes the conversion of carbon dioxide and water to carbonic acid is

_____.
Answer: carbonic anhydrase

Pages: 874
20. Fluid flowing in opposite directions in parallel tubes is called _____ flow.
Answer: countercurrent

Pages: 877
21. Drugs that increase the rate of urine flow are called _____.
Answer: diuretics

Pages: 880
22. Urine is expelled from the bladder by an act called _____, also known as urination or voiding.
Answer: micturition

Pages: 882
23. The product of the catabolism of nucleic acids that is normally present in urine and that may crystallize into kidney stones is _____.
Answer: uric acid

24. The presence of red blood cells in urine is called _____.
    Answer: hematuria

25. The portion of the male urethra that passes through the penis is called the _____ urethra.
    Answer: spongy

26. The polysaccharide used in renal function tests because its clearance equals the glomerular filtration rate is _____.
    Answer: inulin

27. Most tubular reabsorption of water, sodium ions, and potassium ions occurs in the _____.
    Answer: proximal convoluted tubule

28. The renal tubules are least permeable to water in the region known as the _____.
    Answer: thick ascending limb (of the loop of Henle)

29. The deamination of glutamine in the cells of the proximal convoluted tubule results in the secretion of _____ into tubular fluid and the reabsorption of _____ into blood.
    Answer: ammonium ions; bicarbonate ions

30. Secretion of potassium ions by cells of the distal convoluted tubule and the collecting ducts is regulated primarily by the hormone _____.
    Answer: aldosterone

Matching

Choose the item from Column 2 that best matches each item in Column 1.

1. renal clearance	a measure of how effectively the kidneys remove a substance from the blood plasma
2. transport maximum	the upper limit of reabsorption of a substance determined by the number of available carrier molecules
3. renal threshold	the plasma concentration at which a substance starts to spill into the urine
4. tubular secretion	movement of substances from the blood or tubular cells into the tubular fluid

5. tubular reabsorption

    movement of materials from the tubular fluid into the blood of peritubular capillaries or the vasa recta

6. countercurrent mechanism

    the arrangement of juxtamedullary nephrons and the vasa recta that functions to generate and maintain the osmotic gradient in the renal medulla

7. glomerular filtration rate

    the amount of filtrate formed in all the renal corpuscles of both kidneys every minute

8. net filtration pressure

    (GBHP - (CHP + BCOP)), where GBHP = glomerular blood hydrostatic pressure, CHP = capsular hydrostatic pressure, and BCOP = blood colloid osmotic pressure

9. renal autoregulation

    the ability of the kidneys to maintain a constant blood pressure and GFR

10. filtration fraction

    the percentage of plasma entering the nephrons that becomes glomerular filtrate

Essay

Write your answer in the space provided or on a separate sheet of paper.

Pages: 865

1. Describe in detail the renin-angiotensin negative feedback loop that helps regulate blood pressure and glomerular filtration rate.

    Answer: Stress causes decrease in BP, thus GFR. JG cells of JGA sense decreased stretch; macula densa cells detect decreased NaCl and water. JG cells secrete renin, which converts angiotensinogen in blood to angiotensin I, which is converted to angiotensin II by ACE in the lungs. Angiotensin II causes constriction of efferent arterioles, increased thirst, greater ADH secretion from the neurohypophysis, and increased secretion of aldosterone from the adrenal cortex. Blood volume is increased, which increases venous return, stroke volume, cardiac output, thus BP. GFR raised also.

Pages: 870

2. Describe in detail the negative feedback regulation of water reabsorption by antidiuretic hormone.

    Answer: High osmotic pressure of blood detected by osmoreceptors in the hypothalamus. ADH released from neurohypophysis. Principal cells in DCT become more permeable to water, increasing facultative reabsorption. Water added to blood, thus decreasing osmotic pressure.

3. Discuss the importance of countercurrent flow to the functioning of the nephron.
   Answer: Countercurrent flow refers to the flow of fluid in opposite directions in parallel tubing (tubules and blood vessels). The arrangement allows gradients to develop between tubular fluid, blood, and interstitial fluid. Gradients allow for reabsorption of large amounts of water and ions from the tubular fluid.

4. Describe how cells of the renal tubule can work to raise blood pH.
   Answer: To raise blood pH, the kidney can secrete hydrogen ions and reabsorb bicarbonate ions (in the form of carbon dioxide and water). In addition, more hydrogen ions can be excreted when ammonium ions form from the deamination of glutamine. From the same reaction, new bicarbonate ions can be added to the blood.

5. Describe how reabsorption of sodium ions in the proximal convoluted tubule is linked to reabsorption of water and chloride ions and to secretion of hydrogen ions.
   Answer: Sodium ions are reabsorbed by primary active transport. Water accompanies sodium for osmotic balance (obligatory reabsorption). An antiporter links sodium and hydrogen ion transport, such that sodium ions are moved into the cells, and hydrogen ions are moved into the tubular fluid.

6. Describe the structural features of the renal corpuscle that enhance its blood filtering capacity.
   Answer: Endothelial cells of the glomerular are fenestrated. Its basement membrane is part of the filtering mechanism. Podocytes with filtration slits between pedicels wrap the glomerular capillaries. The large surface area also contributes to filtering ability, as does the high glomerular hydrostatic pressure created by the arrangement of the afferent and efferent arterioles (efferent diameter smaller).

7. Predict the effect on reabsorption of sodium ions, bicarbonate ions, and water from the proximal convoluted tubule in an individual who has been given a drug to inhibit carbonic anhydrase activity. Explain your answer.
   Answer: Carbonic anhydrase inhibitors block the conversion of bicarbonate ions to water and carbon dioxide for reabsorption, thus more is excreted. More sodium ions are excreted as they follow bicarbonate ions for electrical balance, and water follows sodium for osmotic balance.

Pages: 862

8. Predict the effect on glomerular filtration rate and net filtration pressure of each of the following: 1) hemorrhage, 2) increased permeability of the endothelial-capsular membrane, 3) constriction of the lumen of the proximal convoluted tubule. Explain your reasoning in each case.

   Answer: 1) Ultimately GFR and NFP decrease as blood volume decreases, since the decreased volume will ultimately decrease BP.

   2) NFP increases as proteins escape into the glomerular capsule, increasing capsular colloid osmotic pressure. Water follows the proteins, ultimately decreasing blood volume and BP, which decreases GFR.

   3) As caspular hydrostatic pressure increases, NFP decreases. As capsular hydrostatic pressure rises, GFR is reduced.

Pages: 870

9. Predict the effects on levels of aldosterone and antidiuretic hormone of each of the following: 1) hemorrhage, 2) increased permeability of the endothelial-capsular membrane, 3) excessive loss of body water by sweating. Explain your answer in each case.

   Answer: 1) Both will increase as BP drops and blood volume decreases.

   2) Both will increase as water is lost with escaping proteins, reducing blood volume, blood concentration, and blood pressure.

   3) Both will increase as both sodium and water are lost from the blood.

Pages: 855

10. Describe the importance of surface area to the functioning of the kidney, and describe the structural features of the kidney that enhance total surface area.

    Answer: A large surface area increases filtering and reabsorptive ability. In the glomerulus, many capillaries are coiled into a small space. Likewise, kidney tubules are highly coiled into small spaces, as are the surrounding peritubular capillaries and vasa recta.

Pages: 848

11. List the functions of the kidneys.

    Answer: 1) regulation of blood volume and composition
    2) regulation of blood pH
    3) regulation of blood pressure
    4) synthesis of calcitriol (active vitamin D)
    5) gluconeogenesis
    6) secretion of erythropoietin to increase RBC production

Chapter 27   Fluid, Electrolyte, and Acid-Base Homeostasis

Multiple-Choice

Choose the one alternative that best completes the statement or answers the question.

Pages: 899
1. ADH saves water by:
    A) promoting the excretion of sodium ions
    B) stimulating the secretion of renin
    C) enhancing passive movement of water out of the collecting ducts
    D) stimulating constriction of the lumen of the distal convoluted tubules
    E) lowering the glomerular filtration rate
    Answer: C

Pages: 893
2. The area that stimulates the conscious desire to drink water is located in the:
    A) adrenal cortex
    B) kidney
    C) medulla oblongata
    D) hypothalamus
    E) lumbar region of the spinal cord
    Answer: D

Pages: 896
3. Levels of sodium ions in the extracellular fluid are regulated primarily by:
    A) ADH
    B) aldosterone
    C) parathyroid hormone
    D) epinephrine
    E) insulin
    Answer: B

Pages: 896
4. Levels of potassium ions in the extracellular fluid are regulated primarily by:
    A) ADH
    B) aldosterone
    C) parathyroid hormone
    D) epinephrine
    E) insulin
    Answer: B

Pages: 893
5. Osmoreceptors are located in the:
    A) hypothalamus
    B) glomerulus
    C) wall of the right atrium
    D) medulla oblongata
    E) lining of the central canal of the spinal cord
    Answer: A

Pages: 873
6. When NH_3 is produced by kidney tubule cells, the effect is to:
    A) help lower the pH of extracellular fluid
    B) promote excretion of hydrogen ions
    C) promote absorption of bicarbonate ions
    D) both B and C are correct
    E) choices A, B and C are all correct
    Answer: D

Pages: 900
7. Which of the following falls within the homeostatic range of pH for extracellular fluid?
    A) 7.30
    B) 7.40
    C) 6.90
    D) 7.50
    E) both A and B are within homeostatic range
    Answer: B

Pages: 901
8. Which of the following would serve to buffer H^+?
    A) NaH_2PO_4
    B) the -COOH end of a protein
    C) HCO_3^-
    D) any strong acid
    E) any weak acid
    Answer: C

Pages: 901
9. Which of the following would serve to buffer OH^-?
    A) NaH_2PO_4
    B) HCO_3^-
    C) the amine end of a protein
    D) a monohydrogen phosphate ion
    E) any strong base
    Answer: A

Pages: 900
10. Carbonic acid is buffered in blood cells primarily by:
    A) carbon dioxide
    B) hemoglobin
    C) sodium monohydrogen phosphate
    D) carbonic anhydrase
    E) oxygen
    Answer: B

11. Bicarbonate ions are reabsorbed from the lumen of the kidney tubule into the tubule cells as:
   A) HCO_3^-
   B) H_2CO_3
   C) CO_2 and H_2O
   D) carbonic anhydrase
   E) a carbamino compound
   Answer: C

12. In order for H+ to be excreted, it must combine with other substances, including:
   A) HCO_3^-
   B) CO_2
   C) $H_2PO_4^-$
   D) hemoglobin
   E) HPO_4^{2-}
   Answer: E

13. In compensating for respiratory alkalosis, the body excretes more:
   A) ammonium ions
   B) bicarbonate ions
   C) dihydrogen phosphate ions
   D) carbonic acid
   E) hydrogen ions
   Answer: B

14. In compensating for metabolic acidosis, the body:
   A) increases respiratory rate
   B) excretes more bicarbonate ions
   C) excretes more monohydrogen phosphate ions
   D) decreases respiratory rate
   E) slows the rate of conversion of ammonia to urea
   Answer: A

15. In human buffer systems, a strong acid is buffered by:
   A) a strong base
   B) a weak base
   C) a weak acid
   D) water
   E) a salt
   Answer: B

Pages: 903
16. If the pH of blood plasma becomes 7.49 due to ingested substances, **ALL** of the following would happen to compensate **EXCEPT**:
    A) respiratory rate decreases
    B) the kidney increases excretion of bicarbonate ions
    C) tubule cells produce more ammonia from glutamate
    D) the partial pressure of carbon dioxide in blood would begin to rise
    E) the kidney excretes fewer dihydrogen phosphate ions
    Answer: C

Pages: 897
17. The kidneys' ability to reabsorb calcium ions is affected mainly by:
    A) ADH
    B) aldosterone
    C) epinephrine
    D) cortisol
    E) parathyroid hormone
    Answer: E

Pages: 896
18. The primary action of aldosterone is to:
    A) lower the glomerular filtration rate
    B) counteract the effects of antidiuretic hormone
    C) promote retention of sodium ions by the distal convoluted tubule
    D) stimulate the release of renin
    E) stimulate constriction of the lumen of the distal convoluted tubule
    Answer: C

Pages: 896
19. Which of the following ions is most abundant in extracellular fluid?
    A) Na^+
    B) K^+
    C) H^+
    D) HPO_4^{2-}
    E) Ca^{2+}
    Answer: A

Pages: 897
20. Hypokalemia is:
    A) low plasma levels of potassium ions
    B) high plasma levels of potassium ions
    C) low plasma level of sodium ions
    D) high plasma levels of sodium ions
    E) low plasma levels of calcium ions
    Answer: A

Pages: 901
21. Which of the following would be able to buffer a strong base?
    A) HCO_3^-
    B) NH_3
    C) $H_2PO_4^-$
    D) hemoglobin
    E) another strong base
    Answer: C

Pages: 893
22. Osmotic diuresis occurs if:
    A) too much water is consumed
    B) too little ADH is present
    C) large amounts of glucose or urea are present in the tubular fluid
    D) aldosterone levels are too high
    E) renal calculi are blocking a kidney tubule
    Answer: C

Pages: 897
23. Which of the following values is within homeostatic range for bicarbonate ions in arterial blood?
    A) 136-142 mEq/liter
    B) 3.8-5.0 mEq/liter
    C) 22-26 mEq/liter
    D) 4.6-5.5 mEq/liter
    E) 1.7-2.6 mEq/liter
    Answer: C

Pages: 896
24. Which of the following falls within homeostatic range for levels of sodium ions in plasma?
    A) 136-142 mEq/liter
    B) 3.8-5.0 mEq/liter
    C) 22-26 mEq/liter
    D) 4.6-5.5 mEq/liter
    E) 1.7-2.6 mEq/liter
    Answer: A

Pages: 897
25. Which of the following falls within homeostatic range for levels of calcium ions in plasma?
    A) 136-142 mEq/liter
    B) 3.8-5.0 mEq/liter
    C) 22-26 mEq/liter
    D) 4.6-5.5 mEq/liter
    E) 1.7-2.6 mEq/liter
    Answer: D

Pages: 894

26. How much sodium chloride is dissolved in one liter of isotonic saline?
    A) To calculate the answer, the molecular weight of sodium chloride is required.
    B) 0.9 moles
    C) 9.0 moles
    D) 0.9 grams
    E) 9.0 grams
    Answer: E

Pages: 894

27. Comparison of the charges carried by ions in different solutions can be made using which of the following units?
    A) %
    B) millimoles/liter
    C) milliequivalents/liter
    D) milliosmoles/liter
    E) mm Hg
    Answer: C

Pages: 894

28. Which of the following units can be used to compare the osmotic activity of substances by expressing the total number of particles in solution?
    A) %
    B) millimoles/liter
    C) milliequivalents/liter
    D) milliosmoles/liter
    E) mm Hg
    Answer: D

Pages: 894

29. Osmotic pressure contributed by one mOsm/liter of a substance can be calculated by multiplying the number of mOsm/liter by:
    A) 19.3 mm Hg
    B) 760 mm Hg
    C) the number of mEq/liter
    D) the number of mmol/liter
    E) the molecular weight of the substance
    Answer: A

Pages: 894

30. Calculate the number of mEq/liter of cations of a 3 mmol/liter solution of Na_2SO_4. Assume complete ionization.
    A) To calculate the answer, the molecular weight is required.
    B) 57.9 mEq/liter
    C) 3 mEq/liter
    D) 6 mEq/liter
    E) 12 mEq/liter
    Answer: D

Pages: 896
31. The cation that is necessary for generation and conduction of action potentials and that contributes nearly half of the osmotic pressure of extracellular fluid is:
    A) sodium ion
    B) potassium ion
    C) calcium ion
    D) chloride ion
    E) phosphate ion
    Answer: A

Pages: 897
32. Which of the following would you expect to see in response to an extracellular fluid calcium ion level of 5.7 mEq/liter?
    A) increased secretion of aldosterone
    B) increased secretion of PTH
    C) increased secretion of CT
    D) decreased secretion of ANP
    E) increased secretion of ADH
    Answer: C

Pages: 896
33. Aldosterone regulates the level of chloride ions in body fluids by:
    A) opening chloride channels in principal cells of distal convoluted tubules
    B) reabsorbing chloride ions for electrical balance as bicarbonate ions are secreted from renal tubules
    C) altering the permeability of glomerular capillaries
    D) regulating secretion from gastric mucosal glands
    E) controlling reabsorption of sodium ions, which chloride ions follow due to electrical attraction
    Answer: E

Pages: 896
34. The primary intracellular cation that plays a role in establishing resting membrane potential and in the repolarization phase of action potentials in nervous and muscle tissue is:
    A) sodium ion
    B) potassium ion
    C) phosphate ion
    D) hydrogen ion
    E) chloride ion
    Answer: B

Pages: 896
35. When bicarbonate ion diffuses out of red blood cells into plasma, it is usually exchanged with which anion?
    A) sodium
    B) potassium
    C) phosphate
    D) hydrogen
    E) chloride
    Answer: E

Pages: 897

36. Why are levels of bicarbonate ion higher in arterial blood than in venous blood?
    A) because the partial pressure of carbon dioxide is higher in arterial blood
    B) because more bicarbonate ions are used up in venous blood to buffer hydrogen ions
    C) because the higher oxygen levels in arterial blood promote dissociation of carbonic acid
    D) because cells in the pulmonary capillaries actively secrete bicarbonate ions into the plasma
    E) because the higher oxygen levels in arterial blood increase the activity of carbonic anhydrase
    Answer: B

Pages: 897

37. Which of the following is **NOT** an important role for calcium ions in the body?
    A) excitability of muscle tissue
    B) blood coagulation
    C) release of neurotransmitters
    D) buffer for metabolic acids
    E) hardness of bone
    Answer: D

Pages: 897

38. Which of the following is **NOT** an effect of increased levels of parathyroid hormone?
    A) increased absorption of calcium ions from the gastrointestinal tract
    B) increased reabsorption of calcium ions by renal tubule cells
    C) increased reabsorption of phosphate ions by renal tubule cells
    D) increased release of calcium ions from mineral salts in bone matrix
    E) increased release of phosphate ions from mineral salts in bone matrix
    Answer: C

Pages: 898

39. The ion that acts as a cofactor for the Na^+/K^+ ATPase is:
    A) chloride
    B) monohydrogen phosphate
    C) magnesium
    D) bicarbonate
    E) calcium
    Answer: C

Pages: 899

40. In studies of fluid balance, the term **water intoxication** refers to:
    A) poisoning of the body's water due to buildup of toxic substances during renal failure
    B) increased blood hydrostatic pressure created by high total blood volume
    C) movement of water from interstitial fluid into intracellular fluid due to osmotic gradients created by ion loss
    D) any situation in which edema develops
    E) failure of the neurohypophysis to secrete sufficient ADH
    Answer: C

41. Hypovolemic shock may occur in patients with renal failure who suffer a decrease in interstitial fluid osmotic pressure because:
    A) water flows from plasma into interstitial fluid along osmotic gradients created as water moves from interstitial fluid to intracellular fluid
    B) lymph channels become blocked by crystalline waste products, so water lost to the interstitial fluid is not returned to blood
    C) renal patients excrete proportionally more ions than water in urine
    D) renal patients do not manufacture enough plasma proteins to keep blood colloid osmotic pressure high enough to balance interstitial fluid osmotic pressure
    E) a drop in interstitial fluid osmotic pressure inhibits secretion of aldosterone
    Answer: A

42. The carboxyl group of an amino acid acts as a buffer for:
    A) excess hydrogen ions
    B) excess hydroxide ions
    C) other carboxyl groups
    D) carbonic acid in red blood cells
    E) hydrochloric acid in gastric juice
    Answer: B

43. The amount of metabolic water produced per day is about:
    A) 200 ml
    B) 800-1000 ml
    C) 1500 ml
    D) 2.5 liters
    E) 5 liters
    Answer: A

44. At a blood pH of 7.4, the two most important amino acid buffers are:
    A) tyrosine and phenylalanine
    B) threonine and methionine
    C) glutamine and lysine
    D) isoleucine and leucine
    E) histidine and cysteine
    Answer: E

45. Hemoglobin picks up a hydrogen ion when:
    A) it releases oxygen to tissues
    B) it binds oxygen in pulmonary capillaries
    C) chloride ions enter red blood cells
    D) the intracellular concentration of monohydrogen phosphate ions is too low to be effective
    E) chloride ions leave red blood cells
    Answer: A

Pages: 901
46. Which of the following does **NOT** act as a weak base?
    A) bicarbonate ion
    B) the -NH_2 end of an amino acid
    C) monohydrogen phosphate ion
    D) deoxyhemoglobin
    E) $H_2PO_4^-$
    Answer: E

Pages: 901
47. The ratio of bicarbonate ions to carbonic acid molecules in extracellular fluid is normally about:
    A) 1:20
    B) 1:1
    C) 2:1
    D) 20:1
    E) 100:1
    Answer: D

Pages: 903
48. Which of the following **cannot** help protect against pH changes caused by respiratory problems in which there is an excess or shortage of carbon dioxide?
    A) plasma protein buffers
    B) hemoglobin
    C) bicarbonate ion/carbonic acid buffers
    D) phosphate buffers
    E) only phosphate buffers can help protect against such pH changes
    Answer: C

Pages: 900
49. Which of the following statements is correct?
    A) A strong acid plus a weak acid yields water plus a weak base.
    B) A strong acid plus a weak base yields a salt plus a weak acid.
    C) A strong acid plus a weak base yields a weak base plus a weak acid.
    D) A strong acid plus a strong base yields a weak acid plus a weak base.
    E) A strong acid plus a weak acid yields a strong base plus a weak base.
    Answer: B

Pages: 900
50. Which of the following statements is correct?
    A) A strong base plus a weak base yields a salt plus a weak base.
    B) A strong base plus a weak acid yields a strong acid and a weak base.
    C) A strong base plus a strong acid yields a weak base plus a weak acid.
    D) A strong base plus a weak acid yields water plus a weak base.
    E) A strong base plus a weak base yields a strong acid plus a weak acid.
    Answer: D

Pages: 901
51. Increasing respiratory rate will:
    A) add more hydrogen ions to the extracellular fluid
    B) result in an increase in excretion of excess bicarbonate ions in urine
    C) result in an increase in excretion of dihydrogen phosphate ions in urine
    D) lower the pH of extracellular fluid
    E) cause a decrease in the affinity of hemoglobin for oxygen
    Answer: B

Pages: 903
52. The inspiratory center in the medulla oblongata triggers more forceful and frequent contractions of the diaphragm if:
    A) a decrease in pCO_2 is detected by peripheral chemoreceptors
    B) a large quantity of an alkaline drug is ingested
    C) levels of ketone bodies become elevated
    D) hydrochloric acid is lost via severe vomiting
    E) blood pressure is increased
    Answer: C

Pages: 902
53. Which of the following values represents a normal concentration of hydrogen ions per liter of arterial blood?
    A) 22-26 mEq
    B) 19-24 mEq
    C) 4.5-5.5 mEq
    D) 7.35-7.45 mEq
    E) 0.035-0.045 mEq
    Answer: E

Pages: 903
54. Which of the following would you expect to see in a patient with mild emphysema?
    A) increased urinary excretion of dihydrogen phosphate ions
    B) pCO_2 less than 35 mm Hg in arterial blood
    C) decreased deamination of glutamine by renal tubule cells
    D) reduced plasma levels of bicarbonate ion
    E) increased affinity of hemoglobin for oxygen
    Answer: A

Pages: 903
55. Aspirin overdose may lead to respiratory alkalosis because aspirin:
    A) is an alkaline drug
    B) depresses the excitability of the diaphragm
    C) induces hyperventilation
    D) stimulates severe vomiting and diarrhea
    E) blocks the action of carbonic anhydrase
    Answer: C

56. Uncontrolled diabetes mellitus may lead to metabolic acidosis because:
    A) high glucose levels depress the respiratory centers in the medulla
    B) glucose is an acidic substance
    C) glucose is osmotically active, and for every water molecule retained, a hydrogen ion is also retained
    D) most diabetics have chronic diarrhea, which leads to excessive loss of bicarbonate ions
    E) increased rates of lipolysis and ketogenesis occur
    Answer: E

57. Which of the following might trigger an increase in the rate of deamination of glutamine by renal tubule cells as a form of compensation for a pH imbalance?
    A) an abrupt move to a high altitude
    B) ingestion of alkaline drugs
    C) plasma levels of bicarbonate ion at 30 mEq/liter
    D) pulmonary edema
    E) severe, prolonged vomiting
    Answer: D

58. A patient whose blood pH is 7.47, whose pCO_2 is 31 mmHg in arterial blood, and whose levels of bicarbonate ion in arterial blood are 23 mEq/liter is in:
    A) compensated metabolic alkalosis
    B) uncompensated respiratory acidosis
    C) uncompensated respiratory alkalosis
    D) uncompensated metabolic acidosis
    E) uncompensated metabolic alkalosis
    Answer: C

59. A patient whose blood pH is 7.47, whose pCO_2 in arterial blood is 40 mm Hg, and whose levels of bicarbonate ion in arterial blood are 28 mEq/liter is in:
    A) compensated metabolic alkalosis
    B) uncompensated respiratory acidosis
    C) uncompensated respiratory alkalosis
    D) uncompensated metabolic acidosis
    E) uncompensated metabolic alkalosis
    Answer: E

60. What percentage of a newborn's total body weight is water?
    A) less than 25%
    B) 25-30%
    C) 40-50%
    D) 55-60%
    E) 75-90%
    Answer: E

61. **ALL** of the following are **TRUE** for newborns **EXCEPT**:
    A) their metabolic rate is higher than that of adults
    B) they have a greater proportion of body volume to surface area than adults
    C) they have a higher respiratory rate than adults
    D) their kidneys do not concentrate urine as well as those of adults
    E) they have a greater percentage of body water in the extracellular fluid than in the
       intracellular fluid
    Answer: B

62. In which of the following persons would you expect to find water constituting the greatest
    percentage of body weight?
    A) a premature infant
    B) an 18-year-old male weight lifter
    C) a 40-year-old obese woman
    D) an 80-year-old male of normal weight
    E) Percentage of water relative to body weight remains constant regardless of age, sex, or
       weight.
    Answer: A

63. At the arterial end of a capillary net movement of water is:
    A) out of the vessel due to high interstitial fluid osmotic pressure
    B) into the vessel due to high interstitial fluid hydrostatic pressure
    C) out of the vessel due to high blood hydrostatic pressure
    D) into the vessel due to high blood colloid osmotic pressure
    E) out of the vessel due to greater capillary permeability
    Answer: C

64. Overexcitability of the central nervous system is symptomatic of:
    A) hyponatremia
    B) alkalosis
    C) hypokalemia
    D) hypercalcemia
    E) all of the above except hyponatremia
    Answer: B

65. On average, daily water loss from all mechanisms totals about:
    A) 200 ml
    B) 800-1000 ml
    C) 1.5 liters
    D) 2.5 liters
    E) 5.0 liters
    Answer: D

True-False

Write T if the statement is true and F if the statement is false.

Pages: 891
1. The largest percentage of body water in the adult is contained within cells.
   Answer: True

Pages: 892
2. Males of average weight usually have a higher percentage of body water than do females of average weight.
   Answer: True

Pages: 894
3. A molecule of glucose creates a greater osmotic pressure than a molecule of sodium chloride because glucose is a much larger molecule.
   Answer: False

Pages: 894
4. Interstitial fluid contains relatively few protein anions.
   Answer: True

Pages: 896
5. The most abundant extracellular ion is sodium ion.
   Answer: True

Pages: 896
6. If the levels of sodium ions are 145 mEq/liter in the plasma, the posterior pituitary stops secreting ADH.
   Answer: False

Pages: 896
7. If sodium ion levels in plasma are 130 mEq/liter, osmosis of water from the extracellular fluid into the intracellular fluid occurs.
   Answer: True

Pages: 897
8. Fibrillation of the heart may result from hyperkalemia.
   Answer: True

Pages: 897
9. The levels of bicarbonate ions in the plasma are regulated primarily by the liver.
   Answer: False

Pages: 897
10. If levels of PTH increase, more calcium ions are reabsorbed through the renal tubule cells into the blood.
    Answer: True

11. If levels of PTH increase, more phosphate ions are deposited into the bone matrix.
Answer: False

12. A person whose blood pH is 7.32 is in alkalosis.
Answer: False

13. A person who has a blood pH of 7.35 has a higher concentration of hydrogen ions in the blood than someone with a blood pH of 7.40.
Answer: True

14. If respiratory rate decreases, the concentration of hydrogen ions in extracellular fluid decreases.
Answer: False

15. A person whose pH imbalance is due to blood levels of HCO_3^- that are less than 22 mEq/liter is in metabolic acidosis.
Answer: True

Short Answer

Write the word or phrase that best completes each statement or answers the question.

1. All body fluids not contained within cells collectively are called _____.
Answer: extracellular fluid

2. Fluid in spaces between cells is called _____.
Answer: interstitial fluid

3. The primary method by which water moves between fluid compartments is _____.
Answer: osmosis

4. As age increases, the percentage of body weight that is water _____; as the amount of adipose tissue increases, the percentage of body weight that is water _____.
Answer: decreases; decreases

5. Increased blood volume stretching the right atrium stimulates the release of _____.
Answer: atrial natriuretic peptide

6. Positively charged ions are called _____; negatively charged ions are called _____.
   Answer: cations; anions

7. An inorganic substance that dissociates into ions when dissolved in water is called a(n)
   _____.
   Answer: electrolyte

8. Isotonic saline is a _____% solution of sodium chloride.
   Answer: 0.9

9. The positive or negative charge equal to the amount of charge in one mole of hydrogen ions is
   called one _____.
   Answer: equivalent

10. Units that express the total number of particles (molecules or ions) in a given volume of
    solution in the body are called _____.
    Answer: milliosmoles/liter

11. In intracellular fluid, the most abundant cation is _____ and the most abundant inorganic
    anion is _____.
    Answer: potassium ion; monohydrogen phosphate

12. The condition in which sodium levels in blood are lower than normal is called _____.
    Answer: hyponatremia

13. The most abundant extracellular anion is _____.
    Answer: chloride ion

14. Plasma levels of potassium ions are regulated primarily by the hormone _____.
    Answer: aldosterone

15. The condition in which levels of potassium ions are higher than normal in the blood is called
    _____.
    Answer: hyperkalemia

16. The inorganic ion that is the major buffer of hydrogen ions in the plasma is _____.
    Answer: bicarbonate ion

Pages: 897
17. As blood passes through the pulmonary capillaries, the plasma level of bicarbonate ion
_____.

Answer: increases

Pages: 897
18. Plasma levels of calcium ions are regulated primarily by the hormones _____ and

_____.

Answer: parathyroid hormone; calcitonin

Pages: 897
19. If levels of aldosterone increase, excretion of potassium ions in urine _____.
Answer: increases

Pages: 897
20. If levels of the hormone _____ increase, excretion of phosphate ions in urine increases.
Answer: parathyroid hormone

Pages: 900
21. The homeostatic range of pH for extracellular fluid is _____.
Answer: 7.35-7.45

Pages: 901
22. When respiratory rate increases, pH of extracellular fluid _____.
Answer: increases

Pages: 902
23. The only way the body can eliminate acids other than carbonic acid is by _____.
Answer: excretion in urine

Pages: 900
24. A strong acid combined with a weak base yields _____ and _____
Answer: water; a weak acid

Pages: 901
25. In the phosphate buffer system, the weak base is _____.
Answer: monohydrogen phosphate

Pages: 900
26. The part of a protein buffer that acts as a weak base is _____.
Answer: the amino group

Pages: 901
27. As a buffer, HCO_3^- acts as a _____.
Answer: weak base

Pages: 903
28. If blood pH becomes 7.32, respiratory rate will _____ as a compensatory mechanism.
Answer: increase

Pages: 903
29. The principal effect of acidosis is _____ of the CNS.
    Answer: depression

Pages: 904
30. By examining pH, _____ , and _____ values, it is possible to determine the cause of an acid-base imbalance.
    Answer: bicarbonate; pCO_2

Matching

Choose the item from Column 2 that best matches each item in Column 1.

Match the following:

1. hyponatremia

    movement of water into cells; muscle weakness, hypotension, tachycardia, shock

2. hypernatremia

    cellular dehydration leading to intense thirst, fatigue, restlessness, agitation

3. hypokalemia

    flaccid paralysis; lengthening of Q-T interval and flattening of T wave on ECG

4. hyperkalemia

    anxiety, paresthesia, cardiac fibrillation

5. hypocalcemia

    hyperactive reflexes, tetany, bone fractures, laryngospasm

6. hypochloremia

    muscle spasms, alkalosis

Match the following...

7. Na^+

    accounts for 50% of osmotic pressure in ECF; important for establishing resting potential and depolarization

8. K^+

    intracellular cation important in establishing resting potential and repolarization

9. Cl^-

    important for anion balance between plasma and red blood cells; secreted by gastric mucosal cells

10. $H_2PO_4^-$

    anion that is an important intracellular and urinary buffer

11. Ca^{2+}

    cation that is an important clotting factor and that is involved in release of neurotransmitters

12. Mg^{2+} important cofactor for functioning of the Na+/K+ pump

13. HCO_3^- important buffer for acids in blood

Essay

Write your answer in the space provided or on a separate sheet of paper.

Pages: 895
1. Identify the functions of electrolytes in the body.
   Answer: 1) control osmosis of water between fluid compartments
   2) help maintain acid-base balance
   3) production of action potential and graded potentials
   4) control secretion of some hormones and neurotransmitters
   5) act as cofactors in enzyme activity

Pages: 893
2. Describe the negative feedback loop that stimulates thirst by dehydration.
   Answer: Dehydration causes 1) decreased flow of saliva, which dries the mouth and pharynx, 2) increased blood osmotic pressure, which stimulates osmoreceptors in the hypothalamus, and 3) decrease blood volume, which lowers BP, increasing release of renin from JG cells, increasing levels of angiotensin II. All of these stimulate the thirst center in the hypothalamus, which increases fluid intake via thirst, thus increasing body water.

Pages: 892
3. Normally, water loss equals water gain within the body. Describe the mechanisms by which water is lost and gained by the body.
   Answer: Water is lost from the GI tract in feces, from the respiratory tract as water vapor in exhaled air, from the skin via insensible perspiration and sweat, and from the excretory system as urine. Water is gained from metabolic reactions, as part of ingested foods, and as simply ingested water.

Pages: 904
4. Identify four reasons why infants experience more fluid and electrolyte and acid-base imbalances than adults.
   Answer: Compared with adults, babies have:
   1) a greater percentage of body water; more water in ECF than ICF; rapid changes in ECF volume due to higher intake/output
   2) a higher metabolic rate generating more metabolic acids & wastes
   3) less efficient kidneys
   4) greater water loss through skin due to relatively high proportion of surface area to volume
   5) a higher infant respiratory rate leading to increased water loss and higher pH
   6) higher levels of K^+ and Cl^- causing a tendency toward acidosis

Pages: 899
5. Explain how water intoxication develops. What is the effect on plasma volume, why does this occur?

Answer: Sodium ions lost via excessive vomiting, diarrhea, or sweating cause interstitial fluid to become hypotonic to ICF. Water enters ICF, causing cells to swell. Osmotic pressure of interstitial fluid rises as water leaves. Water leaves plasma inducing hypovolemic shock.

Pages: 900
6. Explain how buffer systems work using generic formulas.

Answer: A strong acid is buffered by a weak base to yield a salt and a weak acid. A strong base is buffered by a weak acid to form water and a weak base. Conversion of strong acids/bases to weak acids/bases minimizes the stress on the pH of ECF by reducing the number of new hydrogen or hydroxide ions added by the strong acids/bases.

Pages: 900
7. Identify the buffer systems in the body. In each system, identify the principal site of action and the weak acid and weak base.

Answer: The primary buffer system in plasma and body cells is the protein buffer system, in which the weak acid is the carboxyl end of an amino acid, and the weak base is the amino end. Carbonic acid (weak acid)-bicarbonate ion (weak base) buffers are especially important in blood, but work in both ICF and ECF. The phosphate buffers work mostly in ICF and urine. Dihydrogen phosphate is the weak acid; monohydrogen phosphate is the weak base.

Pages: 903
8. What are acidosis and alkalosis, and how do they develop? What are the primary effect of each?

Answer: Acidosis is blood pH < 7.35, which causes depression of the CNS via reduced synaptic transmission. Alkalosis is blood pH > 7.45, which causes overexcitability of the CNS and PNS. Anything that causes pCO_2 to rise in alveoli, or causes loss of bicarbonate ions, or causes buildup of metabolic acids leads to acidosis. Anything that lowers pCO_2, or causes loss of acids, or any excessive ingestion of alkaline substances leads to alkalosis.

Pages: 893
9. A patient has come into the ER with third degree burns over 50% of his body. What aspects of homeostasis are the most likely life-threatening effects of this injury? Explain your answer.

Answer: Excessive water loss from body surface (from loss of stratified epithelium) leading to dehydration and shock; threat of infection from loss of first line of non-specific resistance. [note: Other answers may be acceptable based on student rationale; question is intended to cover multiple chapters.]

Pages: 903
10. Patients with cholera may lose more than five liters of fluid per day from severe diarrhea. What effects on electrolyte and acid/base balance might this produce? What compensatory mechanisms (other than immune function) might be activated as a result of this disease?

Answer: Hyponatremia and loss of bicarbonate ions, leading to metabolic acidosis and depression of CNS; also increase in aldosterone, increased respiratory rate, renal generation of new bicarbonate, increased renal excretion of hydrogen ions via ammonium and dihydrogen phosphate; also all feedback loops to raise BP, which falls due to water loss.

Pages: 901
11. Explain why/how respiratory rate affects the pH of extracellular fluid. When would increases and decreases in respiratory rate be appropriate to alter acid/base balance?

Answer: Carbon dioxide and water combine to form carbonic acid, which dissociates into bicarbonate ion and hydrogen ion. Hydrogen ions lower pH. The direction of the reaction is determined by the pCO_2. To raise pCO_2, thus increasing hydrogen ions and lowering pH, lower respiratory rate to compensate for metabolic alkalosis. To raise pH, increase respiratory rate to compensate for metabolic acidosis by decreasing pCO_2 and consuming hydrogen ions in the reaction.

Pages: 904
12. A patient's blood pH is 7.48. pCO_2 in arterial blood is 32 mm Hg. Levels of bicarbonate ion in blood are 20 mEq/liter. What can you tell about this patient's condition? Explain your answer.

Answer: Patient is in respiratory alkalosis (high pH, low pCO_2), which is partially compensated (low bicarbonate).

Multiple-Choice

Choose the one alternative that best completes the statement or answers the question.

Pages: 915
1. Testosterone is produced by:
   A) spermatozoa
   B) sustentacular cells
   C) interstitial cells
   D) the hypothalamus
   E) all cells in the male
   Answer: C

Pages: 915
2. The acrosome of a sperm cell contains:
   A) the chromosomes
   B) mitochondria for energy production
   C) testosterone
   D) hyaluronidase for egg penetration
   E) the flagellum
   Answer: D

Pages: 939
3. During the menstrual cycle, LH is at its highest levels:
   A) during the menstrual phase
   B) just prior to ovulation
   C) just after ovulation
   D) just before menstruation begins
   E) levels of LH never change
   Answer: B

Pages: 939
4. During the menstrual cycle, progesterone would be at its highest levels:
   A) during the menstrual phase
   B) just prior to ovulation
   C) just after ovulation
   D) late in the postovulatory phase
   E) levels of progesterone never change
   Answer: D

Pages: 941
5. During the menstrual cycle, the endometrium would be at its thickest:
   A) during the menstrual phase
   B) just prior to ovulation
   C) just after ovulation
   D) late in the postovulatory phase
   E) the thickness never changes
   Answer: D

6. Maintenance of the male secondary sex characteristics is the direct responsibility of:
   A) estrogen
   B) testosterone
   C) FSH
   D) progesterone
   E) LH
   Answer: B

7. Which of the following cells are diploid?
   A) secondary oocytes
   B) secondary spermatocytes
   C) primary spermatocytes
   D) spermatids
   E) all of the above except spermatids
   Answer: C

8. A function of FSH in the male is to:
   A) inhibit progesterone
   B) initiate testosterone production
   C) increase protein synthesis
   D) inhibit estrogen
   E) initiate spermatogenesis
   Answer: E

9. Final maturation of sperm cells occurs in the:
   A) epididymis
   B) seminiferous tubules
   C) prostate gland
   D) urethra
   E) female reproductive tract
   Answer: A

10. Seminal vesicles produce:
    A) sperm cells
    B) testosterone
    C) fructose-rich fluid
    D) estrogen
    E) mucus
    Answer: C

Pages: 910
11. Sustentacular cells produce:
    A) testosterone
    B) androgen-binding protein
    C) estrogen
    D) FSH
    E) LH
    Answer: B

Pages: 921
12. The normal number of spermatozoa per milliliter of semen is:
    A) 50-100
    B) fewer than 20,000,000
    C) more than 200,000,000
    D) 50,000,000-150,000,000
    E) about 5 million
    Answer: D

Pages: 939
13. Repair of the endometrium during the preovulatory phase of menstruation is due to rising levels
    of:
    A) FSH
    B) estrogen
    C) hCG
    D) progesterone
    E) inhibin
    Answer: B

Pages: 940
14. During the menstrual cycle, progesterone is produced by:
    A) the secondary oocyte
    B) the corpus luteum
    C) the stroma of the ovary
    D) primary follicles
    E) the endometrium
    Answer: B

Pages: 937
15. Which of the following does **NOT** produce estrogens?
    A) adrenal cortex
    B) placenta
    C) ovarian follicle cells
    D) testes
    E) hypothalamus
    Answer: E

Pages: 913
16. During spermatogenesis, which of the following undergoes a meiotic division to produce haploid cells?
    A) spermatids
    B) secondary spermatocytes
    C) primary spermatocytes
    D) spermatogonia
    E) spermatozoa
    Answer: C

Pages: 913
17. The function of androgen binding protein is to:
    A) provide nourishment for developing spermatocytes
    B) carry testosterone in the bloodstream
    C) keep levels of testosterone high in the testes
    D) nourish mature spermatozoa in semen
    E) prevent chemicals in the female reproductive tract from destroying spermatozoa
    Answer: C

Pages: 926
18. The first meiotic division in oogenesis occurs:
    A) before birth
    B) only if the egg is fertilized
    C) after ovulation
    D) monthly after puberty in response to FSH and LH
    E) when adrenal gonadocorticoids begin to rise at the start of puberty
    Answer: A

Pages: 910
19. The fibrous capsule around the testes is called the:
    A) urogenital diaphragm
    B) germinal epithelium
    C) seminiferous tubules
    D) rete testis
    E) tunica albuginea
    Answer: E

Pages: 915
20. In the male, LH causes:
    A) initiation of spermatogenesis
    B) development of secondary sex characteristics
    C) testosterone production
    D) ejaculation
    E) release of GnRH
    Answer: C

Pages: 939
21. The part of the female reproductive system that is shed during menstruation is the:
    A) myometrium
    B) mucosa of the vagina
    C) tunica albuginea
    D) stratum functionalis of the endometrium
    E) germinal epithelium
    Answer: D

Pages: 941
22. The main function of progesterone during the menstrual cycle is to:
    A) initiate ovulation
    B) initiate menstruation
    C) thicken the endometrium
    D) repair the surface of the ovary after ovulation
    E) stimulate the release of FSH and LH
    Answer: C

Pages: 910
23. The scrotum is the:
    A) tissue covering the end of the penis
    B) accessory organ that produces the largest volume of seminal fluid
    C) duct in which sperm cells mature
    D) structure that supports the testes
    E) part of the penis surrounding the urethra
    Answer: D

Pages: 910
24. The function of the cremaster muscle is to:
    A) elevate the testes during sexual arousal and exposure to cold
    B) generate peristaltic waves in the ductus deferens
    C) control the release of secretions from the seminal vesicles
    D) control the release of sperm cells from the testes into the epididymis
    E) prevent urine from entering the urethra during ejaculation
    Answer: A

Pages: 913
25. The form (stage) of developing male gamete located nearest to the basement membrane of a seminiferous tubule is the:
    A) spermatid
    B) primary spermatocyte
    C) secondary spermatocyte
    D) primordial germ cell
    E) spermatogonium
    Answer: E

Pages: 911
26. Interstitial endocrinocytes are located:
   A) in all the male accessory reproductive organs
   B) interspersed among developing sperm cells in seminiferous tubules
   C) lining the epididymis and ductus deferens
   D) within the tunica albuginea
   E) in spaces between adjacent seminiferous tubules
   Answer: E

Pages: 910
27. The immune system does not normally attack spermatogenic cells because:
   A) they are recognized as "self" structures
   B) they do not have any antigens on their cell membranes
   C) spermatogenic cells are protected by the blood-testis barrier
   D) the acrosome covers any antigens that would be recognized as foreign
   E) spermatogenic cells release chemicals that repel antigen-presenting cells
   Answer: C

Pages: 913
28. The process of crossing-over, or recombination, of genes occurs during:
   A) meiosis I
   B) meiosis II
   C) spermiogenesis
   D) spermiation
   E) fertilization
   Answer: A

Pages: 913
29. Primordial germ cells arise from the:
   A) mesonephric ducts
   B) yolk sac endoderm
   C) genital tubercle
   D) urethral folds
   E) paramesonephric ducts
   Answer: B

Pages: 914
30. The cells that result from the equatorial division of spermatogenesis are called:
   A) spermatogonia
   B) primary spermatocytes
   C) secondary spermatocytes
   D) primordial germ cells
   E) spermatids
   Answer: E

Pages: 914
31. The process of spermiation is the:
   A) reduction division of male gamete production
   B) equatorial division of male gamete production
   C) process producing the liquid portion of semen
   D) release of a sperm cell from its connection to a sustentacular cell
   E) change that occurs in the sperm cell following penetration of the egg
   Answer: D

Pages: 915
32. The principal androgen is:
   A) ABP
   B) FSH
   C) testosterone
   D) hCG
   E) estradiol
   Answer: C

Pages: 917
33. A vasectomy involves removal of a segment of the:
   A) penis
   B) testes
   C) ductus deferens
   D) seminal vesicles
   E) epididymis
   Answer: C

Pages: 917
34. The deep inguinal ring is an opening in the:
   A) transversus abdominis
   B) external oblique
   C) corpus spongiosum penis
   D) cremaster muscle
   E) detrusor muscle
   Answer: A

Pages: 917
35. **ALL** of the following are part of the spermatic cord **EXCEPT** the:
   A) testicular artery
   B) lymphatic vessels
   C) cremaster muscle
   D) ductus deferens
   E) ejaculatory duct
   Answer: E

Pages: 921
36. The part of the penis that surrounds the urethra is the:
    A) corpus spongiosum penis
    B) tunica albuginea
    C) corpora cavernosa penis
    D) crura
    E) prepuce
    Answer: A

Pages: 920
37. The function of fructose in semen is to:
    A) provide an energy source for ATP production by sperm
    B) promote coagulation of semen in the female reproductive tract
    C) buffer acids in the female reproductive tract
    D) inhibit the growth of bacteria in semen and the female reproductive tract
    E) provide an energy source for the zygote
    Answer: A

Pages: 920
38. The seminal vesicles are located:
    A) inferior to the prostate within the urogenital diaphragm
    B) within the lobules of the testes
    C) within the spermatic cord
    D) posterior and inferior to the urinary bladder, in front of the rectum
    E) on the posterior surface of each testis
    Answer: D

Pages: 921
39. Which of the following does **NOT** manufacture products that become part of semen?
    A) seminiferous tubules
    B) bulbourethral glands
    C) penis
    D) seminal vesicles
    E) prostate gland
    Answer: C

Pages: 921
40. The function of seminalplasmin is to:
    A) provide an energy source for sperm
    B) act as an antibiotic to control bacterial numbers
    C) cause semen to coagulate in the female reproductive tract
    D) prevent destruction of spermatogenic cells by the immune system
    E) buffer the acidity of the female reproductive tract
    Answer: B

Pages: 923
41. Circumcision involves the surgical removal of all or part of the:
    A) glans penis
    B) ductus deferens
    C) corpora cavernosa penis
    D) prepuce
    E) testes
    Answer: D

Pages: 924
42. The female structure that is homologous to the testis is the:
    A) ovary
    B) uterus
    C) vagina
    D) clitoris
    E) Bartholin's gland
    Answer: A

Pages: 925
43. The layer of simple epithelium covering the surface of the ovary is the:
    A) cortex
    B) medulla
    C) germinal epithelium
    D) tunica albuginea
    E) stratum basalis
    Answer: C

Pages: 927
44. The glycoprotein layer between the oocyte and the granulosa cells of an ovarian follicle is called the:
    A) theca interna
    B) theca externa
    C) antrum
    D) zona pellucida
    E) corona radiata
    Answer: D

Pages: 927
45. The secretory cells of an ovarian follicle are called the:
    A) theca interna
    B) theca externa
    C) antrum
    D) zona pellucida
    E) corona radiata
    Answer: A

46. Which of the following help move the oocyte into and through the uterine tube?
    A) peristalsis
    B) cilia
    C) flagella
    D) fimbriae
    E) all of the above except flagella
    Answer: E

47. The uterus is located:
    A) between the urinary bladder and the pubic symphysis
    B) between the rectum and the sacrum
    C) between the urinary bladder and the rectum
    D) between the kidneys in the same horizontal plane
    E) along the inferior surface of the urinary bladder
    Answer: C

48. A Pap smear involves removal of cells from the:
    A) ovaries
    B) uterine tubes
    C) fundus of the uterus
    D) cervix of the uterus
    E) clitoris
    Answer: D

49. The opening between the cervical canal and the uterine cavity is called the:
    A) internal os
    B) external os
    C) isthmus
    D) fornix
    E) vagina
    Answer: A

50. The folds of the peritoneum attaching the uterus to either side of the pelvic cavity are called the:
    A) uterosacral ligaments
    B) broad ligaments
    C) cardinal ligaments
    D) round ligaments
    E) suspensory ligaments
    Answer: B

Pages: 930
51. The layer of the uterine wall that is part of the visceral peritoneum is the:
    A) stratum functionalis
    B) stratum basalis
    C) myometrium
    D) perimetrium
    E) mucosa
    Answer: D

Pages: 931
52. Which of the following lists the uterine blood vessels in the correct order of blood flow?
    A) radial arteries, arcuate arteries, spiral arterioles, straight arterioles
    B) spiral arterioles, straight arterioles, radial arteries, arcuate arteries
    C) straight arterioles, radial arteries, arcuate arteries, spiral arterioles
    D) arcuate arteries, radial arteries, straight arterioles, spiral arterioles
    E) arcuate arteries, straight arterioles, spiral arterioles, radial arteries
    Answer: D

Pages: 931
53. The uterine blood vessels that penetrate deep into the myometrium are the:
    A) radial arteries
    B) arcuate arteries
    C) uterine arteries
    D) straight arterioles
    E) spiral arterioles
    Answer: A

Pages: 931
54. The epithelium of the vaginal mucosa is:
    A) simple squamous
    B) simple cuboidal
    C) simple columnar
    D) transitional
    E) stratified squamous
    Answer: E

Pages: 932
55. The female structure that is homologous to the scrotum is the:
    A) mons pubis
    B) labia majora
    C) labia minora
    D) clitoris
    E) hymen
    Answer: B

Pages: 933
56. The perineum is bounded by the:
   A) pubic symphysis, iliac crests, and sacral promontory
   B) anterior and posterior inferior iliac spines, and pubic symphysis
   C) pubic symphysis, ischial tuberosities, and coccyx
   D) pubic symphysis, posterior inferior iliac spines, and coccyx
   E) ischial tuberosities, iliac crests, and anterior inferior iliac spines
   Answer: C

Pages: 936
57. Milk production is stimulated primarily by the hormone:
   A) FSH
   B) oxytocin
   C) DHT
   D) relaxin
   E) PRL
   Answer: E

Pages: 937
58. **ALL** of the following are functions of estrogens **EXCEPT**:
   A) help control fluid and electrolyte balance
   B) promote protein anabolism
   C) help regulate secretion of FSH
   D) promote development and maintenance of female secondary sex characteristics
   E) raise blood cholesterol
   Answer: E

Pages: 940
59. If fertilization does not occur, the corpus luteum:
   A) is expelled into the pelvic cavity
   B) begins to secrete low levels of FSH
   C) degenerates into the corpus albicans
   D) continues to secrete progesterone until the next ovulation
   E) both A and C are correct
   Answer: C

Pages: 942
60. The process by which sperm are deposited into the vagina is called:
   A) fertilization
   B) spermiation
   C) spermiogenesis
   D) coitus
   E) capacitation
   Answer: D

61. **ALL** of the following are **sympathetic** responses during sexual intercourse **EXCEPT**:
    A) peristalsis in the ductus deferens
    B) increased blood pressure
    C) contraction of perineal muscles
    D) ejaculation of semen
    E) erection of the penis/clitoris
    Answer: E

62. Oral contraceptives for women typically contain:
    A) human chorionic gonadotropin
    B) progestin and estrogen
    C) low levels of both FSH and LH
    D) nonoxynol-9
    E) testosterone
    Answer: B

63. Oral contraceptives for women work by:
    A) binding to the estrogen receptors on target cells
    B) killing sperm cells by increasing the permeability of their cell membranes
    C) reducing the percentage of body fat to levels too low to support menstruation
    D) providing negative feedback to decrease levels of FSH, LH, and GnRH
    E) forming a physical barrier around the oocyte that sperm cannot penetrate
    Answer: D

64. Onset of puberty in both sexes is signaled by sleep-associated increases in levels of:
    A) estradiol
    B) DHT
    C) oxytocin
    D) PRL
    E) LH
    Answer: E

65. The uterus and vagina are derived from the embryonic:
    A) paramesonephric ducts
    B) mesonephric ducts
    C) genital tubercle
    D) urethral groove
    E) urethral folds
    Answer: A

True-False

Write T if the statement is true and F if the statement is false.

Pages: 911
1. Spermatogonia produce testosterone.
   Answer: False

Pages: 913
2. Secondary spermatocytes contain 23 chromosomes.
   Answer: True

Pages: 913
3. Crossing-over refers to exchange of genes between homologous maternal and paternal chromosomes
   Answer: True

Pages: 914
4. Spermatids are formed by the first nuclear division (reduction division) of spermatogenesis.
   Answer: False

Pages: 914
5. The DNA of a sperm cell is contained in the acrosome.
   Answer: False

Pages: 915
6. GnRH regulates secretion of both FSH and LH.
   Answer: True

Pages: 915
7. The receptors for testosterone and DHT are found within the nuclei of target cells.
   Answer: True

Pages: 920
8. The seminal vesicles are located along the posterior border of each testis.
   Answer: False

Pages: 925
9. The layer of cells covering the surface of the ovary is called the germinal epithelium.
   Answer: True

Pages: 927
10. The form (stage) of the female sex cell that is ovulated is called an ovum.
    Answer: False

Pages: 927
11. The reduction division of oogenesis is completed before birth.
    Answer: False

Pages: 929
12. The uterus is located between the urinary bladder and the pubic symphysis.
Answer: False

Pages: 940
13. During the postovulatory phase of the menstrual cycle, new cells are added to the myometrium.
Answer: False

Pages: 939
14. Spiral arterioles of the endometrium undergo changes during the menstrual cycle.
Answer: True

Pages: 932
15. The term **vulva** refers to the external genitalia of the female.
Answer: True

Short Answer

Write the word or phrase that best completes each statement or answers the question.

Pages: 910
1. The blood-testis barrier is formed just internal to the basement membrane of the seminiferous tubule by tight junctions between _____ cells.
Answer: sustentacular

Pages: 911
2. The cells in the seminiferous tubule that secrete testosterone are the _____.
Answer: interstitial endocrinocytes

Pages: 913
3. The process of cell division in which diploid cells are converted to haploid gametes is called

_____.
Answer: meiosis

Pages: 913
4. Division of each primary spermatocyte eventually produces _____ spermatids.
Answer: four

Pages: 915
5. Release of LH and FSH is regulated by the hormone _____ produced by the hypothalamus.
Answer: GnRH

Pages: 915
6. A hormone secreted by sustentacular cells that targets the anterior pituitary to inhibit secretion of FSH is _____.
Answer: inhibin

Pages: 917

7. A tightly coiled tube, 6 m X 1 mm, that lies along the posterior border of the testis is the

    _____.

    Answer: epididymis

Pages: 917

8. The testicular artery, veins, autonomic nerves, lymphatic vessels, and the cremaster muscle together constitute the _____.

    Answer: spermatic cord

Pages: 920

9. An accessory gland lying inferior to the urinary bladder and surrounding the urethra in the male is the _____.

    Answer: prostate gland

Pages: 921

10. The average volume of semen in an ejaculation is _____ with a sperm count of _____ per milliliter.

    Answer: 2.5-5 ml; 50-150 million

Pages: 921

11. An antibiotic present in semen is _____.

    Answer: seminalplasmin

Pages: 921

12. The paired dorsolateral masses of erectile tissue of the penis are called the _____; the smaller midventral mass is called the _____.

    Answer: corpora cavernosa penis; corpora spongiosum penis

Pages: 921

13. The vascular changes resulting in an erection are the result of a _____ reflex.

    Answer: parasympathetic

Pages: 922

14. Covering the glans of an uncircumcised penis is the foreskin, also known as the _____.

    Answer: prepuce

Pages: 918

15. The _____ is formed by the union of the duct from the seminal vesicle and the ductus deferens.

    Answer: ejaculatory duct

Pages: 925

16. The white capsule of dense, irregular connective tissue that is immediately deep to the germinal epithelium of the ovary is called the _____.

    Answer: tunica albuginea

Pages: 926

17. During early female fetal development, primordial germ cells migrate from the endoderm of the yolk sac to the ovaries, where they differentiate into _____.

    Answer: oogonia

Pages: 926
18. Degeneration of primary germ cells during female fetal development is called _____.
Answer: atresia

Pages: 927
19. Finger-like projections of the infundibulum of the uterine tubes are called _____.
Answer: fimbriae

Pages: 929
20. The inferior narrow portion of the uterus that opens into the vagina is called the _____.
Answer: cervix

Pages: 930
21. The layer of the endometrium nearest the uterine cavity that is shed during menstruation is the

_____.
Answer: stratum functionalis

Pages: 931
22. A hysterectomy is surgical removal of the _____.
Answer: uterus

Pages: 933
23. The female structure that is homologous to the penis of the male is the _____.
Answer: clitoris

Pages: 939
24. The form (stage) of the egg that is ovulated is the _____.
Answer: secondary oocyte

Pages: 940
25. During the postovulatory phase of the menstrual cycle, the endometrium is prepared to receive a fertilized ovum principally by the hormone _____ produced by the corpus luteum.
Answer: progesterone

Pages: 944
26. One of the most widely used spermicides in contraceptive creams, foams, sponges, etc., is

_____.
Answer: nonoxynol-9 (or octoxynol-9)

Pages: 946
27. Onset of the first menses is called _____; permanent cessation of menses as part of the aging process is called _____.
Answer: menarche; menopause

Pages: 947
28. The male pattern of differentiation of primitive gonads in the embryo depends on the presence of a master gene on the Y chromosome called _____.
Answer: SRY

Pages: 950

29. A chancre at the point of contact is the symptom of the primary stage of the sexually transmitted disease _____.
    Answer: syphilis

Pages: 952

30. Infection of the uterine tubes is called _____.
    Answer: salpingitis

Matching

Choose the item from Column 2 that best matches each item in Column 1.

1. oxytocin

    stimulates ejection of milk from mammary glands

2. GnRH

    produced by the hypothalamus; promotes secretion of FSH and LH

3. LH

    stimulates secretion of testosterone

4. FSH

    stimulates spermatogenesis

5. testosterone

    promotes protein anabolism; promotes musculoskeletal growth resulting in wide shoulders and narrow hips; contributes to libido in both sexes

6. beta-estradiol

    helps control fluid and electrolyte balance; lowers blood cholesterol; promotes fat distribution to abdomen, breasts, and hips

7. progesterone

    works synergistically with estrogens to prepare the endometrium for implantation of the fertilized egg and the mammary glands for milk secretion

8. hCG

    produced by chorion of embryo to maintain activity of corpus luteum

9. inhibin

    secreted by Sertoli cells and corpus luteum; inhibits secretion of FSH

10. relaxin

    secreted by corpus luteum; inhibits uterine contractions; softens pelvic connective tissues

Essay

Write your answer in the space provided or on a separate sheet of paper.

Pages: 910

1. State the functions of sustentacular cells. Where are they located?

   Answer: Support & protect developing spermatogenic cells; form blood-testis barrier; nourish developing sperm cells; phagocytize excess cytoplasm during sperm development; mediate effects of testosterone and FSH; control movement of spermatogenic cells & release into lumen of seminiferous tubule; secrete inhibin; produce fluid for sperm transport

Pages: 913

2. Compare and contrast the processes of spermatogenesis and oogenesis.

   Answer: Both processes result in formation of gametes (haploid cells). One female stem cell yields one functional gamete plus 2-3 polar bodies; one male stem cell yields four functional gametes. Once spermatogenesis begins at puberty, it continues throughout life, producing 300 million sperm daily. Oogenesis begins during fetal life, resumes at puberty, and produces one mature gamete monthly until menopause.

Pages: 941

3. Describe the development and possible fates of an ovarian follicle in a woman of reproductive age.

   Answer: A primordial follicle is a single layer of cells around an oocyte. A primary follicle is 6-7 layers of cuboidal epithelial cells around the oocyte. A secondary follicle has the theca externa, theca interna, and follicular fluid filling antrum. The mature follicle (larger) ovulates the secondary oocyte, then collapses to form the corpus hemorrhagicum (clot inside). The clot is absorbed and the cells enlarge to form the corpus luteum, which either degenerates into the corpus albicans or is maintained by hCG depending on whether fertilization occurred.

Pages: 921

4. Identify four factors examined in semen analysis, and explain how each may contribute to male infertility.

   Answer: Low volume, low count or delayed liquefaction prevent sufficient numbers from reaching the egg. Poor motility and abnormal cell shape prevent the sperm from swimming well enough to reach egg. Low pH prevents proper buffering of vaginal secretions. Low fructose reduces ATP production.

Pages: 930

5. Describe the role of cervical mucus in the female reproductive tract.

   Answer: Mucus forms a plug to impede penetration of sperm except at ovulation. It also supplements the energy needs of the sperm, serves as a sperm reservoir, and protects the sperm from pH damage and phagocytosis. Mucus may also play a role in capacitation.

Pages: 929

6. Explain how bacteria entering the vagina could ultimately cause peritonitis.

   Answer: Motile bacteria can move unaided into the reproductive tract. Others can be carried by sperm cells. Because the uterine tubes are open to the pelvic cavity, bacteria can easily enter the pelvic cavity and come in contact with folds of the peritoneum. Some bacteria may enter the blood or lymph and travel directly to the peritoneum.

7. Identify the factors that increase a woman's risk of developing breast cancer.
   Answer: Family history (especially mother or sister); no children or first child after age 34; previous cancer in one breast; exposure to ionizing radiations; excess intake of fat and/or alcohol; cigarette smoking

8. Describe the hormonal changes that occur during a normal menstrual cycle.
   Answer: Menstrual phase - low estrogen and progesterone causing sloughing of the stratum functionalis, rising FSH and slightly rising LH promote follicular development; Preovulatory phase - FSH and LH surge just prior to ovulation, estrogen rises steeply as follicle develops, progesterone still low, inhibin rising; Postovulatory phase - FSH, LH, and estrogen fall sharply after ovulation, estrogen rises midway through phase then falls off, progesterone rises sharply (from corpus luteum), then falls off just prior to menstruation

9. Name the three predominant estrogens, and state the functions of estrogens in the non-pregnant female.
   Answer: Beta estradiol, estrone, and estriol are major estrogens. They promote development and maintenance of female reproductive structures and secondary sex characteristics and breasts, help control F&E balance, increase protein anabolism, and lower blood cholesterol.

10. Describe the positive feedback loop involved in ovulation.
    Answer: FSH and LH promote follicular development, thus increasing estrogen production. High levels of estrogen during late preovulatory phase stimulate release of GnRH from hypothalamus. GnRH promotes release of more FSH and LH from anterior pituitary.

11. Describe the secondary sex characteristics of males and females.
    Answer: Female: fat distribution to breasts, abdomen, mons pubis, and hips; higher pitched voice; broad pelvis; axillary and pubic hair Male: musculoskeletal development to widen shoulders and narrow hips; deepened voice via enlargement of larynx; thickened skin; increased sebaceous gland secretions; axillary, pubic, facial, and chest hair

12. How can sexually transmitted diseases be prevented?
    Answer: Abstinence from sexual contact; use of condoms (male or female); chemical contraceptives that also inhibit microorganisms; education; tracking and treating sexual contacts of infected persons (other answers may be acceptable)

Chapter 29   Development and Inheritance

Multiple-Choice

Choose the one alternative that best completes the statement or answers the question.

Pages: 978
1. The observable characteristics of a person's genetic makeup are known as the:
    A) genotype
    B) phenotype
    C) gene pool
    D) karyotype
    E) autosomes
    Answer: B

Pages: 960
2. During early pregnancy, the main function of hCG is to:
    A) nourish the embryo
    B) stimulate placental growth
    C) maintain the corpus luteum
    D) depress estrogen production
    E) prevent passage of testosterone to the developing embryo
    Answer: C

Pages: 958
3. How many days after fertilization does implantation of the blastocyst occur?
    A) 2 days
    B) 6 days
    C) 14 days
    D) 28 days
    E) 120 days
    Answer: B

Pages: 964
4. The chorion develops from the:
    A) inner cell mass
    B) blastocele
    C) trophoblast
    D) yolk sac
    E) corpus luteum
    Answer: C

Pages: 963
5. The embryonic disc develops from the:
    A) inner cell mass
    B) blastocele
    C) trophoblast
    D) yolk sac
    E) corpus luteum
    Answer: A

Pages: 963
6. The fetus is protected from mechanical injury by fluid contained within the:
    A) allantois
    B) yolk sac
    C) blastocele
    D) amnion
    E) decidua
    Answer: D

Pages: 966
7. Human chorionic gonadotropin is produced by the:
    A) corpus luteum
    B) embryo
    C) endometrium
    D) trophoblast cells of the chorion
    E) blastocele
    Answer: D

Pages: 970
8. Human chorionic gonadotropin is at its highest levels during:
    A) ovulation
    B) fertilization
    C) implantation
    D) the ninth month of pregnancy
    E) the third month of pregnancy
    Answer: E

Pages: 980
9. How many pairs of autosomes does a normal human have?
    A) 22
    B) 23
    C) 44
    D) 46
    E) one
    Answer: A

Pages: 966
10. **ALL** of the following are **TRUE** for the placenta **EXCEPT**:
    A) it makes estrogen
    B) it allows for mixing of maternal and fetal blood
    C) it makes progesterone
    D) it allows for exchange of nutrients, wastes, and gases
    E) it develops in part from the trophoblast
    Answer: B

11. An individual whose alleles for a particular trait are the same is said to be:
    A) dominant
    B) recessive
    C) homologous
    D) homozygous
    E) heterozygous
    Answer: D

12. What percentage of sperm cells introduced into the vagina normally reach the oocyte?
    A) less than 1%
    B) about 10%
    C) 25-30%
    D) 50%
    E) close to 100%
    Answer: A

13. One function of the enzyme **acrosin** is to:
    A) neutralize the acidity of vaginal secretions
    B) open calcium ion channels in the fertilized oocyte
    C) stimulate cleavage of the zygote
    D) stimulate sperm motility and migration within the female reproductive tract
    E) liquefy cells of the endometrium to allow implantation to occur
    Answer: D

14. The term **capacitation** refers to:
    A) union of the male and female pronuclei
    B) equatorial division of the secondary oocyte following penetration by a sperm cell
    C) functional changes that sperm undergo in the female reproductive tract that allow them to fertilize the secondary oocyte
    D) functional changes in the zona pellucida caused by release of calcium ions
    E) a sperm cell's penetration of the zona pellucida and entry into a secondary oocyte
    Answer: C

15. The deepest layer of follicular cells around the secondary oocyte is the:
    A) acrosome
    B) trophoblast
    C) corona radiata
    D) morula
    E) zona pellucida
    Answer: C

Pages: 958
16. The term **syngamy** refers to the:
   A) union of male and female pronuclei
   B) equatorial division of the secondary oocyte following penetration by a sperm cell
   C) functional changes that sperm undergo in the female reproductive tract that allow them to fertilize a secondary oocyte
   D) functional changes in the zona pellucida caused by release of calcium ions
   E) a sperm cell's penetration of the zona pellucida and entry into a secondary oocyte
   Answer: E

Pages: 959
17. At day 4 after fertilization, the solid ball of cells that has formed is called the:
   A) zygote
   B) blastocyst
   C) gastrula
   D) morula
   E) embryo
   Answer: D

Pages: 958
18. What is the next event following syngamy?
   A) penetration of the zona pellucida by a sperm cell
   B) depolarization and release of calcium ions by the oocyte
   C) cleavage
   D) meiosis II
   E) implantation
   Answer: B

Pages: 960
19. Implantation usually occurs in the:
   A) uterine tube
   B) myometrium
   C) cervix adjacent to the internal os
   D) posterior fornix
   E) posterior wall of the body or fundus of the uterus
   Answer: E

Pages: 960
20. The enzymes that allow implantation to occur are produced by the:
   A) syncytiotrophoblast
   B) cytotrophoblast
   C) blastocele
   D) inner cell mass
   E) acrosome
   Answer: A

Pages: 960
21. The reason the corpus luteum is maintained in early pregnancy is to:
    A) continue production of hCG
    B) produce ATP for the developing embryo
    C) keep levels of estrogen and progesterone high enough to maintain the endometrium
    D) produce the enzymes necessary for implantation
    E) serve as the connection between the developing embryo and the endometrium
    Answer: C

Pages: 962
22. The embryonic period of development covers what time period?
    A) the time between fertilization and implantation
    B) the first two months following fertilization
    C) the first trimester of pregnancy
    D) the first two trimesters of pregnancy
    E) the time between the appearance of the primary germ layers and the development of major
       abdominal organs
    Answer: B

Pages: 963
23. The process by which an embryonic structure composed of the primary germ layers is formed is
    called:
    A) meiosis II
    B) blastogenesis
    C) capacitation
    D) gastrulation
    E) syngamy
    Answer: D

Pages: 963
24. The amnion forms from the:
    A) syncytiotrophoblast
    B) blastocele
    C) cytotrophoblast
    D) decidua
    E) inner cell mass
    Answer: C

Pages: 963
25. The yolk sac forms from the:
    A) ectoderm of the inner cell mass
    B) fluid within the amniotic cavity
    C) allantois
    D) decidua
    E) endoderm of the inner cell mass
    Answer: E

Pages: 963
26. The extraembryonic coelom develops between the:
    A) two mesodermal layers of the embryonic disc
    B) endoderm and ectoderm prior to the appearance of the mesodermal layer
    C) amnion and chorion
    D) chorion and decidua
    E) syncytiotrophoblast and cytotrophoblast
    Answer: A

Pages: 963
27. The extraembryonic coelom becomes the:
    A) ectoderm
    B) endoderm
    C) amniotic cavity
    D) ventral body cavity
    E) blastocele
    Answer: D

Pages: 964
28. The "water" referred to when a woman's "water breaks" prior to delivery is:
    A) amniotic fluid released when the amnion ruptures
    B) maternal plasma leaking from weakened blood vessels
    C) maternal urine expelled as compression of the bladder occurs
    D) interstitial fluid released by separation of the layers of the placenta
    E) mucus from stimulated glands in the cervix and vagina
    Answer: A

Pages: 965
29. **ALL** of the following are derived from ectoderm **EXCEPT**:
    A) brain
    B) epidermis
    C) adrenal medulla
    D) lens
    E) skeletal muscle
    Answer: E

Pages: 965
30. **ALL** of the following are derived from mesoderm **EXCEPT**:
    A) middle ear
    B) pituitary gland
    C) blood
    D) dermis
    E) bone
    Answer: B

Pages: 966
31. The embryonic membrane that forms the vascular portion of the umbilical cord is the:
    A) allantois
    B) yolk sac
    C) amnion
    D) chorion
    E) decidua
    Answer: A

Pages: 966
32. Exchange of gases, nutrients, and wastes between maternal and fetal blood takes place between the:
    A) amnion and chorion
    B) decidua capsularis and chorion
    C) decidua basalis and yolk sac
    D) decidua basalis and chorionic villi
    E) decidua capsularis and decidua basalis
    Answer: D

Pages: 966
33. The portion of the endometrium that covers the embryo and is located between the embryo and the uterine cavity is the:
    A) decidua parietalis
    B) decidua basalis
    C) decidua capsularis
    D) stratum basalis
    E) amnion
    Answer: C

Pages: 966
34. The portion of the endometrium that becomes the maternal portion of the placenta is the:
    A) decidua parietalis
    B) decidua basalis
    C) decidua capsularis
    D) stratum basalis
    E) amnion
    Answer: B

Pages: 966
35. What fills intervillous spaces?
    A) amniotic fluid
    B) maternal blood only
    C) fetal blood only
    D) interstitial fluid
    E) both fetal and maternal blood
    Answer: B

36. Most materials cross the placenta by:
    A) primary active transport
    B) filtration
    C) phagocytosis
    D) diffusion
    E) exocytosis
    Answer: D

37. Deoxygenated fetal blood is carried to the placenta via the:
    A) uterine arteries
    B) uterine veins
    C) umbilical arteries
    D) umbilical vein
    E) decidua basalis
    Answer: C

38. Peak secretion of hCS occurs:
    A) between fertilization and implantation
    B) during the second trimester
    C) late in the third trimester
    D) just after delivery
    E) at about the ninth week of pregnancy
    Answer: C

39. Once hCG levels decrease, estrogen and progesterone are secreted mainly by the:
    A) placenta
    B) embryo
    C) corpus luteum
    D) hypothalamus
    E) stratum basalis
    Answer: A

40. Which of the following has occurred by the end of the first month of development?
    A) The placenta has developed.
    B) The eyes have formed and are open.
    C) Ossification has begun.
    D) Urine has begun to form.
    E) The heart has formed and begun beating.
    Answer: E

41. **ALL** of the following have occurred by the end of the third month of development **EXCEPT**:
    A) The limbs have fully formed.
    B) The eyes have opened.
    C) Ossification has begun.
    D) Urine has begun to form.
    E) The heart has formed and begun beating.
    Answer: B

42. Softening of the connective tissue around pelvic joints is stimulated by:
    A) estrogen
    B) progesterone
    C) inhibin
    D) hCS
    E) relaxin
    Answer: E

43. In a pregnant woman, decreased utilization of glucose and increased release of fatty acids from adipose tissue are promoted by the hormone:
    A) estrogen
    B) progesterone
    C) inhibin
    D) hCS
    E) relaxin
    Answer: D

44. Early pregnancy tests are based on detection of what substance in the urine?
    A) amniotic fluid
    B) blastomeres
    C) hCG
    D) high levels of progesterone
    E) hCS
    Answer: C

45. The human gestation period is about:
    A) 9 weeks
    B) 24 weeks
    C) 32 weeks
    D) 38 weeks
    E) 48 weeks
    Answer: D

46. **ALL** of the following are **increased** in pregnant women **EXCEPT**:
    A) cardiac ouput
    B) blood volume
    C) tidal volume
    D) glomerular filtration rate
    E) expiratory reserve volume
    Answer: E

47. Hypertension associated with impaired renal function at about the 20th week of pregnancy is called:
    A) emesis gravidarum
    B) preeclampsia
    C) placenta previa
    D) dystocia
    E) Turner's syndrome
    Answer: B

48. The major problem for premature infants is an insufficient level of:
    A) red blood cells
    B) antibodies
    C) epinephrine
    D) surfactant
    E) cerebrospinal fluid
    Answer: D

49. Colostrum is different from true milk because it contains less lactose and virtually no:
    A) fat
    B) protein
    C) sodium
    D) iron
    E) antibodies
    Answer: A

50. The two alternative forms of a gene that code for the same trait and are at the same locus on homologous chromosomes are called:
    A) alleles
    B) autosomes
    C) Barr bodies
    D) blastomeres
    E) chromatids
    Answer: A

Pages: 978
51. A person is heterozygous for a particular trait if he/she has:
    A) more than two copies of a particular gene
    B) genes for the trait on both sex chromosomes and autosomes
    C) one dominant allele and one recessive allele for the trait
    D) two dominant alleles for the trait
    E) two recessive alleles for the trait
    Answer: C

Pages: 978
52. One person is homozygous dominant for a particular trait, and another person is heterozy
    for the same trait. Which of the following statements is **TRUE** regarding these two people'
    A) The homozygous person is female and the heterozygous person is male.
    B) The homozygous person exhibits the trait, but the heterozygous person does not.
    C) The heterozygous person exhibits the trait, but the homozygous person does not.
    D) Neither person exhibits the trait.
    E) The two people both exhibit the trait.
    Answer: E

Pages: 979
53. If a person's phenotype is intermediate between homozygous dominant and homozygou
    then inheritance of this trait is an example of:
    A) codominance
    B) sex-linked inheritance
    C) incomplete dominance
    D) nondisjunction
    E) lyonization
    Answer: C

Pages: 980
54. Inheritance of the ABO blood type is an example of:
    A) codominance
    B) sex-linked inheritance
    C) incomplete dominance
    D) nondisjunction
    E) lyonization
    Answer: A

Pages: 980
55. What would be the possible blood phenotypes of the offspring of parents whose genoty
    ii and $I^A I^B$?
    A) type O only
    B) type AB only
    C) types A or B only
    D) types A, B, or O only
    E) all ABO types are possible
    Answer: C

56. A person who expresses the SRY gene is:
    A) exhibiting sickle-cell anemia
    B) expressing a recessive trait
    C) expressing DNA that was originally part of a virus
    D) phenotypically female
    E) phenotypically male
    Answer: E

57. Teratogens are:
    A) traits carried only on sex chromosomes
    B) agents that induce physical defects in developing embryos
    C) cells with an abnormal number of chromosomes
    D) all of the alleles that contribute to a particular trait
    E) genes that are inactivated during fetal development
    Answer: B

58. Nondisjunction of chromosome 21 results in:
    A) Klinefelter's syndrome
    B) Turner's syndrome
    C) sickle-cell anemia
    D) Down syndrome
    E) hemophilia
    Answer: D

59. A person whose sex chromosome genotype is XO has:
    A) Klinefelter's syndrome
    B) Turner's syndrome
    C) sickle-cell anemia
    D) Down syndrome
    E) hemophilia
    Answer: B

60. Which of the following is a sex-linked disorder?
    A) SIDS
    B) PKU
    C) sickle-cell anemia
    D) Down syndrome
    E) hemophilia
    Answer: E

61. What event marks the beginning of the stage of expulsion in true labor?
    A) when the woman's water breaks
    B) appearance of lochia
    C) complete cervical dilation
    D) severing of the umbilical cord
    E) completion of two hours of rhythmic uterine contractions
    Answer: C

62. A major advantage of chorionic villus sampling over amniocentesis is that chorionic villus sampling:
    A) can be done earlier in the pregnancy
    B) has less risk of inducing spontaneous abortion
    C) is more accurate
    D) can detect conditions that amniocentesis cannot
    E) is non-invasive
    Answer: A

63. A couple who are both phenotypically normal have a child who expresses a sex-linked recessive trait. Which of the following represents this child's genotype? [Let the trait be designated T (dominant) or t (recessive).]
    A) Tt
    B) tt
    C) $X^t X^t$
    D) $X^t Y$
    E) both C and D could be correct
    Answer: D

64. A child expresses an autosomal recessive trait. Which of the following are **NOT** possible genotypes for the parents?
    A) Both father and mother are homozygous recessive.
    B) Both father and mother are heterozygous.
    C) The father is heterozygous and the mother is homozygous recessive.
    D) The mother is heterozygous and the father is homozygous recessive.
    E) The father is heterozygous and the mother is homozygous dominant.
    Answer: E

65. By doing karyotyping, one can determine the gender of the child because:
    A) the external genitalia can be seen on the screen
    B) the total number of chromosomes would be different between the sexes
    C) the Y chromosome is much smaller than the X chromosome
    D) levels of testosterone are higher in the amniotic fluid of male embryos
    E) the embryonic membranes are thicker around male embryos
    Answer: C

True-False

Write T if the statement is true and F if the statement is false.

Pages: 978

1. A person who is heterozygous for a particular trait may express either the dominant or recessive allele.
   Answer: False

Pages: 979

2. A zygote produced from the union of a sperm cells and an oocyte in which nondisjunction occurred would be described as aneuploid.
   Answer: True

Pages: 980

3. The alleles involved in inheritance of ABO blood types are all codominant.
   Answer: False

Pages: 982

4. A female exhibits red-green color blindness only if both X chromosomes contain the recessive gene.
   Answer: True

Pages: 982

5. In female cells, one of the X chromosomes is randomly and permanently inactivated early in development.
   Answer: True

Pages: 958

6. The change that occurs in the zona pellucida that prevents entry of multiple sperm into an oocyte is called capacitation.
   Answer: False

Pages: 959

7. Monozygotic twins develop from separate oocytes.
   Answer: False

Pages: 963

8. The embryo develops from the inner cell mass of the blastocyst.
   Answer: True

Pages: 965

9. The extraembryonic coelom develops into the ventral body cavity.
   Answer: True

Pages: 965

10. All nervous tissue is derived from endoderm.
    Answer: False

11. Human chorionic gonadotropin mimics the action of progesterone.
Answer: False

12. More human chorionic somatomammotropin is secreted at the 28th week of pregnancy tha
12th week of pregnancy due to an increase in placental mass.
Answer: True

13. Venous return is decreased in pregnancy due to a 20% decrease in cardiac output.
Answer: False

14. During true labor, the cervix is considered to be completely dilated at about 10 cm.
Answer: True

15. A normal stimulus for release of oxytocin is the sucking action of the infant at the breast.
Answer: True

Short Answer

Write the word or phrase that best completes each statement or answers the question.

1. _____ is the term for the functional changes that sperm undergo in the female reproduct
tract that allow them to fertilize a secondary oocyte.
Answer: Capacitation

2. The glycoprotein layer internal to the corona radiata surrounding the oocyte is called the

_____.
Answer: zona pellucida

3. The fertilized ovum is called a(n) _____.
Answer: zygote

4. The development of an embryo or fetus outside the uterine cavity is called a(n) _____.
Answer: ectopic pregnancy

5. The part of the female reproductive tract in which fertilization normally occurs is the

_____.
Answer: uterine tube

Pages: 959
6. By the end of the third day after fertilization, the fertilized egg has become a solid ball of cells called the _____.
Answer: morula

Pages: 959
7. The hollow ball of cells that is implanted into the uterine wall is called the _____.
Answer: blastocyst

Pages: 962
8. During the first two months of development, the developing human is called a(n) _____.
Answer: embryo

Pages: 963
9. The primary germ layers are the _____, the _____, and the _____.
Answer: ectoderm; endoderm; mesoderm

Pages: 963
10. The process by which the two-layered inner cell mass is converted into a structure composed of the primary germ layers is called _____.
Answer: gastrulation

Pages: 963
11. The fetal membrane that serves as an early site of blood formation and that is the source of the cells that differentiate into primitive germ cells is the _____.
Answer: yolk sac

Pages: 964
12. The structure derived from the trophoblast of the blastocyst that becomes the principal embryonic part of the placenta is the _____.
Answer: chorion

Pages: 966
13. Development of the placenta is accomplished by the _____ month of pregnancy.
Answer: third

Pages: 966
14. The portion of the endometrium that becomes modified following implantation is known as the _____.
Answer: decidua

Pages: 970
15. The chorion of the placenta secretes the hormone _____, which mimics the action of LH.
Answer: human chorionic gonadotropin

Pages: 971
16. The time a developing human is carried in the female reproductive tract between fertilization and birth is called _____, which is normally about _____ weeks.
Answer: gestation; 38

Pages: 972
17. A condition that appears after the 20th week of gestation in which hypertension results from impaired renal function is called _____.
Answer: preeclampsia

Pages: 976
18. The principal hormone promoting lactation is _____.
Answer: prolactin

Pages: 977
19. The initial low-lactose, low-fat fluid produced by the mammary glands during late pregnancy and for the first few days following delivery is called _____.
Answer: colostrum

Pages: 978
20. The complete genetic makeup of an organism is called the _____.
Answer: genome

Pages: 978
21. The two alternative forms of a gene that code for the same trait and are at the same locus on homologous chromosomes are called _____; an individual in whom the two forms are the same is said to be _____, while an individual in whom the two forms are different is said to be _____.

Answer: alleles; homozygous; heterozygous

Pages: 978
22. The physical or outward expression of a gene is called the _____.
Answer: phenotype

Pages: 979
23. An error in meiosis called _____ occurs when homologous chromosomes fail to separate properly during anaphase of the reduction division.
Answer: nondisjunction

Pages: 980
24. A diploid human cell contains _____ pair(s) of autosomes and _____ pair(s) of sex chromosomes.
Answer: 22; one

Pages: 982
25. Traits inherited on the X and/or Y chromosomes are referred to as _____ traits.
Answer: sex-linked

Pages: 983
26. The dark-staining inactivated X chromosome seen in the nuclei of female mammalian cells is called a _____.
Answer: Barr body

Pages: 984
27. Nondisjunction of chromosome 21 results in _____.
Answer: Down syndrome

Pages: 970
28. The hormone that serves as a basis for early pregnancy tests is _____.
    Answer: human chorionic gonadotropin

Pages: 963
29. The blastocyst has three portions - the outer _____, the _____, and the internal, fluid-filled cavity called the _____.
    Answer: trophoblast; inner cell mass; blastocele

Pages: 963
30. The embryo develops from the layer of the blastocyst called the _____.
    Answer: inner cell mass

Matching

Choose the item from Column 2 that best matches each item in Column 1.

1.	zygote	a segmentation nucleus, cytoplasm, and the zona pellucida
2.	morula	solid mass of cells called blastomeres
3.	blastocyst	hollow ball of cells that implants into the uterine wall
4.	trophoblast	part of blastocyst that secretes hCG
5.	amnion	embryonic membrane nearest to the embryo
6.	chorion	embryonic membrane that becomes the principal embryonic part of the placenta
7.	decidua basalis	portion of the endometrium that becomes the maternal part of the placenta
8.	allantois	embryonic membrane that forms the vascular portion of the umbilical cord
9.	decidua capsularis	portion of the endometrium that covers the embryo
10.	inner cell mass	part of the blastocyst that becomes the embryo

Essay

Write your answer in the space provided or on a separate sheet of paper.

Pages: 960

1. Describe the morphology of the blastocyst, and describe the fates of the different cell regions.

   Answer: The blastocyst is a hollow ball of cells consisting of the fluid-filled blastocele, outer trophoblast cells, and inner cell mass. The trophoblast consists of the cytotrophoblast and syncytiotrophoblast layers. Part of the former becomes the amnion, and both layers become the chorion. The inner cell mass becomes the embryo.

Pages: 970

2. Name and describe the roles of the hormones involved in pregnancy.

   Answer: Estrogen and progesterone maintain the lining of the uterus and prepare the mammary glands to secrete milk. hCG mimics the action of LH to maintain the corpus luteum to keep levels of estrogen and progesterone high until the placenta is large enough to produce adequate amounts. Relaxin softens connective tissues at joints. hCS increases protein synthesis, prepares mammary glands for lactation, decreases maternal glucose utilization, and increases release of fatty acids from adipose tissue.

Pages: 983

3. Describe the potential hazards to the embryo and fetus associated with alcohol consumption and cigarette smoking.

   Answer: Acetaldehyde, a metabolic product of alcohol, causes fetal alcohol syndrome, which is characterized by slow growth, small head, unusual facial features, defective heart and other organs, malformed limbs, and CNS abnormalities that may lead to behavioral problems. Cigarette smoking leads to low birth weight and increased risk of fetal/infant mortality, cardiac problems, anencephaly, and cleft lip and cleft palate.

Pages: 976

4. Describe the cardiovascular adjustments that occur in the infant at birth.

   Answer: The foramen ovale closes first (between atria), then the ductus arteriosus closes. Both divert blood to lungs. After severing of the umbilical cord, blood no longer travels through the ductus venosus (bypassing liver). Pulse rate is high (120-160 bpm) then slows. Production of red blood cells and hemoglobin is increased. Very high leukocyte count decreases after about one week.

Pages: 976

5. Describe the respiratory adjustments that occur in the infant at birth.

   Answer: Fetal lungs are collapsed or filled with amniotic fluid. After severing of umbilical cord, the partial pressure of carbon dioxide rises, thus stimulating inspiratory center. Respiratory muscles contract, and the infant draws a very deep breath. Vigorous exhalation and spontaneous crying occur. Respiratory rate is rapid for about two weeks.

Pages: 958-960
6. Describe the events of early development from fertilization through implantation.
   Answer: At fertilization, the male and female pronuclei fuse into the segmentation nucleus, which with cytoplasm and zona pellucida, form the zygote in the uterine tube. The zygote undergoes cleavage to form blastomeres, which are arranged as a solid ball of cells called the morula. By the fifth day after fertilization, the hollow blastocyst has formed (trophoblast, inner cell mass, and blastocele), which implants into the uterine wall oriented toward the endometrium.

Pages: 963
7. Identify the embryonic/fetal membranes, and describe the location and functions of each.
   Answer: Allantois   outpouching from yolk sac; early site of blood formation; becomes umbilical cord
   Yolk sac - extends as sac between amnion and chorion; early site of blood cell formation and source of cells that become primitive germ cells
   Amnion - over embryonic disc, eventually surrounds embryo/fetus; holds amniotic fluid (shock absorber, etc.)
   Chorion - between amnion and decidua - villi grow into decidua; principal embryonic part of placenta for exchange of materials between mother and fetus

Pages: 976
8. Describe the hormonal control of lactation.
   Answer: After delivery, PRH and OT secretion are stimulated by the sucking action at the nipple, and PIH secretion is inhibited. PRL is released to stimulate milk production, and OT stimulates contraction of myoepithelial cells for milk ejection.

Pages: 980
9. A child is born who is blood type O. The mother's blood type is A. The man the mother claims is the father of the baby is blood type B. Because he is blood type B, this man claims he cannot be the baby's father. Is he correct? Explain your answer. What, if anything, can you tell about the blood phenotypes and genotypes of the maternal grandparents? Explain your answer.
   Answer: The man is not correct. The woman could be genotype $I^A i$, and he could be $I^B i$. A Punnett square would show the possibility of an ii genotype. The woman's parents could have been phenotypes A ($I^A I^A$ or $I^A i$) and O (ii), A ($I^A i$) and A ($I^A i$), or A ($I^A I^A$ or $I^A i$) and B ($I^B i$).

Pages: 982
10. A woman who is a carrier of a sex(X)-linked trait marries a man who does not express the trait. What are the possible phenotypes and genotypes for male and female offspring? Explain your answer.
    Answer: All girls will have a normal phenotype, because all will possess at least one dominant allele. Half will be homozygous dominant, half will be heterozygous carriers. Half of the boys will be of normal phenotype, because they will have the dominant allele on the X chromosome. Half of the boys will express the trait because they will have only one X chromosome, which has the recessive allele.